Metasurface-Driven Electronic Warfare

Metasurface-Driven Electronic Warfare

Rafael Gonçalves Licursi de Mello
Ponteiro, France

IEEE Press
WILEY

Published by John Wiley & Sons, Inc., Hoboken, New Jersey.
Published simultaneously in Canada.

Library of Congress Cataloging-in-Publication Data
Names: Licursi de Mello, Rafael Gonçalves, author.
Title: Metasurface-driven electronic warfare / Rafael Gonçalves Licursi de Mello.
Description: Hoboken, New Jersey : Wiley, [2025] | Includes bibliographical references and index.
Identifiers: LCCN 2024042391 (print) | LCCN 2024042392 (ebook) | ISBN 9781394226672 (hardback) | ISBN 9781394226696 (adobe pdf) | ISBN 9781394226689 (epub)
Subjects: LCSH: Metasurfaces. | Electronics in military engineering.
Classification: LCC TK7871.15.M48 L53 2025 (print) | LCC TK7871.15.M48 (ebook) | DDC 623/.043–dc23/eng/20241030
LC record available at https://lccn.loc.gov/2024042391
LC ebook record available at https://lccn.loc.gov/2024042392

Cover Design: Wiley
Cover Image: © Olena.07/Shutterstock

Set in 9.5/12.5pt STIXTwoText by Straive, Pondicherry, India

Contents

About the Author *xi*
Foreword *xiii*
Preface *xv*
About the Companion Website *xvii*

1 Introduction to Metasurface-Driven Electronic Warfare *1*
1.1 From Static Radios to a Metasurface-Driven EW *3*
1.1.1 Early Days *4*
1.1.2 The Modern Era *6*
1.1.3 To the Future *8*
1.2 Can We Actually Rely on Electromagnetic Waves? *9*
1.3 The (Super) Natural Powers of Metasurfaces *10*
1.3.1 Metamaterials: A Proof of Concept *12*
1.3.2 The Need for a Two-Dimensional Version *13*
1.4 How Do Metasurfaces Fit EW? *15*
1.5 Reconfigurability at the Tactical Level *18*
1.6 Book Presentation and Organization *19*
1.7 Final Remarks of the Chapter *20*
References *21*

2 Electromagnetics: A Review on Pertinent Points *33*
2.1 Introduction to Electromagnetics *34*
2.1.1 Fundamental Concepts *34*
2.1.2 Complex Representation *37*
2.1.3 Traveling and Standing Waves *38*
2.1.4 Reflection, Transmission, and the Concept of Impedance *41*
2.1.5 Oblique Incidence and the Refraction of Waves *46*
2.1.6 Reflection and Transmission at Oblique Incidence *48*
2.1.7 Surface and Evanescent Waves *50*
2.1.8 Takeaways *52*
2.2 Unusual Properties for the Propagation of Waves *52*

2.2.1 Negative Refraction *52*
2.2.2 Negative Phase Velocity and Backward Waves *54*
2.2.3 Amplification of Evanescent Waves *55*
2.3 Demystifying the Break of Fundamental Laws *56*
2.3.1 Mechanisms Behind Constitutive Parameters of Media *57*
2.3.2 Typical Resonant Behaviors of Material *59*
2.4 Final Remarks of the Chapter *64*
References *64*

3 Mechanisms Behind the Operation of Metasurfaces *67*
3.1 Obtaining Arbitrary Medium Properties *68*
3.2 Obtaining Arbitrary Boundary Conditions *72*
3.2.1 The Mushroom-Like Metasurface *73*
3.2.2 Surface Impedances *74*
3.2.3 Transmission-Line Equivalent Schematic *76*
3.2.4 Arbitrary Reflection Phase *78*
3.2.5 Huygens' Principle *79*
3.2.6 Arbitrary Transmission Phase *81*
3.2.7 Arbitrary Reflection and Refraction Directions *84*
3.3 Diversity of Approaches *87*
3.4 Metasurfaces' Limitations *88*
3.4.1 Metasurfaces' Frequency Dependence *88*
3.4.2 Metasurfaces' Angle/Polarization Dependence *89*
3.4.3 Low Profile and Conformal Shape in Metasurfaces *90*
3.5 Final Remarks of the Chapter *91*
References *92*

4 Passive Stealth with Metasurfaces *99*
4.1 The Radar Equation and the Concept of Radar Cross-Section *100*
4.2 Traditional Stealth Technologies *103*
4.3 Frequency Selective Surfaces *104*
4.4 Scattering Metasurfaces *106*
4.4.1 Checkerboard Metasurfaces *106*
4.4.2 Phase-Gradient Metasurfaces *109*
4.4.3 Polarization-Conversion Metasurfaces *112*
4.4.4 Time-Varying Metasurfaces *116*
4.4.5 Space–Time-Modulated Metasurfaces *118*
4.4.6 Limitation of the Scattering Approach *120*
4.5 Metasurface Absorbers *121*
4.5.1 Conventional Metasurface Absorbers *121*
4.5.2 Rasorbers *125*

4.6 Challenges for Stealth with Passive Metasurfaces *127*
4.6.1 Low Frequencies *127*
4.6.2 Deployment Rate *131*
4.7 Final Remarks of the Chapter *133*
References *134*

5 Active-Cancelation Stealth with Metasurfaces *143*
5.1 The LPI Radar *145*
5.2 A Brief Introduction to DRFMs *148*
5.2.1 Generic DRFM Architecture *149*
5.2.2 Complexity in the Analog Domain *150*
5.2.3 Complexity in the Digital Domain *151*
5.3 Working Principle of Active-Cancelation Stealth *153*
5.3.1 Impact of the DOA Estimation *156*
5.3.2 Impact of Complicated RCS Patterns *157*
5.3.3 Impact of the Waveform *157*
5.3.4 Impact of the Time Delay *159*
5.4 Power-Amplifying Metasurfaces *163*
5.4.1 Modeling of Power-Amplifying Metasurfaces *163*
5.4.2 Miniaturization of Active Circuits *165*
5.4.3 Cross-Polarized, Power-Amplified Reflections *166*
5.4.4 Co-polarized, Power-Amplified Reflections *167*
5.4.5 Roadmap to a Metasurface-Driven Active Stealth *169*
5.5 Introduction to the EW Scenario Simulator in Matlab *171*
5.5.1 Simulation Modeling *171*
5.5.2 Simulation Algorithm *172*
5.6 Simulation: Comparison Between Stealth Techniques *174*
5.7 Exercises *177*
5.7.1 Exploring the EW Scenario Simulator *178*
5.7.2 Simulating a Realistic DRFM *178*
5.7.3 Simulating a Realistic Active-Cancelation Stealth *179*
5.8 Final Remarks of the Chapter *180*
References *181*

6 Metasurfaces for Electronic Support Applications *189*
6.1 The One-Way Equation and ES Detection *192*
6.2 Introduction to Frequency-Independent Antennas *193*
6.2.1 Scaling Principle *194*
6.2.2 Self-Complementariness Principle *201*
6.3 Traditional Methods to Obtain Unidirectional Patterns *204*
6.3.1 Lossy Cavities *204*

6.3.2 PEC Antenna Reflectors *205*
6.4 Obtaining Unidirectional Patterns with Metasurfaces *207*
6.4.1 Single-Band AMC Reflectors *207*
6.4.2 Frequency-Reconfigurable AMCs *209*
6.4.3 Multiband-AMC Reflection Scheme *210*
6.4.4 Extending the Number of Bands of Antenna Reflectors *216*
6.4.5 Roadmap to Powerful Metasurface-Based ES Antennas *219*
6.4.6 Composite Surfaces and Inhomogeneous AMCs *221*
6.5 Simulation: Impact of Radiation Pattern in Collected Signals *225*
6.6 Exercises *226*
6.6.1 Simulating the Impact of Gain in Direction Finding *226*
6.6.2 Simulating a DOA-Driven DRFM *228*
6.6.3 Simulating the Impact of DOA Estimation in Active Stealth *228*
6.6.4 Simulating the Impact of Gain in LPI Signal Classification *229*
6.7 Chapter Final Remarks *229*
References *230*

7 Metasurface-Driven Electronic Attack *239*
7.1 The Monopulse Radar *241*
7.2 The Jamming Equation and the Burn-Through Range *245*
7.3 Cross-Polarization Jamming with Metasurfaces *247*
7.3.1 Working Principle with Conventional EA Systems *248*
7.3.2 Working Principle with Metasurfaces *250*
7.4 Simulation: Comparative Study on Cross-Polarization Jamming *253*
7.4.1 Initial Results for Conventional EA Systems and Metasurfaces *254*
7.4.2 Burn-Through Range Mitigation *256*
7.4.3 Radar Reaction *258*
7.4.4 Impact of Polarization-Conversion Ratio/Deployment Rate *258*
7.5 Blinking Jamming with Metasurfaces *260*
7.5.1 Working Principle with Conventional EA Systems *262*
7.5.2 Working Principle with Metasurfaces *265*
7.5.3 Implementation with Reconfigurable Metasurface Absorbers *266*
7.5.4 Implementation with Power-Amplifying Metasurfaces *268*
7.6 Simulation: Comparative Study on Blinking Jamming *269*
7.6.1 Initial Results and Radar Reaction *270*
7.6.2 Burn-Through Range Mitigation *272*
7.7 Envision of Generalized EA with Metasurfaces *274*
7.8 Exercises *276*
7.8.1 Modeling the Missile Guidance *277*
7.8.2 Impact of DRFM on the Proposed EA Measures *277*
7.8.3 Simulating the Generalized EA Measure *277*

7.9 Chapter Final Remarks *278*
References *279*

8 Metasurfaces for Electronic Protection and Multifunctionality *285*
8.1 Electronic Protection Against Directed-Energy Weapons *287*
8.1.1 A Brief Introduction to Directed-Energy Weapons *287*
8.1.2 Countering Directed-Energy Weapons with Metasurfaces *287*
8.2 Electronic Protection Against Cross-Polarization Jammers *289*
8.2.1 Countering Cross-Polarization Jammers with Metasurfaces *289*
8.3 Exercise: Simulating EP Against Cross-Polarization Jammers *290*
8.4 Metasurfaces for Radar Polarimetry *291*
8.5 Metasurfaces as an Alternative to Phased Arrays *292*
8.5.1 Working Principle of Phased-Array Antennas *293*
8.5.2 Beamsteering with Phase-Gradient Metasurfaces *297*
8.5.3 Beam-Steerable Antennas with Reflective Metasurfaces *298*
8.5.4 Beam-Steerable Antennas with Transmissive Metasurfaces *299*
8.5.5 Fabry–Perot Resonances *300*
8.5.6 Beamsteering with Radiating Element Surrounded by Metasurfaces *305*
8.6 Multifunctional Antennas with Metasurfaces *308*
8.6.1 Challenges in Multiband/Multifunctional Antennas with Metasurfaces *309*
8.6.2 Obtaining Stable Radiation Patterns in Multiple Bands *311*
8.6.3 Example of Multiband/Multifunctional Antenna with Metasurfaces *312*
8.6.4 Roadmap to New Metasurface-Based Multifunctional Antennas *317*
8.7 Chapter Final Remarks *318*
References *319*

9 Emerging Use Cases of Metasurfaces in EW *327*
9.1 Metasurface-Driven ES Capabilities in Drone Swarms *329*
9.1.1 Direction Finding with Metasurfaces *329*
9.1.2 Challenges to Bring the Technology to the Real World *332*
9.1.3 Functionalities Unlocked by ES-Capable Drone Swarms *332*
9.2 Metasurface-Driven Cognitive Jamming in Drone Swarms *333*
9.2.1 Cognitive Jamming with Reinforcement Learning *334*
9.2.2 Cognitive Jamming with Drone Swarms *336*
9.2.3 The Role of Metasurfaces in Cognitive Jamming *337*
9.3 Metasurface-Controlled Spectrum Access *338*
9.3.1 Multipath Propagation Channel *339*
9.3.2 Introduction to Reconfigurable Intelligent Surfaces *341*
9.3.3 Improving the Radio Link Between Assets of Friendly Forces *343*
9.3.4 Improving Signal Interception by Friendly Forces *346*

9.3.5 Avoiding Signal Interception by Adversary Forces *348*
9.3.6 Denying Spectrum Access to Adversary Forces *351*
9.3.7 Further Ideas for the Application of RISs in an EW Context *353*
9.4 Exercises *354*
9.4.1 ES-Solving a Dense Scenario *354*
9.4.2 ES-Solving a Multifunctional Radar *355*
9.4.3 Simulating Cognitive Jamming with a Drone Swarm *356*
9.5 Chapter Final Remarks *356*
References *357*

10 Summary and Final Remarks *369*

Index *373*

About the Author

Dr. Rafael Gonçalves Licursi de Mello earned his PhD in Electronics from the Institut Polytechnique de Paris, with a focus on reconfigurable metasurfaces and multiband/wideband antennas. Experienced in both the hardware and software aspects of electronic warfare (EW), he has worked on metasurfaces, antennas, signal processing, and signals intelligence at various deep-tech companies. He started his career as a pilot, radar operator, and EW officer in Maritime Patrol Aviation, where he conducted real EW missions and identified the needs for future scenarios.

Foreword

Arthur C. Clarke's third law states "Any sufficiently advanced technology is indistinguishable from magic."

Rafael's description of metasurfaces is a prime example of this concept. He describes the electromagnetic properties and math that gives metamaterials and metasurfaces almost supernatural powers, including the ability to make an object invisible. These unusual properties caused quite a stir in academia because they seem to break some of the laws of Physics and moreover cannot be usually found in nature. It was only after the first real-world demonstrations in the late 1990s that the concept was considered reasonable.

Many practical demonstrations of this magic came in the beginning of the 2000s. The absorptive properties of metasurfaces led to excitement among electronic warfare (EW) professionals as a method to accomplish stealth: a metasurface coating would dramatically reduce the radar cross section of an object.

Rafael shows in this book how to expand from this core passive-absorption concept into other EW applications with reflective and transmissive features. The applications include active cancellation, electronic support (ES), electronic attack (EA), and electronic protection (EP). He also describes the simplicity, compactness, power efficiency, and low cost of metasurfaces, making them suitable for more settings, such as lightweight, cheap drones, where digital radio frequency memories (DRFM) are too large and expensive. Metasurfaces are even elected as a central element for commercial 6G systems for cheap, energy-efficient, secure communications.

Importantly, metasurfaces add another layer of dynamism to EW systems, at the physical layer. In a cognitive EW system, we use artificial intelligence (AI) to control all aspects of the EW loop: ES, EP, and EA in a closed-loop, interactive setting. Using feedback from the environment, machine learning models can choose how to reconfigure the underlying EW system. Metasurfaces enhance the cognitive EW's ability to accomplish mission objectives.

Moreover, we are seeing increased blending of EW applications across military and civilian settings. Jamming of police radios and drone incursions at airports have the same characteristics that appear in traditionally military EW problem spaces. Military systems must leverage the commercial advancements in metasurfaces, especially 6G, at the speed of relevance.

EW has determined the winners of wars. Since the first long-distance RF transmission in 1899, EW has been a part of war. The ability to manage the electromagnetic spectrum–communications or radar–determines the outcomes of future battles. Each subsequent technology advance has changed the nature of warfare. During the Russo-Japanese War of 1904–1905, the Russians wanted to jam the Japanese signals; denied permission to perform the jamming, the Japanese won the decisive Battle of Tsushima, ending the war in Japan's favor. By World War II, EW surveillance and jamming was a core battlefield domain. The Ukraine war has seen continuous use of EW and swarming drone technology. Metasurfaces can and will be another leap forward.

—Karen Z. Haigh, author of *Cognitive Electronic Warfare: An Artificial Intelligence Approach*, 2021.

Preface

Writing *Metasurface-Driven Electronic Warfare* was harder than I expected but more rewarding than I could have imagined. But why did I do it?

Tons of papers about metasurfaces for defense exist in the open literature. However, each of them focuses on a very specific problem. Putting it differently, an organization of the many potentials unlocked by the whole plethora of existing metasurfaces in the EW space was lacking.

What allowed me to go through this hard process was, besides extensive academic research, a clear vision of the next configuration of EW. To this end, the deeply technical roles I performed in the industry was of utmost importance, as well as my operational experience in the maritime patrol aviation.

Many interpretations exist for the EW activity. One of them, EW consists of a cat-and-mouse game where opponents innovate as fast as they can to ensure their safe usage of the electromagnetic spectrum while denying it to adversaries. Although this perspective is highly adaptable to increasingly complex scenarios, this dynamic is not a privilege of EW in defense. In essence, the EW game is nothing but another manifestation of the three fundamental interactions described by Clausewitz a couple of centuries ago in his masterpiece *On War*: more specifically, the conflict escalation reasoning, the reciprocal fear of annihilation, and the full exploitation of peoples' military capacity, defined by their *tangible means* and *will*.

After a 20th-century full of large-scale conflicts, the cat-and-mouse EW game is still manifesting such interactions. The relentless implementation of deep technologies, like AI, lightweight drones, and many others, is creating an increasingly dynamic theater of operation. This book anticipates how metasurfaces will add to the dynamism.

EW is an extremely interdisciplinary activity. One challenge in writing this book was to accommodate all the main reader profiles. Generally, every individual able to understand engineering concepts can follow the book. Chapter 1 provides readers of all profiles with the big picture of the EW dynamism and the advent of metasurfaces. Chapters 2 and 3 equip nonexperts with everything they need to understand the working principles of metasurfaces. Chapters 4–8 discuss metasurface applications in stealth, ES, EA, and EP. Although the described use cases are many, most are just clear examples of the applicability of the technology *today*. As it will become obvious to the reader as they go through these chapters, the exotic capabilities that metasurfaces provide will soon unlock many new ways of doing EW. Chapter 9 is where we gather some of them per the latest trends.

I thank people who directly or indirectly contributed to this book. Mentioning everyone would be impossible. Nonetheless, I can't let two of them go unnoticed. About a decade ago, Professor José Edimar Barbosa Oliveira conveyed to me his unequalled enthusiasm in developing research of excellence in EW. More recently, Dr. Karen Zita Haigh also inspired me with her capacity in organizing a complex domain, AI in EW.

In conclusion, having worked with both software and hardware parts of the EW problem, I often get questions about which one is more important. Actually, I tend to see EW forces as octopuses whose brains are composed of software but whose tentacles are manifested through hardware. In the upcoming scenario, AI will be what empowers their brains, and metasurfaces what *extend* their tentacles – or *increase their number*. True, we want to develop the *smartest* octopuses but who can also act over *many, strong, agile, long* tentacles. Thus, software and hardware are different, interdependent instances of a same problem to which we should prepare right now – the EW of the future:

> Victory smiles upon those who anticipate the change in the character of war, not upon those who wait to adapt themselves after the changes occur.
>
> —Giulio Douhet, air power theorist.

About the Companion Website

This book is accompanied by a companion website:

www.wiley.com/go/licursidemello/metasurfacedrivenelectronicwarfare1

The website includes:

MATLAB codes.

1

Introduction to Metasurface-Driven Electronic Warfare

Modern Electronic Warfare (EW) is a highly dynamic activity where many reconfigurable systems can optimize specific tasks on the fly. One may consider the EW problem as a huge scenario where systems exchange information by means of electromagnetic waves that interact with platforms and the environment they propagate through. Today, reconfigurability is essentially located at the extremity of the problem, that is, in software modules (the processing) and the hardware they control (mainly the antenna and RF front end), all depicted in Fig. 1.1.

This book describes how metasurfaces can add extra layers of dynamism to the core of the EW problem. By tailoring electromagnetic waves, metasurfaces can make platforms and even the environment itself reconfigurable. The reader does not need to master advanced concepts on electromagnetics, whose pertinent points we will concisely explain. However, we expect some familiarity with the radio and EW terminologies. As such, the reader will easily understand the possibilities and challenges related to the application of metasurfaces in EW. In some

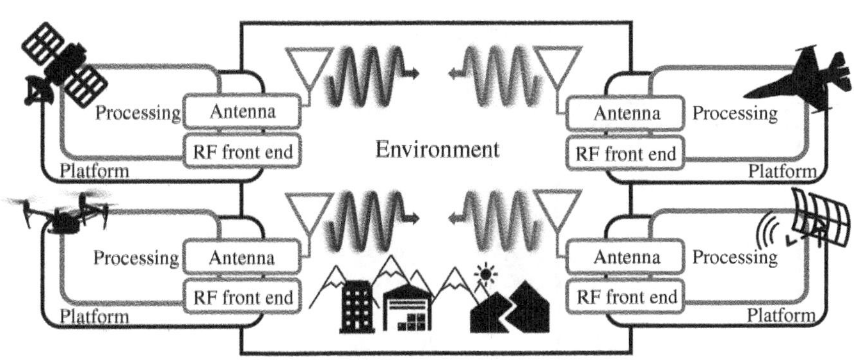

Fig. 1.1 Modern EW: today, reconfigurability is located at the processing, RF-front-end, and antenna levels

Metasurface-Driven Electronic Warfare, First Edition. Rafael Gonçalves Licursi de Mello.

Published 2025 by John Wiley & Sons, Inc.
Companion website: www.wiley.com/go/licursidemello/metasurfacedrivenelectronicwarfare1

years, cities will deploy metasurfaces everywhere, and the ordinary person will get used to them. Unavoidably, defense applications will also exploit them extensively.

Metasurfaces are the convenient, two-dimensional version of metamaterials, which in turn may receive the following, classical definition:

> Metamaterials are man-made materials with physical properties superior to natural materials. 'Meta' is a Greek prefix meaning 'beyond' or 'a level above', as in metaphysical or metalogical.
>
> —André de Lustrac, Emeritus Professor at Université Paris Sud [1].

These artificially engineered surfaces are made of tiny unit cells that one can implement by etching, drilling, and deploying electronic components over dielectric laminates, rather like conventional printed circuit boards. The appropriate design of the unit cells provides an arbitrary electromagnetic response. Figure 1.2 exemplifies how a metasurface looks like by illustrating a structure for the absorption of waves at low frequencies [2]. Such metasurface is particularly interesting for stealth applications, and may present an ultra-thin conformal shape, as shown in the upper inset.

Although researchers have thoroughly studied metasurfaces in the last decades [3], to date few products have reached the market, and the maturity level of the

Fig. 1.2 Example of metasurface absorber: the upper inset highlights the conformal shape, and the lower one is the unit cell. "Reproduced from Khuyen et al. [2]/Springer Nature/CC BY 4.0."

technology is not among the highest. The main reason is that the arrival of metamaterials and metasurfaces divided researchers into spirited discussions about some physics fundamentals considered untouchable so far, such as the law of refraction [4, 5]. Indeed, we had to review this and other concepts [6].

Despite the initial controversies, today, academia acknowledges the exotic properties of metasurfaces. In a civilian context, the community regards today as the time when the technology will start to become popular and commercially successful [7, 8]. In particular, regulatory bodies recently elected metasurfaces as a central element for the physical layer of the next generation of mobile communications, the 6G [9]. In Defense, however, if many isolated papers are available in the open literature, the application of these academic ideas is still unclear for many.

Metasurface-driven electronic warfare fills this gap, connecting the dots between metasurfaces and EW. The book aims at clarifying to students, engineers, researchers, and EW practitioners how metasurfaces work, what they can do, and how they would perform in EW problems. The book covers not only the so-called metasurface-based stealth technology, but also applications in Electronic Support (ES), Electronic Attack (EA), and Electronic Protection (EP). Moreover, it discusses the use of metasurfaces in radars.

Metasurfaces are promoting a true revolution in electromagnetics. The community have already envisioned them as a way of creating smart electromagnetic environments [10–14]. Of course, malicious applications are also showing up [15, 16]. In parallel, in the EW context, one-on-one combats are increasingly uncommon, with current trends reinforcing the idea of battle between EW networks [17–23]. Hence, this book also discusses the perspectives on the use of metasurface-driven drone swarms.

Most of the applications studied here are immediate. However, some of them envision the near future because metasurfaces also have limitations that the community must address yet, depending on the application. Among them, they are inherently frequency-dependent, and their electromagnetic response typically varies with the angle of incidence and polarization of waves. Besides, although metasurfaces can present low-profile and conformal shape, some applications are extremely demanding in this matter. Thus, this book also makes an analysis of the challenges to solve before metasurfaces reach the theater of operation.

1.1 From Static Radios to a Metasurface-Driven EW

Unlike metasurfaces, radio technology has already reached a remarkable maturity level, with quite elaborated products in the market for both civilian and military purposes. Nowadays, digital radios covering instantaneous bandwidths (BW) of

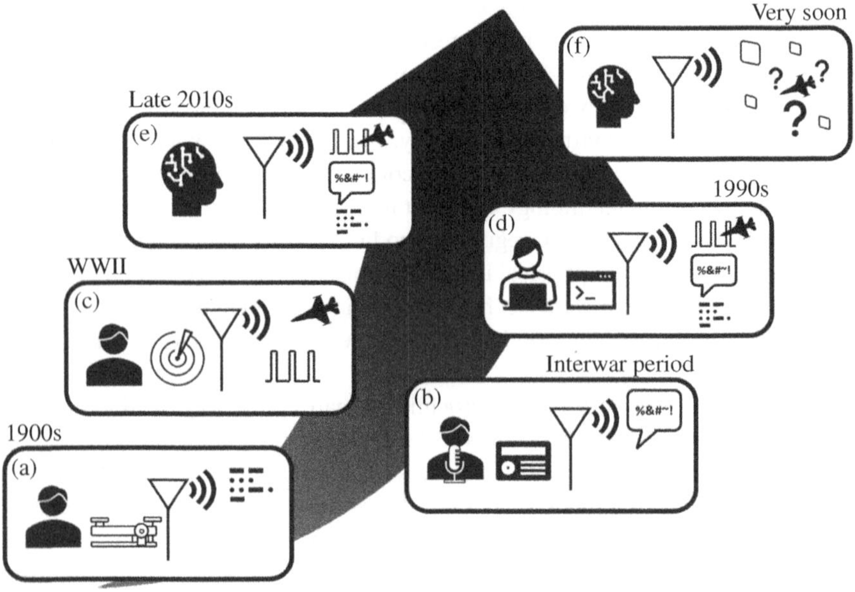

Fig. 1.3 Radio evolution from an EW perspective: (a) static radios and Morse codes; (b) tunable radios; (c) radars and EW; (d) software-defined radio; (e) cognitive EW; (f) metasurface-driven EW

1.0 GHz and 16 bits of quantization levels no longer shock anyone. Flexible circuits enable fast agility in the tuning frequency, a wide control of gain and sampling rate, and even an onboard signal processing. The most interesting, everything ruled by software, allowing for the implementation of cognitive systems [24]. But in the past things were not like that. Figure 1.3 depicts the evolution of radio since its creation.

1.1.1 Early Days

The first functional radios made use of static circuits. Their application was the exchange of Morse code pulses in commercial and diplomatic activities, as in Fig. 1.3a. They quickly attracted many amateurs who started intercepting whatever signals they could. As such, they had already laid the foundations for a wide, military activity of signals intelligence (SIGINT) by the beginning of the First World War which recruited many of these amateurs [25]. The First World War made clear the military significance of radio. While some Generals wanted to fight the traditional war, others did not leave the advantages of the new technology aside. The interception and location of radio signals are behind the unexpected outcomes of several battles that changed the course of that conflict [26].

When the Second World War (WWII) broke out, radios were much more elaborated. Continuous wave transmitters and crystal receivers allowed the exchange of amplitude-modulated (AM) sound messages at frequencies tuned through a knob, as in Fig. 1.3b. The development of automatic gain controllers (AGC) added another degree of reconfigurability [27]. The arrival of the magnetron [28] allowed the use of higher frequencies, unlocking a series of technologies that made possible applications other than communications, like the radar, depicted in Fig. 1.3c.

By the time the first radars entered into service, it was clear that radio technology would shape the way humans make war. Many consider that moment as the birth of EW [29]. The beauty of such activity begins at the very first moment one decides to ensure to their forces the safe usage of the spectrum while denying it to adversaries, ultimately taking advantages in the process. To this end, researchers developed a series of systems to detect and counter radars. Further, they introduced a new branch of SIGINT, called Electronic Intelligent (ELINT), to build libraries to classify and identify radars through their signals [30]. As the spectrum became denser with many radar signals together, advanced pulse deinterleaving [31] and jamming techniques [32] became increasingly required. Consequently, a cat-and-mouse game where opponents had to adopt new systems and actions to counter one another was established. Fig. 1.4a depicts this dispute, still in place

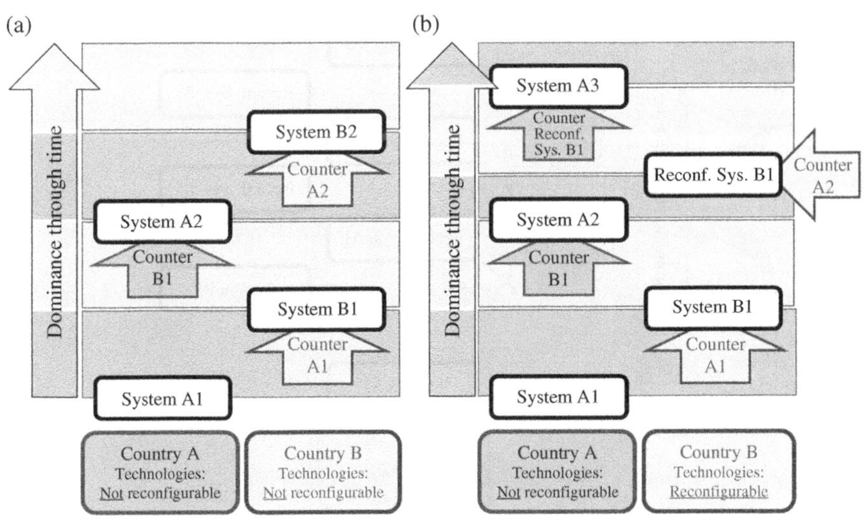

Fig. 1.4 Cat-and-mouse EW game: (a) none of the countries use reconfigurability; (b) country B presents a longer dominance through time due to reconfigurability

today. In the figure, Country A achieves dominance of the electromagnetic spectrum by deploying System A1 in the theater of operation. After a while, Country B reacts by releasing System B1, aiming at countering System A1 and taking dominance back. Country A must develop a new system to counter System B1. This cycle repeats in the cat-and-mouse EW game, and dominance passes from one country to another during the course of the conflict.

The cat-and-mouse game made reconfigurability an invaluable EW asset because simple reconfigurations of an existing system can counter many new adversary systems. Fig. 1.4b illustrates this situation, where System B1 optimizes the dominance through time of Country B by means of reconfigurability.

1.1.2 The Modern Era

After WWII, radars with frequency hopping and reconfigurable baseband signals were developed, demanding increasingly elaborated counter systems. Back in the day, engineers implemented a rather limited reconfigurability with analog circuits. In that context, the nature of EW, based on measures and countermeasures, made the rate of innovation a crucial factor. Rate of innovation is how fast one can deploy new systems or reconfigure existing ones. Figure 1.5 represents the cat-and-mouse EW game when both countries have reconfigurable technologies. In

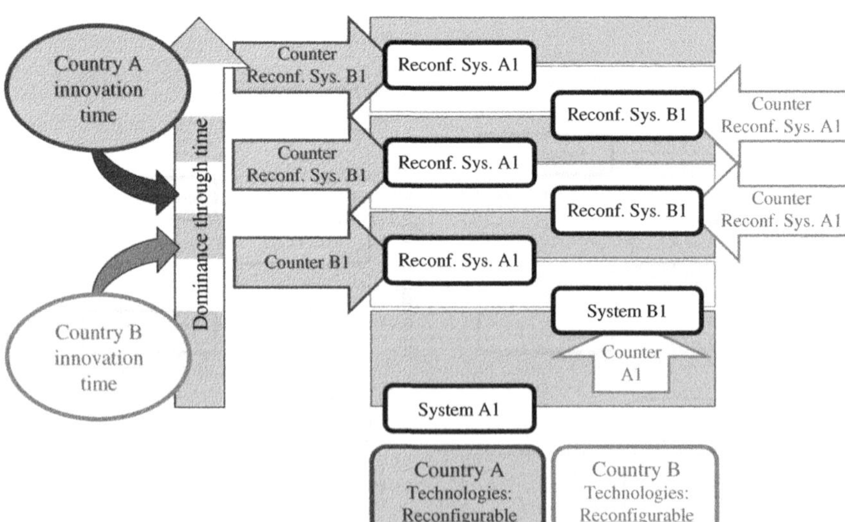

Fig. 1.5 Cat-and-mouse EW game when both countries have reconfigurable technologies: the innovation rate defines which country will own dominance through a longer time

this case, the country that will hold dominance the longest is the one that can reconfigure its systems the fastest for effective counters.

Parallel to the spread of analog reconfigurability, the arrival of the transistor after WWII [33] and significant advances in computer theory considerably boosted the development of digital technology [34]. By the 1980s, a single, tiny integrated circuit one could pick up at the fingertips could accommodate tens of thousands of reconfigurable, logic circuits. The development of powerful microcontrollers (MCU) and analog-to-digital (ADC) and digital-to-analog converters (DAC) for use with radiofrequency (RF) front ends unlocked the digital radio and computer-based techniques. In ES, this approach showed great potential, enabling digital deinterleaving techniques both by the clustering of radar pulses [35] and the recognition of time-of-arrival (TOA) patterns [36]. These innovations also allowed the development of digital radiofrequency memories (DRFM) [37]. The latter represented a major advance in terms of EA, mainly in front of newly developed low-probability-of-intercept (LPI) radars, which could transmit coded pulses and filter the target skin returns accordingly [38].

In the 1990s, J. Mitola proposed a couple of groundbreaking concepts. In the first, flexible hardware allied to digital signal processors (DSP) allows software to define functions traditionally carried out by analog circuitry [39]. An ideal software-defined radio (SDR) converts signals to the digital domain right after reception, opening doors for a code to perform the whole signal processing. As such, the SDR can perform a specific function for a while (for example, communications) soon after readapting to work in another role (such as a radar), as illustrated in Fig. 1.3d. In practice, the SDR hardware has constraints in terms of operating frequency range, BW, antenna performance, and so forth, which limits the applications.

Mitola also proposed the concept of cognitive radio [40]. The implementation of radios in the software domain allows an artificial intelligence (AI) model to observe the environment and learn from experience. Next, the model can orient, plan, decide, and act to optimize the allocation of radio resources and users in the spectrum. Reconfigurability reaches the network as a whole. In 2006, researchers extended this concept to radars [41, 42], where AI optimizes transmitted signals and filters, pursuing specific performances. Similar to what happened when radars became reconfigurable, their entry in the cognitive world dragged all EW with them.

In the cognitive EW, represented in Fig. 1.3e, systems interact with the environment and learn from experience. An in-mission learning can provide real-time reconfigurability, pushing to the limit the innovation rate and dominance of the electromagnetic spectrum. However, the community must overcome some challenges to consolidate cognitive EW in the theater of operations. Among them, the access to high-quality data, how much of the environment is observable, and the management of computational resources [43].

The transition of much of the radio architecture from the hardware to the software domain represented a watershed in EW, making many legacy systems receive an updated, digital version with extended capabilities [44–47]. Recently, reconfigurability achieved levels previously unimaginable. By the mid-2010s, a cognitive radar could tune its carrier frequency in a range of 4.0 GHz, vary its BW from 10.0 kHz to 1.0 GHz, use a pulse repetition frequency (PRF) of up to 30 kHz, and change its waveform in a pulse-to-pulse basis [48]. The wideness of these parameter ranges can make many signals overlap in ELINT libraries, leading to countless ambiguities in traditional ES tasks for radar classification and identification. Thus, strong research on AI exists today to accomplish these tasks based on complex signal parameters instead of traditional libraries [49–52]. The aforesaid versatility also allows radars to switch between multiple modes according to the environment responses and a task schedule. AI can also conduct these radars optimally from the most to the least dangerous modes by EA means [53]. In this context, one can use RF systems on chips (RFSoC) to implement powerful, digital EW systems based on field-programmable gate arrays (FPGA) with several gigahertz of BW and up to 16 bits of quantization levels [54–57]. A single RFSoC contains all the necessary digital circuits to this end, which considerably reduces the development time and enhances the innovation rate even more.

The extensive use of software also facilitated the integration of EW in the context of Joint All Domain Command and Control (JADC2) [58] and Information Warfare [59]. Modern conflicts take place in a Combat Cloud where Land, Sea, Air, Space, and Cyber meet [60]. There, EW networks allocate resources to enable efficient command, control, communications, computers, intelligence, surveillance, target acquisition, and reconnaissance (C4ISTAR) in a scenario populated by both manned and unmanned platforms [17]. From the radar perspective, the exchange and processing of information gathered by distributed networks improve robustness to traditional jammers [18–21]. On the other side, jammer networks equipped with modern DRFMs carry out resource allocation tasks in terms of available power, antenna beams, and processing capabilities. As such, the number of jammed radars and the communication therebetween is optimized, as well as between unmanned platform networks [61–67].

1.1.3 To the Future

So attractive these software capabilities are, the fact that radio systems are relentlessly constrained by hardware and the physical limitations imposed by electromagnetics must not be ignored. Here, metasurface technology becomes an invaluable asset. In the metasurface-driven EW, represented in Fig. 1.3f, hardware components receive extended capabilities. Further, the arbitrary tailoring of electromagnetic waves adds an extra reconfigurability layer in the physics level.

An analysis of recent conflicts, in particular, the wars of Nagorno-Karabakh, Ukraine, and Gaza, indicates a deep change in the character of warfare. The main trends suggest that lightweight, cheap, highly specialized, autonomous drone swarms will lead to a distributed EW instead of its concentration on major platforms [68–73]. In the news, great expectations also exist for the use of AI as an EP measure in lightweight drones [74]. Of course, new measures to counter these new threats, like directed-energy weapons, are on the road. Originally envisioned to physically damage the circuits of adversary radars and communication systems [75–78], directed-energy weapons are a strong candidate to counter drone swarms [79–83].

Metasurface's features of simplicity, lightweight, and low costs are an enabler for many functionalities in this new EW that forms. Their impact in the EW context is not only the insertion of new capabilities in the scenario. Most importantly, new phenomena and behaviors in the theater of operation will require EW operators and AI models to retrain – or be replaced.

1.2 Can We Actually Rely on Electromagnetic Waves?

Homing pigeons were once responsible for the noble task of transmitting messages in a conflict. Despite the outstanding ability to find their way back home over extremely long distances, these birds could not always succeed in their missions. Sometimes, a tragedy could happen in the trip, like the interference of a natural predator or even the opponent. For this reason, one could not be sure that the message would reach its destination, as in Fig. 1.6a.

The arrival of radio changed this reality. The idea was that, besides transmitting messages at the speed of light, electromagnetic waves could do so reliably. A technology fully supported by science would not fail. Once one had the right equipment and conditions, radio waves would propagate through the communication channel and do the job, as in Fig. 1.6b, whether the adversary wanted it or not. A special case would be if the radio receivers were under an EA measure. However, in this case, electromagnetic waves would have done their jobs, carrying the message through the channel, even though the receivers experienced interference.

In this context, metasurfaces represent a paradigm shift because they can arbitrarily shape electromagnetic waves. They can even tailor the channel itself. Metasurfaces can absorb [84, 85] and control the polarization [86, 87] of waves. They allow an asymmetric transmission [88–90], imposing different effects on the waves depending on the direction they are traversed. They can steer [6, 91], focus [92, 93], and even shape [6, 91–94] electromagnetic beams. Consequently, if metasurfaces

Fig. 1.6 Paradigm shift in EW: (a) low reliability with homing pigeons; (b) increased reliability with radios; (c) low reliability again with the introduction of metasurfaces

are deployed in the theater of operation, as in Fig. 1.6c, waves will not necessarily arrive where opponents intend them to. In the metasurface-driven EW scenario, adversaries cannot be sure anymore whether an electromagnetic beam carrying a message will arrive at its destination or not. Also, the certainty that a radar pulse will hit a target is not absolute anymore.

1.3 The (Super) Natural Powers of Metasurfaces

Occasionally, news about metamaterials and metasurfaces gives the impression that they have supernatural powers, including the ability to make an object invisible. This idea comes from the fact that metasurfaces can provide properties not readily found in nature. However, their capabilities are justified by physics, in spite of the fact that they have tricked even for the most renowned researchers at a first moment. Chapter 3 will detail the operation principle of metamaterials and metasurfaces, and this section provides next a brief explanation.

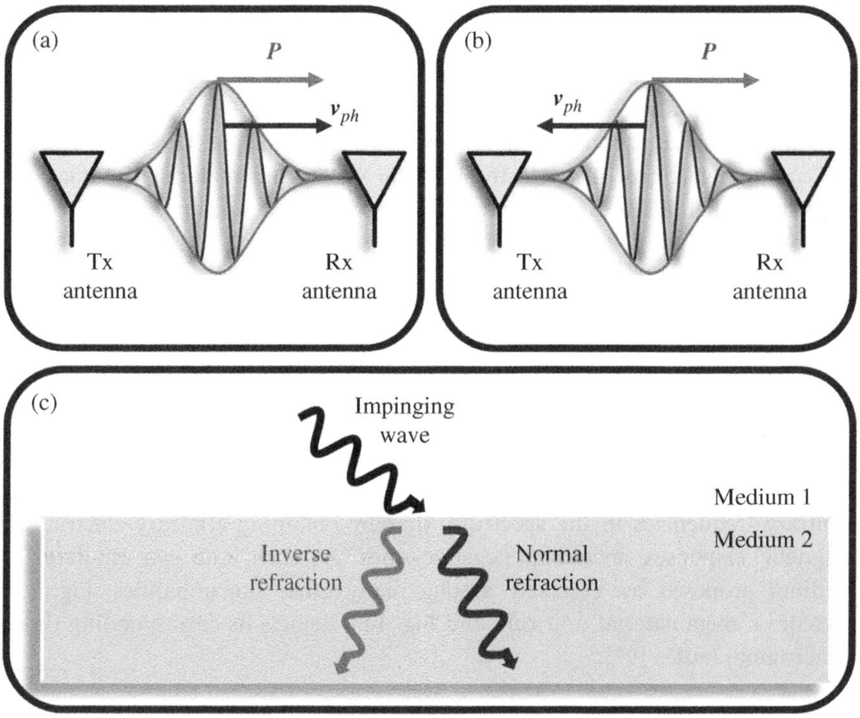

Fig. 1.7 Unusual phenomena in Veselago's medium: (a) example of wave propagation in conventional media; (b) propagation in Veselago's medium; (c) inverse refraction

In 1968, V. Veselago [95] proposed a deliberation on the propagation of waves in a hypothetical medium whose both electric permittivity and magnetic permeability would be negative. Applying such quantities to time-harmonic Maxwell's equations [96] would result in a series of unusual phenomena, such as

- the existence of backward waves, whose phase velocity $\boldsymbol{v}_{ph}$, unlike that of conventional media (Fig. 1.7a), would be in the opposite direction of the Poynting vector $\boldsymbol{P}$ (Fig. 1.7b);
- the amplification of evanescent waves instead of their usual vanishing as the observation point distances from a reference point;
- the refraction in an interface with a conventional medium would occur in an inverted direction, as indicated in Fig. 1.7c.

At first glance, some researchers concluded that the implementation of such hypothetical materials would be impossible since they would break some physics fundamentals. From their understanding, the existence of backward waves would imply that an envelope of energy would propagate from a transmitting (Tx) antenna to a receiving (Rx) antenna, whereas the wavefronts within the envelope propagate from the Rx to the Tx antenna, as in Fig. 1.7b, calling into question the principle of causality. Also, the amplification of evanescent waves in a passive medium cast doubt upon the energy conservation principle. However, through the comprehension of how electromagnetic waves and materials interact, the mechanisms behind the unusual properties become clear, and one can realize that Veselago's medium does not break any of these principles.

1.3.1 Metamaterials: A Proof of Concept

Metamaterials are three-dimensional lattices composed of tiny unit cells designed to provide a specific electromagnetic response. The unit cell subwavelength dimensions allow us to regard the structure as an effectively continuous medium. By inserting the proper metal inclusions, one can make the unit cells resonate at arbitrary frequencies in the spectrum, thereby obtaining arbitrary electric and magnetic responses, including negative ones. As such, one can emulate the medium proposed by Veselago, among many other functionalities. Fig. 1.8a sketches a metamaterial unit cell, and Fig. 1.8b depicts its corresponding three-dimensional lattice [97].

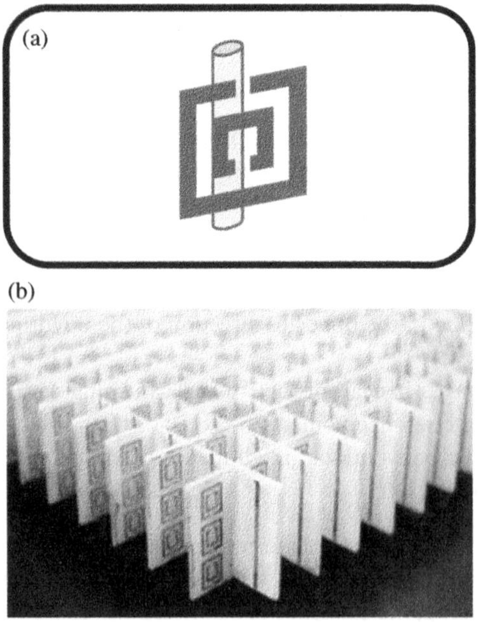

Fig. 1.8 Metamaterial providing simultaneous negative permeability and permittivity: (a) unit cell composed of split-ring resonators and wire; (b) composite medium fabricated with the unit cell of (a). "Reproduced with permission from Houck et al. [97]/American Physical Society."

By the late 1990s, that is, 30 years after Veselago's work, J.B. Pendry et al. implemented a structure composed of thin wires arranged in a cubic lattice that provided a negative electric permittivity at arbitrary microwave frequencies [98]. Right after, Pendry et al. demonstrated another cubic lattice but now composed of split-ring resonators to provide a negative magnetic permeability [99].

In 2000, D. R. Smith et al. succeed in arranging in a same lattice wires and split-ring resonators to obtain simultaneous negative electric permittivity and magnetic permeability at microwave frequencies [100]. A similar structure (the one of Fig. 1.8b) experimentally validated that metamaterials can indeed perform a negative refraction [97]. Besides, experiments also confirmed the existence of backward waves [101], as well as the amplification of evanescent waves [102] and the so-called invisibility cloak [103].

1.3.2 The Need for a Two-Dimensional Version

The first works on metamaterials [97–103] provided evidence that such structures could really offer arbitrary electromagnetic properties albeit the controversies in academia. However, the first useful applications initially were not clear even to the leading researchers in the field:

> Somewhere along the way John (Pendry) decided 'hey, we can make an invisibility cloak out of this'. (...) Once it was demonstrated that we could make something that didn't really exist in nature it kind of took off, but people really didn't know what to make of it at first.
>
> —David R. Smith, Professor, Duke University [104].

And, indeed, a series of questions associated to the operation of metamaterials limits their use in real applications. The three-dimensional lattice makes metamaterials bulky, costly, and hard to fabricate, also leading to reproducibility challenges. Furthermore, their resonance-based principle of operation causes an inherently frequency-dependent behavior. For instance, the reduction of scattering of a metal cylinder using a multilayer cloak presented a $BW = 2.4\%$ in the best of the cases [105].

One way of circumventing the frequency-dependence problem would exploit reconfigurability. As such, metamaterials could operate in a wide frequency range, even if not instantaneously. However, the possible reconfigurability methods impose limitations since a complex, if not unfeasible, feeding system should individually address each of the unit cells in the three-dimensional lattice.

The implementation of metamaterials in a two-dimensional disposition, when they are called metasurfaces, yields a series of advantages. Figure 1.9 illustrates the difference in the geometrical disposition of metamaterials and metasurfaces.

out of band) [110]. We can perfectly apply the concept of different behaviors depending on the operating frequency to metasurfaces. For instance, rasorbers allow the transmission of most of the energy in one frequency band and absorb it out of the band [111]. Other metasurfaces can control the transmission phase in one band without changing it in others [112, 113], allowing the steering of the antenna beam in one band without bothering the radiation patterns in others.

In metasurface absorbers, lossy materials or components are placed in strategic positions within the unit cells to dissipate energy guided thereto [84, 85]. The straightforward way of using such metasurfaces in EW is to coat the most reflective parts of a platform therewith for stealth purposes. However, one can also apply metasurface absorbers in the design of antennas, considerably reducing their radar cross-section (RCS). Besides, they can also perform a passive blinking jamming by exploiting reconfigurability [114]. Earlier, Fig. 1.2 depicted a metasurface absorber [2].

Reflective metasurfaces usually present a sheet made of a good electrical conductor whose behavior is close to the one of a perfect electric conductor (PEC). The metal sheet has the function of a ground plane, which does not allow the transmission of impinging waves. The unit cells carry out specific functions in the wave reflection, such as converting the polarization [115, 116] or imposing a specific phase shift [91, 117]. One can use reflective metasurfaces in stealth applications by scattering waves at multiple directions, polarizations, and frequencies or by making use of reconfigurability to active-cancel echoes of radar signals. One can also use them to estimate the direction of arrival (DOA) of signals in ES applications with space-modulated unit cells [118]. Besides, they allow the design of multiband reflectors for antenna applications [119–121], enabling a high gain covering different parts of the spectrum. Moreover, reconfigurable reflective metasurfaces can work in EA applications like blinking jamming [114] or cross-polarization jamming when associated with power amplifiers [122, 123]. Figure 1.11 exemplifies a reflective metasurface.

Transmissive metasurfaces usually do not have a ground plane. Their unit cells perform specific functions in the transmission of waves, such as converting the polarization [89, 90], imposing a specific phase shift [112, 124], or performing an asymmetric transmission, when different effects are imposed on waves depending on the direction they traverse the surface [88–90]. The asymmetric transmission allows waves to enter antenna radomes but not to leak back from it, which is useful to reduce the RCS of Rx antennas. Reconfigurable transmissive metasurfaces can perform the role of dynamic lenses for antennas, by steering [6], focusing [92, 93], and shaping [6, 92–94] electromagnetic beams. For this reason, they can work as an alternative to phased-array antennas. Lastly, one can exploit reconfigurable polarization-conversion transmissive metasurfaces both in polarimetric

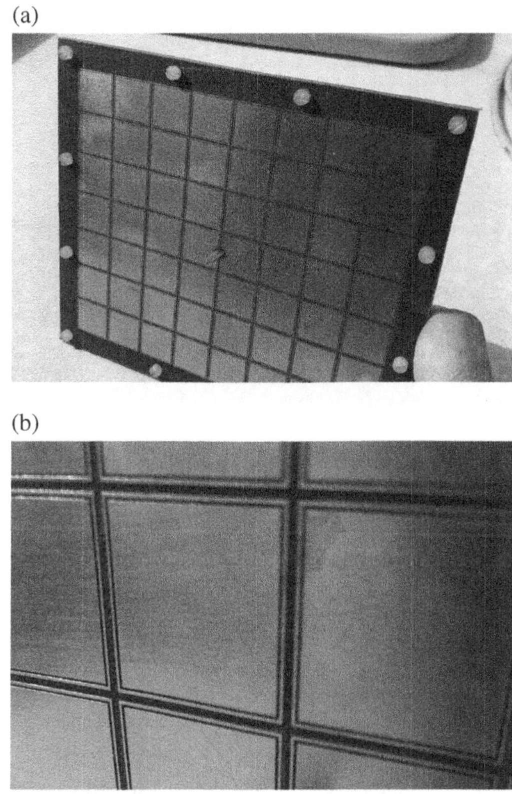

Fig. 1.11 Example of reflective metasurface: (a) full structure; (b) detailed unit cells

radars and in EP measures that detect the presence of cross-polarization jamming, avoiding thereby the use of multiple receivers, and reducing the system complexity and cost. Figure 1.12 exemplifies a transmissive metasurface [125].

The advent of metasurfaces also made it possible to glimpse a new concept in the physical layer of wireless systems. Reconfigurable intelligent surfaces (RIS) are an abstract concept where a surface may create additional, controllable propagation paths to optimize a previously uncontrollable channel [10–14]. The aforementioned types of metasurfaces can populate an EW environment to tailor the propagation of waves therein in a dynamic manner. In this way, they can control the access to the electromagnetic spectrum.

Figure 1.13 relates the three major groups of metasurfaces and their possible applications in EW. The applications listed here are clear for the current scenario and metasurface technology state. In the future, other possibilities will certainly arise.

Fig. 1.12 Example of transmissive metasurface; "Reproduced from Li [125]/IntechOpen/CC BY 3.0."

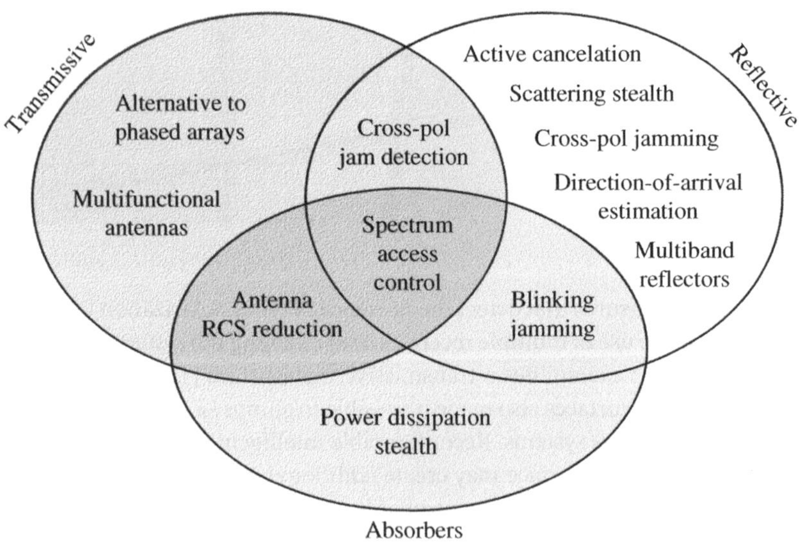

Fig. 1.13 Possible EW applications of the three major types of metasurfaces

1.5 Reconfigurability at the Tactical Level

In Section 1.1.1, we discussed how reconfigurability can counter new opponents' systems and actions. Reconfigurability, in that case, is an invaluable asset at the strategic level. However, one can also make use of reconfigurability at the tactical level to generate behaviors never seen before whereby tricking operators, tracking systems, and even cutting-edge AI models.

We provide next an easy-understanding case. Let us exploit a simple metasurface capability: the tunable absorber. Suppose that tunable metasurfaces that can absorb waves at specific, desired frequencies coat an aircraft. In this way, the metasurfaces are able to turn the aircraft skin returns "on and off" from the perspective of a radar by alternately tuning and detuning the absorption at the radar's operating frequency. At this point, the reader may wonder why one would do that rather than remaining totally invisible to the radar. And here is the answer: to create a totally new behavior in a scenario composed of diverse platforms.

From an EW point of view, radar signals can provide much information about their operation, but the particular logic behind radar tracking systems, formulated by its designer, is usually unknown [126]. Nonetheless, tracking rules (either performed by a mechanical tracking system or an AI model – or the operator's eyes) usually consider time information because to each platform a reasonable range of velocities can be associated. Playing with the skin returns of diverse platforms denies such information to the radar and may thus create a real trouble in the tracking dynamics.

Today, there seem to be countless possibilities for exploiting reconfigurability at the tactical level. Active-cancelation stealth and blinking jamming are just examples. Further, using multiple unmanned platforms covered by metasurfaces allows for a huge number of combinations, opening doors for many ideas. In summary, much room for innovation exists in this regard.

1.6 Book Presentation and Organization

We made an active effort throughout the whole writing of this book to make it accessible to non-electromagnetics professionals – keeping at the same time the level of detail the experts are interested in.

After providing a good theoretical basis for those who need it, we start from the simplest and most intuitive applications to the most complex ones. Before approaching the use of metasurfaces thereon, the book will firstly explain each application in its traditional form. Further, EW scenarios composed of radars illuminating aircraft will be the main exemplification instrument in this book. In this way, we avoid defining many different situations, which would actually only make the reader tired (and perhaps confused). Nevertheless, the core of the concepts presents a high degree of generality, and the reader requires little effort to extend the examples to ships, satellites, missiles, and lightweight drones using comms or lidars either in the VHF, UHF, infrared or visible ranges, among many others.

We organized the remainder of this book not only to provide the reader with a complete view of applications of metasurfaces in the current EW scenario but also to allow a glimpse of future possibilities:

- *Chapter 2* provides a pertinent review on electromagnetics for readers lacking a contact with it for a long time (or having the very first right now); this is not the most enticing introduction for the metasurface expert. Chapter 4 is the place for them to start;
- *Chapter 3* provides a straightforward explanation of the working principle of metasurfaces; it also makes it explicit limitations that can become challenges in some EW applications;
- *Chapter 4* studies the metasurface-based passive stealth application with different principles like wave absorption and wave scattering in space, frequency, and polarization;
- *Chapter 5* proposes an active-cancelation stealth using power-amplifying metasurfaces instead of DRFMs;
- *Chapter 6* approaches ES applications, such as the improvement of the antenna gain and the reduction of its thickness with wideband and multiband metasurface reflectors;
- *Chapter 7* explores EA applications, like cross-polarization and blinking jamming, using passive and active metasurfaces; it also envisions a generalized EA measure;
- *Chapter 8* studies EP applications against directed-energy weapons and cross-polarization jamming; it also discusses the application of metasurfaces in polarimetric radars or as an alternative to phased-array antennas; last, it covers the design of metasurface-based multifunctional antennas;
- *Chapter 9* is the most conceptual chapter of this book; it proposes metasurface solutions enabling the use of drone swarms in the implementation of distributed EW networks; in this way, it envisions ES- and EA-capable drone swarms; last, it discusses the tailoring of electromagnetic waves in urban and open environments for the ultimate control of the spectrum access;
- *Chapter 10* concludes the book.

1.7 Final Remarks of the Chapter

In the last two decades, researchers have thoroughly studied metasurfaces due to the many functionalities they can provide for the interaction with electromagnetic waves. Such functionalities match current EW trends, providing a whole plethora of new possibilities to the EW scenario, which in turn is becoming each day more and more dynamic.

The latest advances in EW consisted mainly of achieving reconfigurability through software means. Metasurfaces can add an extra layer of dynamism by extending hardware capabilities and by unlocking reconfigurability in the propagation of electromagnetic waves. However, their application in practical EW problems is until unclear for many.

In the remaining of this book, we will address this gap. We will take into account all the advantages metasurfaces can provide, as well as their limitations and the difficulty in connecting theory and practice in such a complex problem.

This book will show that the metasurface-driven EW is the continuation of the cat-and-mouse EW game in its purest essence, and in a world where an unparalleled innovation rate is not asking permission to impose itself:

> To win, you must decide what you want your tomorrow to be, and then make it happen faster than the rate of change in your competitive environment.
>
> —John A. Warden III, air power theorist.

References

1 Lustrac, A. (2016). Métamatériaux électromagnétiques et optiques. *Techniques de l'Ingénieur* AF3713 (V2): 1–21. https://doi.org/10.51257/a-v2-af3713

2 Khuyen, B.X., Tung, B.S., Kim, Y.J. et al. (2018). Ultra-subwavelength thickness for dual/triple-band metamaterial absorber at very low frequency. *Scientific Reports* 8 (11632): 1–9. https://doi.org/10.1038/s41598-018-29896-4

3 Bukhari, S.S., Vardaxoglou, J., Whittow, W. (2019). A metasurfaces review: definitions and applications. *MDPI Applied Sciences* 9 (2727): 1–14. https://doi.org/10.3390/app9132727

4 Valanju, P.M., Walser, R.M., Valanju, A.P. (2002). Wave refraction in negative-index media: always positive and very inhomogeneous. *Physics Review Letters* 88, (187401): 1–4. https://doi.org/10.1103/PhysRevLett.88.187401

5 Munk, B.A. (2009). *Metamaterials: critique and alternatives.* Hoboken, NJ: Wiley.

6 Yu, N., Genevet, P., Kats, M.A. et al. (2011). Light propagation with phase discontinuities: generalized laws of reflection and refraction. *Science* 334 (6054): 333–337. https://doi.org/10.1126/science.1210713

7 Pile, D. (2023). Metamaterials for the masses. *Nature Photonics* 17: 2–3. https://doi.org/10.1038/s41566-022-01124-6

8 Gu, T., Kim, H.J., Rivero-Baleine, C. et al. (2023). Reconfigurable metasurfaces towards commercial success. *Nature Photonics* 17: 48–58. https://doi.org/10.1038/s41566-022-01099-4

9 Wang, C.-X., You, X., Gao, X. et al. (2023). On the road to 6G: visions, requirements, key technologies, and testbeds. *IEEE Communications Surveys & Tutorials* 25 (2): 905–974. https://doi.org/10.1109/COMST.2023.3249835

10 Barbuto, M., Zarghani, Z.H., Longhi, M. et al. (2022). Metasurfaces 3.0: a new paradigm for enabling smart electromagnetic environments. *IEEE Transactions on Antennas and Propagation* 70 (10): 8883–8897. https://doi.org/10.1109/TAP.2021.3130153

11 Martini, E., Maci, S. (2022). Theory, analysis, and design of metasurfaces for smart radio environments. *Proceedings of the IEEE* 110 (9): 1227–1243. https://doi.org/10.1109/JPROC.2022.3171921

12 Gong, S., Lu, X., Hoang, D.T. et al. (2020). Toward smart wireless communications via intelligent reflecting surfaces: a contemporary survey. *IEEE Communications Surveys & Tutorials* 22 (4): 2283–2314. https://doi.org/10.1109/COMST.2020.3004197

13 Di Renzo, M., Zappone, A., Debbah, M. et al. (2020). Smart radio environments empowered by reconfigurable intelligent surfaces: how it works, state of research, and the road ahead. *IEEE Journal on Selected Areas in Communications* 38 (11): 2450–2525. https://doi.org/10.1109/JSAC.2020.3007211

14 Björnson, E., Wymeersch, H., Matthiesen, B. et al. (2022). Reconfigurable intelligent surfaces: a signal processing perspective with wireless applications. *IEEE Signal Processing Magazine* 39 (2): 135–158. https://doi.org/10.1109/MSP.2021.3130549

15 Lin, Z., Niu, H., An, K. et al. (2024). Pain without gain: destructive beamforming from a malicious RIS perspective in IoT networks. *IEEE Internet of Things Journal* 11 (5): 7619–7629. https://doi.org/10.1109/JIOT.2023.3316830

16 Rivetti, S., Demir, O.T., Bjornson, E. et al. (2024). Malicious reconfigurable intelligent surfaces: how impactful can destructive beamforming be? *ArXiv (preprint)*. https://doi.org/10.48550/arXiv.2402.13351

17 Varghese, V., Nene, M.J. (2017). Battlefield-of-things and its implications in modern day battlefield. *Proceedings of the IEEE International Conference on Computational Intelligence and Computing Research (ICCIC)*, Coimbatore, India (14–16 December 2017), New York, US: IEEE. https://doi.org/10.1109/ICCIC.2017.8524515

18 Berry, P.E., Dahal, N., Venkataraman, K. (2022). On the design of an optimal coherent multistatic radar network configuration. *IET Radar, Sonar & Navigation* 16 (5): 869–884. https://doi.org/10.1049/rsn2.12226

19 Li, Z., Xie, J., Liu, W. et al. (2022). Transmit antenna selection and power allocation for joint multi-target localization and discrimination in MIMO radar with distributed antennas under deception jamming. *MDPI Remote Sensing* 14 (3904): 1–17. https://doi.org/10.3390/rs14163904

20 Shi, C., Wang, F., Sellathurai, M. et al. (2020). Low probability of intercept-based optimal power allocation scheme for an integrated multistatic radar and communication system. *IEEE Systems Journal* 14 (1): 983–994. https://doi.org/10.1109/JSYST.2019.2931754

21 Zhang, S., Zhou, Y., Zhang, L. et al. (2021). Target detection for multistatic radar in the presence of deception jamming. *IEEE Sensors Journal* 21 (6): 8130–8141. https://doi.org/10.1109/JSEN.2021.3050008

22 Zhang, P., Huang, Y., Jin, Z. (2023). An electronic jamming method based on a distributed information sharing mechanism. *MDPI Electronics* 12 (2130): 1–18. https://doi.org/10.3390/electronics12092130

23 Ren, Y., Li, B., Wang, H. et al. (2019). A novel cognitive jamming architecture for heterogeneous cognitive electronic warfare networks. *Information Science and Applications* 621: 97–104. https://doi.org/10.1007/978-981-15-1465-4_11

24 Manco, J., Dayoub, I., Nafkha, A. (2022). Spectrum sensing using software defined radio for cognitive radio networks: a survey. *IEEE Access* 10: 131887–131908. https://doi.org/10.1109/ACCESS.2022.3229739

25 West, N. (1986). *The SIGINT secrets: the signals intelligence war, 1900 to today*, Stroud, UK: The History Press.

26 Matthews, P. (2018). *SIGINT: the secret history of signals intelligence in the world wars*, Stroud, UK: The History Press.

27 Küpfmüller, K. (1928). On the dynamics of automatic gain controllers. *Elektrische Nachrichtentechnik* 5 (11): 459–467.

28 Okabe, K. (1927). Production of extra short electromagnetic waves by split-anode magnetron. *Proceedings of Imperial Academy* 3 (8): 514–515. https://doi.org/10.2183/pjab1912.3.514

29 Price, A. (2022). *Instruments of darkness: the history of electronic warfare 1939–1945*, Yorkshire, UK: Frontline Books.

30 Wiley, R.G. (2006). *ELINT: the interception and analysis of radar signals*. Norwood, MA: Artech House.

31 Adamy, D. (2001). *EW 101: a first course in electronic warfare*. Norwood, MA: Artech House, Section 5.2.4, pp. 80.

32 Lothes, R.N., Szymanski, M.B., Wiley, R.G. (1990). *Radar vulnerability to jamming*. Boston, MA: Artech House.

33 Riordan, M. (2004). The lost history of the transistor. *IEEE Spectrum* 41 (5): 44–49. https://doi.org/10.1109/MSPEC.2004.1296014

34 Turing, A.M. (1937). On computable numbers, with an application to the entscheidungsproblem. *Proceedings of the London Mathematical Society* 2–42 (1): 230–265. https://doi.org/10.1112/plms/s2-42.1.230

35 Mardia, H.K. (1988). Adaptive multidimensional clustering for ESM. *Proceedings of IEE Colloquium on Signal Processing for ESM Systems*, London, UK (26 April 1988), New York, US: IEEE.

36 Mardia, H.K. (1989). New techniques for the deinterleaving of repetitive sequences. *Proceedings of Institution of Electrical Engineering* 136 (4): 149–154. https://doi.org/10.1049/ip-f-2.1989.0025

37 Roome, S.J. (1990). Digital radio frequency memory. *Electronics & Communication Engineering Journal* 2 (4): 147–153. https://doi.org/10.1049/ecej:19900035

38 Schleher, D.C. (1985). Low probability of intercept radar. *Proceedings of the IEEE Radar Conference*, Washington, DC, USA (6–9 May 1985), New York, US: IEEE.

39 Mitola, J. (1992). Software radios-survey, critical evaluation and future directions. *Proceedings of NTC-92: National Telesystems Conference*, Washington, DC, USA (19–20 May 1992), New York, US: IEEE. https://doi.org/10.1109/NTC.1992.267870

40 Mitola, J., Maguire, G.Q. (1999). Cognitive radio: making software radios more personal. *IEEE Personal Communications* 6, (4): 13–18. https://doi.org/10.1109/98.788210

41 Haykin, S. (2006). Cognitive radar: a way of the future. *IEEE Signal Processing Magazine* 23 (1): 30–40. https://doi.org/10.1109/MSP.2006.1593335

42 Farina, A., De Maio, A., Haykin, S. (2017). *The impact of cognition on radar technology*. London, UK: Scitech Publishing.

43 Haigh, K.Z., Andrusenko, J. (2021). *Cognitive electronic warfare: an artificial intelligence approach*, Boston, MA: Artech House.

44 Neri, F. (2018). *Introduction to electronic defense systems*, 3rd ed. Boston, MA: Artech House.

45 Gonçalves Licursi de Mello, R., De Sousa, F.R., Junqueira, C. (2017). SDR-based radar-detectors embedded on tablet devices. *Proceedings of the SBMO/IEEE MTT-S International Microwave and Optoelectronics Conference (IMOC)*, Aguas de Lindoia, Brazil (27–30 August 2017), New York, US: IEEE. https://doi.org/10.1109/IMOC.2017.8121126

46 Gonçalves Licursi de Mello, R., De Sousa, F.R. (2018). Precise techniques to detect superimposed radar pulses on ESM systems. *IET Radar, Sonar & Navigation* 12 (7): 735–741. https://doi.org/10.1049/iet-rsn.2017.0563.

47 Gonçalves Licursi de Mello, R., De Sousa, F.R. (2019). Non-analytical direction-finding method as a key step in pursuing low size, weight, costs, and computational power in the deinterleaving of radar pulses. *IET Radar, Sonar & Navigation* 13 (11): 1876–1882. https://doi.org/10.1049/iet-rsn.2018.5011

48 Smith, G.E., Cammenga, Z., Mitchell, A. (2016). Experiments with cognitive radar. *IEEE Aerospace and Electronic Systems Magazine* 31 (12): 34–46. https://doi.org/10.1109/MAES.2016.150215

49 Qu, Z., Hou, C., Hou, C. et al. (2020). Radar signal intra-pulse modulation recognition based on convolutional neural network and deep Q-learning network. *IEEE Access* 8: 49125–49136. https://doi.org/10.1109/ACCESS.2020.2980363

50 Xiao, Z., Yan, Z. (2021). Radar emitter identification based on novel time-frequency spectrum and convolutional neural network. *IEEE Communications Letters* 25 (8): 2634–2638. https://doi.org/10.1109/LCOMM.2021.3084043

51 Bhatti, S.G., Bhatti, A.I. (2022). Radar signals intrapulse modulation recognition using phase-based STFT and BiLSTM. *IEEE Access* 10: 80184–80194. https://doi.org/10.1109/ACCESS.2022.3195273

52 Yuan, S., Xu, T., Zhang, M. (2022). Extraction and sequential recognition of MFR pulse groups in intercepted pulse trains. *IEEE Access* 10: 106988–106998. https://doi.org/10.1109/ACCESS.2022.3211938

53 Feng, L.-W., Liu, S.-T., Xu, H.-Z. (2022). Multifunctional radar cognitive jamming decision based on dueling double deep Q-network. *IEEE Access* 10: 112150–112157. https://doi.org/10.1109/ACCESS.2022.3214842

54 Feng, W., Wang, X., Shi, Z. et al. (2024). Research on high-speed detection method of radar signal based on RFSoC. *International Conference on Wireless Communications and Signal Processing (WCSP)*, Hangzhou, China (02–04 November 2023), New York, US: IEEE. https://doi.org/10.1109/WCSP58612.2023.10405269

55 Michalak, F., Zabołotny, W., Podkalicki, Ł. et al. (2022). Universal RFSoC-based signal recorder for radar applications. *23rd International Radar Symposium (IRS)*, Gdansk, Poland (12–14 September 2022), New York, US: IEEE. https://doi.org/10.23919/IRS54158.2022.9905032

56 Javaid, A., Ahmed, T., Ali, S. (2022). Performance evaluation of Xilinx Zynq UltraScale+ RFSoC device for low latency applications. *19th International Bhurban Conference on Applied Sciences and Technology (IBCAST)*, Islamabad, Pakistan (16–20 August 2022), New York, US: IEEE. https://doi.org/10.1109/IBCAST54850.2022.9990470

57 De Martino, A. (2018). *Introduction to modern EW systems*, 2nd ed. Norwood, MA: Artech House, Section 5.3, pp. 300.

58 Lingel, S., Hagen, J., Hastings, E. et al. (2020). Joint all domain command and control for modern warfare: an analytic framework for identifying and developing artificial intelligence applications. https://apps.dtic.mil/sti/citations/AD1106119 (accessed 21 Aug. 2024).

59 Poisel, R.A. (2013). *Information warfare and electronic warfare systems*. Boston, MA: Artech House.

60 Hess, J., Kiser, A., Bouhafa, E.M. et al. (2019). *The combat cloud: enabling multidomain command and control across the range of military operations*. Montgomery, AL: Air University Press.

61 Xing, H., Xing, Q., Wang, K. (2023). A joint allocation method of multi-jammer cooperative jamming resources based on suppression effectiveness. *MDPI Mathematics* 11 (826): 1–25. https://doi.org/10.3390/math11040826

62 Zheng, G., Na, S., Huang, T. et al. (2019). A barrage jamming strategy based on CRB maximization against distributed MIMO radar. *MDPI Sensors* 19 (2453): 1–13. https://doi.org/10.3390/s19112453

63 Wang, Q., Yan, X., Liu, F. et al. (2022). An optimal allocation method for cooperative jamming resources based on multi-agent genetic algorithm. *Proceedings of the 2nd International Conference on Frontiers of Electronics, Information and Computation Technologies (ICFEICT)*, sWuhan, China (19–21 August 2022), New York, US: IEEE. https://doi.org/10.1109/ICFEICT57213.2022.00056

64 Ke, M., Zhang, H., Wu, H. et al. (2021). Game model in multi-target jamming task allocation. *Proceedings of the 2nd International Conference on Electronics, Communications and Information Technology (CECIT)*, Sanya, China (27–29 December 2021), New York, US: IEEE. https://doi.org/10.1109/CECIT53797.2021.00114

65 Shao, S., Zhang, T., Ye, F. (2017). Jamming resource allocation aimed to data link based on simulant annealing algorithm. *Proceedings of the Progress in Electromagnetics Research Symposium - Fall (PIERS-FALL)*, Singapore, Singapore (19–22 November 2017), New York, US: IEEE. https://doi.org/10.1109/PIERS-FALL.2017.8293276

66 Sun, J., Zhang, D., Yi, W. (2021). Jamming resource allocation for multi-jammer cooperatively suppressing netted radar systems. *Proceedings of the CIE International Conference on Radar (Radar)*, Haikou, Hainan, China (15–19 December 2021), New York, US: IEEE. https://doi.org/10.1109/Radar53847.2021.10028342

67 Guo, J., Wang, H., Cheng, H. et al. (2022). Multiple methods based resource allocation optimization of jamming system for UAV clusters. *Proceedings of the IEEE International Conference on Unmanned Systems (ICUS)*, Guangzhou, China (28–30 October 2022), New York, US: IEEE. https://doi.org/10.1109/ICUS55513.2022.9986780

68 Dixon, R. (2020). Azerbaijan's drones owned the battlefield in Nagorno-Karabakh — and showed future of warfare. *The Washington Post* (11 November 2020). https://www.washingtonpost.com/world/europe/nagorno-karabkah-drones-azerbaijan-aremenia/2020/11/11/441bcbd2-193d-11eb-8bda-814ca56e138b_story.html (accessed 08 August 2023).

69 Hecht, E. (2022). Drones in the Nagorno-Karabakh war: analyzing the data. *Military Strategy Magazine* 7 (4): 31–37. https://www.militarystrategymagazine.com/article/drones-in-the-nagorno-karabakh-war-analyzing-the-data (accessed 08 August 2023).

70 Pietralunga, C. (2023). Russia and Ukraine take drone warfare to unprecedented scale. *Le Monde* (18 June 2023). https://www.lemonde.fr/en/international/article/2023/06/18/russia-and-ukraine-take-drone-warfare-to-unprecedented-scale_6033281_4.html (accessed 08 August 2023).

71 Cook, E. (2024). Ukraine drones learning to dodge Russian EW signal jammers. *Newsweek* (04 March 2024). https://www.newsweek.com/ukraine-fpv-drones-russia-electronic-warfare-1875577 (accessed 11 March 2024).

72 Tegler, E. (2023). A new micro kind of electronic warfare may be unfolding in Gaza. *Forbes* (03 November 2023). https://www.forbes.com/sites/erictegler/2023/11/03/a-new-micro-kind-of-electronic-warfare-may-be-unfolding-in-gaza (accessed 03 November 2023).

73 Watling, J., Reynolds, N. (2023). Stormbreak: fighting through Russian defences in Ukraine's 2023 offensive. *Special Report*. Royal United Services Institute. https://www.rusi.org/explore-our-research/publications/special-resources/stormbreak-fighting-through-russian-defences-ukraines-2023-offensive (accessed 10 October 2023).

74 Hudson, J., Khudov, K. (2023). The war in Ukraine is spurring a revolution in drone warfare using AI. *The Washington Post* (26 July 2023). https://www.washingtonpost.com/world/2023/07/26/drones-ai-ukraine-war-innovation (accessed 08 August 2023).

75 Weinberg, G.V. (2022). Performance prediction of directed energy weapons. *Progress in Electromagnetics Research Letters* 108: 79–88. http://doi.org/10.2528/PIERM21111201

76 Neri, F. (2018). *Introduction to electronic defense systems*, 3rd ed. Boston, MA: Artech House, Section 7.5, pp. 532–535.

77 De Martino, A. (2018). *Introduction to modern EW systems*, 2nd ed. Norwood, MA: Artech House, Section 1.4, pp. 20–22.

78 Adamy, D. (2001). *EW 101: a first course in electronic warfare*. Norwood, MA: Artech House, Section 9.2.3, pp. 185–186.

79 Weinberg, G.V., Kracman, M. (2023). Swarm UAV defeat modelling through lifetime distribution analysis. *Progress in Electromagnetics Research Letters* 111: 71–78. http://doi.org/10.2528/PIERL23022808

80 Weinberg, G.V. (2022). Prediction of UAV swarm defeat with high-power radio frequency fields. *IEEE Transactions on Electromagnetic Compatibility* 64 (6): 2157–2162. https://doi.org/10.1109/TEMC.2022.3193881

81 Zhao, M., Chen, Y., Zhou, X. et al. (2022). Investigation on falling and damage mechanisms of UAV illuminated by HPM pulses. *IEEE Transactions on Electromagnetic Compatibility* 64 (5): 1412–1422. https://doi.org/10.1109/TEMC.2022.3187017

82 Zhang, J., Han, K., Zhang, P. et al. (2023). A survey on joint-operation application for unmanned swarm formations under a complex confrontation environment. *Journal of Systems Engineering and Electronics* 34 (6): 1432–1446. https://doi.org/10.23919/JSEE.2023.000162

83 Esteves, J.L., Cottais, E., Kasmi, C. (2018). Unlocking the access to the effects induced by IEMI on a civilian UAV. *International Symposium on Electromagnetic*

Compatibility (EMC EUROPE), Amsterdam, Netherlands (27–30 August 2018), New York, US: IEEE. https://doi.org/10.1109/EMCEurope.2018.8484990

84 Amiri, M., Tofigh, F., Shariati, N. et al. (2020). Wide-angle metamaterial absorber with highly insensitive absorption for TE and TM modes. *Scientific Reports* 10 (13638): 1–13. https://doi.org/10.1038/s41598-020-70519-8

85 Begaud, X., Lepage, A.C., Varault, S. et al. (2018). Ultra-wideband and wide-angle microwave metamaterial absorber. *MDPI Materials* 11 (10): 1–9. https://doi.org/10.3390/ma11102045

86 Liu, C., Bai, Y., Zhou, J. et al. (2019). High-performance bifunctional polarization switch chiral metamaterials by inverse design method. *npj Computational Materials* 5 (93): 1–8. https://doi.org/10.1038/s41524-019-0230-z

87 Yuan, Y., Zhang, K., Ratni, B. et al. (2020). Independent phase modulation for quadruplex polarization channels enabled by chirality-assisted geometric-phase metasurfaces. *Nature Communications* 11 (4186): 1–9. https://doi.org/10.1038/s41467-020-17773-6

88 Chen, K., Feng, Y., Cui, L. et al. (2017). Dynamic control of asymmetric electromagnetic wave transmission by active chiral metamaterial. *Scientific Reports* 7 (42802): 1–10. https://doi.org/10.1038/srep42802

89 Ji, W., Cai, T., Wang, G. et al. (2019). High-efficiency and ultra-broadband asymmetric transmission metasurface based on topologically coding optimization method. *Optics Express* 27 (3): 2844–2854. https://doi.org/10.1364/OE.27.002844

90 Stephen, L., Yogesh, N., Subramanian, V. (2018). Broadband asymmetric transmission of linearly polarized electromagnetic waves based on chiral metamaterial. *Journal of Applied Physics* 123 (3): 1–7. https://doi.org/10.1063/1.5008614

91 Ratni, B., de Lustrac, A., Piau, G.-P. et al. (2018). Reconfigurable meta-mirror for wavefronts control: applications to microwave antennas. *Optics Express* 26: 2613–2624. https://doi.org/10.1364/OE.26.002613

92 Pfeiffer, C., Grbic, A. (2020). Metamaterial Huygens' surfaces: tailoring wave fronts with reflectionless sheets. *Physics Review Letters* 110: 1–5, 2013. https://doi.org/10.1103/PhysRevLett.110.197401

93 Raeker, B.O., Grbic, A. (2019). Compound metaoptics for amplitude and phase control of wave fronts. *Physics Review Letters* 122: 1–6. https://doi.org/10.1103/PhysRevLett.122.113901

94 Li, H., Wang, G., Cai, T. et al. (2019). Wideband transparent beam-forming metadevice with amplitude- and phase-controlled metasurface. *Physics Review Letters* 11: 1–9. https://doi.org/10.1103/PhysRevApplied.11.014043

95 Veselago, V.G. (1968). The electrodynamics of substances with simultaneously negative values of ε and μ. *Physics Uspekhi* 10 (4): 509–514.

96 Balanis, C.A. (2012). *Advanced engineering electromagnetics*, 2nd ed. New York, NY: Wiley, Table 1.4, pp. 23.

97 Houck, A.A., Brock, J.B., Chuang, I.L. (2003). Experimental observations of a left-handed material that obeys Snell's law. *Physics Review Letters* 90 (137401): 1–4. https://doi.org/10.1103/PhysRevLett.90.137401

98 Pendry, J.B., Holden, A.J., Robbins, D.J. (1998). Low frequency plasmons in thin-wire structures. *Journal of Physics: Condensed Matter* 10: 4785–4809. https://doi.org/10.1088/0953-8984/10/22/007

99 Pendry, J.B., Holden, A.J., Robbins, D.J. et al. (1999). Magnetism from conductors and enhanced nonlinear phenomena. *IEEE Transactions on Microwave Theory and Techniques* 47 (11): 2075–2084. https://doi.org/10.1109/22.798002

100 Smith, D.R., Padilla, W.J., Vier, D.C. (2003). Composite medium with simultaneously negative permeability and permittivity. *Physics Review Letters* 84 (4184): 1–4. https://doi.org/10.1103/PhysRevLett.84.4184

101 Hrabar, S., Bartolic, J. (2003). Backward wave propagation in waveguide filled with negative permeability meta material. *Proceedings of the IEEE Antennas and Propagation Society International Symposium on Digest Held in conjunction with: USNC/CNC/URSI North American Radio Sci. Meeting (Cat. No.03CH37450)*, Columbus, OH, USA (22–27 June 2003), New York, US: IEEE. https://doi.org/10.1109/APS.2003.1217413

102 Popa, B.-I., Cummer, S.A. (2006). Direct measurement of evanescent wave enhancement inside passive metamaterials. *Physics Review E* 73 (016617): 1–5. https://doi.org/10.1103/PhysRevE.73.016617

103 Schurig, D., Mock, J.J., Justice, B.J. et al. (2006). Metamaterial electromagnetic cloak at microwave frequencies. *Science* 314 (5801): 977–980. https://doi.org/10.1126/science.1133628

104 Kingery, K. (2018). Beyond Materials: from invisibility cloaks to satellite communications. *DukeStories*, https://stories.duke.edu/beyond-materials-from-invisibility-cloaks-to-satellite-communications (accessed 08 March 2023).

105 Ivsic, B., Sipus, Z., Hrabar, S. (2009). Analysis of uniaxial multilayer cylinders used for invisible cloak realization. *IEEE Transactions on Antennas and Propagation* 57 (5): 1521–1527. https://doi.org/10.1109/TAP.2009.2016695

106 Viaene, S., Ginis, V., Danckaert, J. (2016). Metamaterials open up entirely new possibilities in optics. https://phys.org/news/2016-11-metamaterials-possibilities-optics.html (accessed 08 March 2023).

107 Li, Z., Liu, W., Cheng, H. et al. (2015). Realizing broadband and invertible linear-to-circular polarization converter with ultrathin single-layer metasurface. *Scientific Reports* 5 (18106): 1–9. https://doi.org/10.1038/srep18106

108 Brewitt-Taylor, C.R. (2007). Limitation on the bandwidth of artificial perfect magnetic conductor surfaces. *IET Microwaves, Antennas & Propagation* 1 (1): 255–260. https://doi.org/10.1049/iet-map:20050309

109 Shao, L., Zhu, W. (2021). Electrically reconfigurable microwave metasurfaces with active lumped elements: a mini review. *Frontiers in Materials* 8 (689665): 1–7. https://doi.org/10.3389/fmats.2021.689665

110 Munk, B.A. (2000). *Frequency selective surfaces: theory and design*, Hoboken, NJ: Wiley.

111 Han, Y., Chang, Y., Che, W. (2022). Frequency-selective rasorbers: a view of frequency-selective rasorbers and their application in reducing the radar cross sections of antennas. *IEEE Microwave Magazine* 23 (2): 86–98. https://doi.org/10.1109/MMM.2021.3125463

112 Gonçalves Licursi de Mello, R., Lepage, A.C., Begaud, X. (2021). Metasurface de Huygens multibande avec dépointage du faisceau dans la bande 5G 3.4–3.8 GHz. *Proceedings of the 9ème Conférence Plénière du GDR ONDES*, Lille, France (1 December 2021).

113 Gonçalves Licursi de Mello, R., Lepage, A.C., Begaud, X. (2023). Taming Fabry–Pérot resonances in a dual-metasurface multiband antenna with beam steering in one of the bands. *Scientific Reports* 13 (9871): 1–10. https://doi.org/10.1038/s41598-023-36828-4

114 Gonçalves Licursi de Mello, R. (2023). Metasurfaces and blinking jamming: convergent study, comparative analysis, and challenges. *MDPI Micromachines* 14 (7): 1405. https://doi.org/10.3390/mi14071405

115 Lin, B., Guo, J., Lv, L. et al. (2019). Ultra-wideband and high-efficiency reflective polarization converter for both linear and circular polarized waves. *Applied Physics A* 126 (76): 1–8. https://doi.org/10.1007/s00339-018-2368-9

116 Lončar, J., Grbic, A., Hrabar, S. (2018). A reflective polarization converting metasurface at X-band frequencies. *IEEE Transactions on Antennas and Propagation* 66 (6): 3213–3218. https://doi.org/10.1109/TAP.2018.2816784

117 Sievenpiper, D.F., Schaffner, J.H., Song, H.J. (2003). Two-dimensional beam steering using an electrically tunable impedance surface. *IEEE Transactions on Antennas and Propagation* 51 (10): 2713–2722. https://doi.org/10.1109/TAP.2003.817558

118 Fang, X., Li, M., Han, J. et al. (2022). Accurate direction–of–arrival estimation method based on space–time modulated metasurface. *IEEE Transactions on Antennas and Propagation* 70 (11): 10951–10964. https://doi.org/10.1109/TAP.2022.3184556

119 Gonçalves Licursi de Mello, R., Lepage, A.C., Begaud X. (2023). A low-profile, triple-band, and wideband antenna using dual-band AMC. *MDPI Sensors* 23 (4): 1920. https://doi.org/10.3390/s23041920

120 Gonçalves Licursi de Mello, R., Lepage, A.C., Begaud X. (2022). Réflecteur multibande à métamatériaux. *Speech presented at the Journées Scientifiques URSI France 2022* in Palaiseau, France (22–23 March 2022).

121 Gonçalves Licursi de Mello, R., Lepage, A.C., Begaud X. (2023). A Compact, quad-band, and wideband antenna using triple-band AMC. *Proceedings of 13th International Conference on Metamaterials, Photonic Crystals and Plasmonics (Meta2023)*, Paris, France (18–21 July 2023), Paris, France: Metaconferences. https://metaconferences.org/META23/files/meta23_proceedings.pdf

122 Taravati, S., Eleftheriades, G.V. (2021). Full-duplex reflective beamsteering metasurface featuring magnetless nonreciprocal amplification. *Nature Communications* 12 (4414): 1–11. https://doi.org/10.1038/s41467-021-24749-7

123 Lončar, J., Grbic, A., Hrabar, S. (2019). Ultrathin active polarization-selective metasurface at X-band frequencies. *Physics Review B* 100 (7): 075131-1–075131-7. https://doi.org/10.1103/PhysRevB.100.075131

124 Afzal, M.U., Esselle, K.P., Koli, M.N.Y. (2022). A beam-steering solution with highly transmitting hybrid metasurfaces and circularly polarized high-gain radial-line slot array antennas. *IEEE Transactions on Antennas and Propagation* 70 (1): 365–377. https://doi.org/10.1109/TAP.2021.3111522

125 Li, W., Yu, Q., Qiu, J.H. et al. (2024). Intelligent wireless power transfer via a 2-bit compact reconfigurable transmissive-metasurface-based router. *Nature Communications* 15 (2807): 1–10. https://doi.org/10.1038/s41467-024-46984-4

126 Neri, F. (2018). *Introduction to electronic defense systems*, 3rd ed. Boston, MA: Artech House, Section 5.4.20, pp. 449.

2

Electromagnetics: A Review on Pertinent Points

Metamaterials and metasurfaces can both provide arbitrary properties for the interaction with electromagnetic waves. While metamaterials have proven to enable unusual phenomena, like a refraction in an inverse direction [1] (see Fig. 1.7c), the impact of metasurfaces was so that even Snell's laws for the refraction of light, accepted by scientists since the 17th century, had to be adjusted [2].

Although metasurfaces present advantages of low profile, weight, and costs, besides an ease of fabrication and reconfigurability, one can understand them easier by starting from metamaterials. The main reason is that the exercise of understanding the properties of metamaterials provides deep insight on atypical features of electromagnetic waves.

Veselago's hypothetical medium [3] implies the existence of backward waves and the amplification of evanescent waves besides the negative refraction. One may get confused with the physics behind these phenomena and arrive at misconceptions concerning the break of some fundamental laws depending on the form of Maxwell's equations used in the analysis. However, understanding how electromagnetic waves interact with matter makes the mechanisms behind such unusual properties become clear, ultimately making explicit that a medium like Veselago's one does not contradict physics.

This chapter is for readers lacking a good review on Electromagnetics–or having a first contact with it. It covers those points needed to understand the operation principles behind metasurfaces, which Chapter 3 will address. This is not the most enticing introduction nor it is necessary for the metasurface expert, who can jump to Chapter 4. If that is not your case, we will start here from the basics, and next introduce advanced concepts with the support of figures. We also derive the mathematical formulation. Following it is not a must for understanding the rest of the book, but getting aware of the main points (summarized in Section 2.1.8) is highly recommended.

Metasurface-Driven Electronic Warfare, First Edition. Rafael Gonçalves Licursi de Mello.

Published 2025 by John Wiley & Sons, Inc.
Companion website: www.wiley.com/go/licursidemello/metasurfacedrivenelectronicwarfare1

2.1 Introduction to Electromagnetics

Classical Electromagnetics relies over four equations named nowadays after scientists that made major contributions to each one of them. Experimental observations and analogies with the Newtonian theory of fields were what mainly allowed the derivation of such equations. James Clerk Maxwell added one term that united them in a closed set making clear the relationship between electricity and magnetism.

From Maxwell's equations, we can derive the mathematical formulation that support electromagnetic waves, as well as the reflection, transmission, and refraction thereof across the interface between different media. Moreover, the notion of impedance can be understood. All these concepts are crucial for a good visualization of the operation principles of metasurfaces.

2.1.1 Fundamental Concepts

Electromagnetics mathematics starts from the foundation of Maxwell's equations. Maxwell's equations are a set of coupled partial differential equations that model electric and magnetic quantities as a function of charges, currents, and their changes. We can express Maxwell's equations in their instantaneous differential form as [4]:

$$\nabla \times \mathcal{H} = \mathcal{J}_i + \mathcal{J}_c + \frac{\partial \mathcal{D}}{\partial t} \text{ (Ampère's law)} \tag{2.1}$$

$$\nabla \times \mathcal{E} = -\mathcal{M}_i - \frac{\partial \mathcal{B}}{\partial t} \text{ (Faraday–Lenz's law)} \tag{2.2}$$

$$\nabla \cdot \mathcal{B} = q_{mv} \text{ (Gauss's law for the magnetic flux)} \tag{2.3}$$

$$\nabla \cdot \mathcal{D} = q_{ev} \text{ (Gauss's law for the electric flux)} \tag{2.4}$$

where

- $\nabla\times$ and $\nabla\cdot$ respectively denote the curl and divergent operations;
- $\mathcal{H}$ is the magnetic field, in amperes per meter [A/m];
- $\mathcal{J}_i$ is the impressed electric current density [A/m^2];
- $\mathcal{J}_c$ is the conduction electric current density [A/m^2];
- $\mathcal{D}$ is the electric flux density, in coulombs per square meter [C/m^2];
- $\mathcal{E}$ is the electric field, in volts per meter [V/m];
- $\mathcal{M}_i$ is the impressed magnetic current density [V/m^2];
- $\mathcal{B}$ is the magnetic flux density, in webers per square meter [Wb/m^2];
- q_{mv} is the magnetic charge density [Wb/m^3];
- q_{ev} is the electric charge density [C/m^3].

All quantities in (2.1)–(2.4) are instantaneous. Boldface quantities refer to vectors, which in the rectangular coordinate system can be associated to their scalar components as follows:

$$\mathcal{A} = \mathcal{A}_x\hat{\mathbf{a}}_x + \mathcal{A}_y\hat{\mathbf{a}}_y + \mathcal{A}_z\hat{\mathbf{a}}_z \tag{2.5}$$

where $\hat{\mathbf{a}}_x$, $\hat{\mathbf{a}}_y$, and $\hat{\mathbf{a}}_z$ are unit vectors in the respective directions x, y, and z. Consider also the following constitutive relations [5]:

$$\mathcal{B} = \hat{\mu} * \mathcal{H} \text{ (medium magnetic response)} \tag{2.6}$$

$$\mathcal{D} = \hat{\varepsilon} * \mathcal{E} \text{ (medium electric response)} \tag{2.7}$$

$$\mathcal{J}_c = \hat{\sigma} * \mathcal{E} \text{ (medium conductivity)} \tag{2.8}$$

where

- $*$ denotes the operation of convolution;
- $\hat{\mu}$, $\hat{\varepsilon}$, $\hat{\sigma}$, are respectively the impulse response of the medium for the magnetic permeability, in henries per meter [H/m], the electric permittivity, in farads per meter [F/m], and the electric conductivity, in siemens per meter [S/m].

In anisotropic media, the permeability $\hat{\mu}$, permittivity $\hat{\varepsilon}$, and conductivity $\hat{\sigma}$ are 3×3 tensors representing the relationship between each component of the quantities they associate in the three directions of the coordinate system. For example, (2.7) in matrix form is:

$$\begin{bmatrix} \mathcal{D}_x \\ \mathcal{D}_y \\ \mathcal{D}_z \end{bmatrix} = \begin{bmatrix} \varepsilon_{xx} & \varepsilon_{xy} & \varepsilon_{xz} \\ \varepsilon_{yx} & \varepsilon_{yy} & \varepsilon_{yz} \\ \varepsilon_{zx} & \varepsilon_{zy} & \varepsilon_{zz} \end{bmatrix} * \begin{bmatrix} \mathcal{E}_x \\ \mathcal{E}_y \\ \mathcal{E}_z \end{bmatrix} \tag{2.9}$$

and in expanded form is:

$$\begin{aligned} \mathcal{D}_x &= \varepsilon_{xx} * \mathcal{E}_x + \varepsilon_{xy} * \mathcal{E}_y + \varepsilon_{xz} * \mathcal{E}_z \\ \mathcal{D}_y &= \varepsilon_{yx} * \mathcal{E}_x + \varepsilon_{yy} * \mathcal{E}_y + \varepsilon_{yz} * \mathcal{E}_z \\ \mathcal{D}_z &= \varepsilon_{zx} * \mathcal{E}_x + \varepsilon_{zy} * \mathcal{E}_y + \varepsilon_{zz} * \mathcal{E}_z \end{aligned} \tag{2.10}$$

In the free space, $\hat{\mu}$ reduces to $\mu_0 = 4\pi \times 10^{-7}$ H/m and $\hat{\varepsilon}$ to $\varepsilon_0 = 8.854 \times 10^{-12}$ F/m for all directions. The features of other media can be referred to by using a relative permeability μ_r and permittivity ε_r in relation to μ_0 and ε_0.

Equations (2.1) through (2.4) are general. We can use them to describe and relate the associated quantities at any point in space at any time. Equation (2.1) means that the presence of impressed and conducted electric current densities ($\mathcal{J}_i$ and $\mathcal{J}_c$) as well as a time-varying electric flux density $\mathcal{D}$ at a specific point in space implies in the establishment of a magnetic field $\mathcal{H}$ according to a curl at the same point. Likewise, (2.2) shows that the existence of an impressed magnetic current

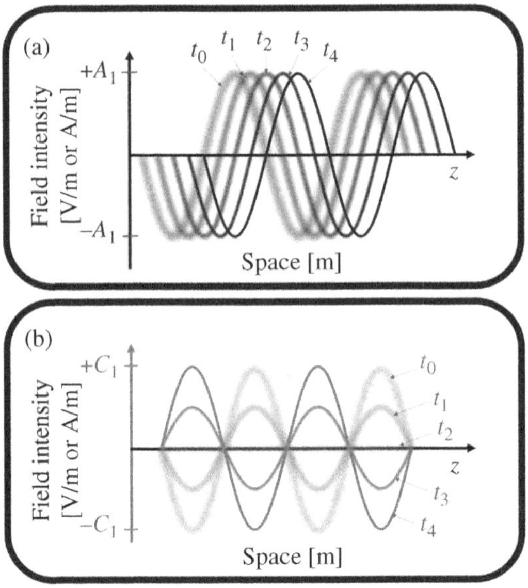

Fig. 2.1 Different types of waves: (a) traveling wave; (b) standing wave

the first term of (2.32) represents a traveling wave of amplitude A_1 and wavenumber k propagating through space with phase velocity v_{ph} in the $+\hat{\mathbf{a}}_z$ direction. Fig. 2.1a represents this wave at increasing instants t_0–t_4.

At this moment, it is important to understand that, at a fixed point in space, the phase of (2.34) is exclusively defined by the angular frequency ω as a function of the time t. On the other hand, at a fixed instant t, it is the wavenumber k that exclusively defines the phase as a function of space. Equation (2.34) can, for example, represent the electric field of an x-polarized plane wave traveling at the $+\hat{\mathbf{a}}_z$ direction with uniform fields in the x and y dimensions. Figure 2.2 illustrates such a wave on the xz plane ($y = 0$) at a fixed instant t. The arrows indicate the direction of propagation. For the sake of visualization, the oscillations in the surface plot

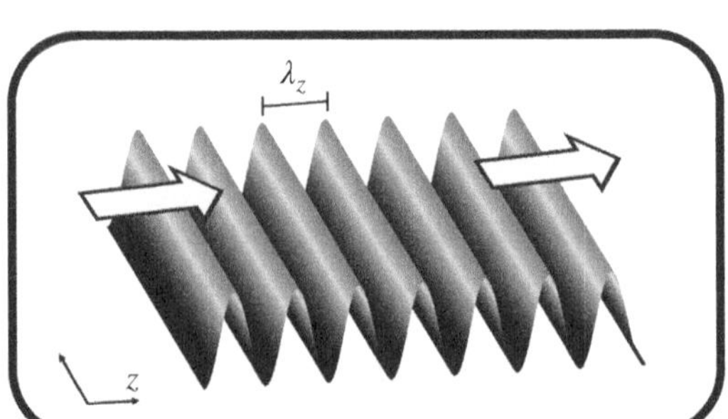

Fig. 2.2 Representation of plane wave at a fixed instant t

indicate variations of the electric field, not a displacement in the spatial dimension y nor to the wave polarization. In the exemplified plane wave, the distribution of electric field is the same for every plane of constant y.

We can also draw the above conclusions by making $A_1 = 0$ instead of $B_1 = 0$ in (2.32). In this case, the argument of the instantaneous form of $g_1(z)$ would be $\omega t + kz$, keeping constant for decreasing values of z in space when the time t increases. That means that the second term of (2.32) is also a traveling wave of wavenumber k, but with amplitude B_1 and propagation toward the opposite direction, $-\hat{\mathbf{a}}_z$.

We can also investigate the first term of (2.33). To this end, we make $D_1 = 0$ and use the relation (2.16) to obtain the instantaneous form of $g_2(z)$, as:

$$\mathscr{g}_2(z;t) = \mathrm{Re}\left[C_1 \cos(kz) e^{j\omega t}\right] = C_1 \cos(kz) \cos(\omega t) \tag{2.37}$$

In (2.37), two cosines modulate space and time independently. That means that the first term of (2.33) represents a standing wave of amplitude C_1 and wavenumber k. Fig. 2.1b sketches such wave at increasing instants t_0–t_4, which is distributed through the dimension z in space but does not propagate. We obtain similar conclusions for the second term of (2.33), with differences in the amplitude and phase distribution in space due to the exchange of C_1 for D_1 and the cosine for a sine.

Note that the x, y, and z variables are exchangeable in the derivation of (2.29) and (2.31). That is why we know such equations as wave equations for the rectangular coordinate system. Actually, wave equations in arbitrary directions projected over the x, y, and z dimensions present the same form as (2.29) and (2.31) [10]. Therefore, we can project all the features of the traveling and standing waves presented here from an arbitrary direction to the x, y, and z dimensions. The wavenumber k_u projected from an arbitrary direction to another can be calculated as:

$$k_u = k \cos \theta_u \tag{2.38}$$

where θ_u is the angle between the original direction and the direction of interest $\hat{\mathbf{a}}_u$. Similarly, using (2.35) and (2.38), the wavelength λ_u projected from an arbitrary direction to another is:

$$\lambda_u = \frac{2\pi}{k_u} = \frac{\lambda}{\cos \theta_u} \tag{2.39}$$

2.1.4 Reflection, Transmission, and the Concept of Impedance

A final solution for an electromagnetic wave distributed in space for any time includes correct values assigned to A_1, B_1, C_1, and D_1 in (2.32) and (2.33). That is where boundary conditions (2.11)–(2.14) come into play.

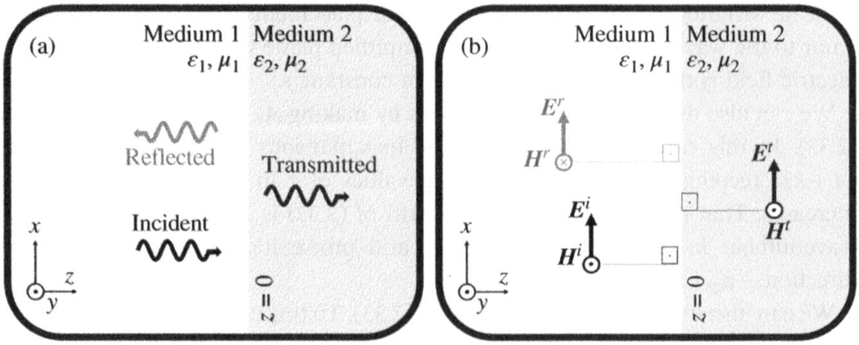

Fig. 2.3 Transmission and reflection of waves across interface: (a) disposition of waves; (b) associated vectors

Consider in Fig. 2.3a an interface between two isotropic, source-free ($\boldsymbol{J}_i = q_{ev} = 0$, and $\boldsymbol{M}_i = q_{mv} = 0$), and lossless ($\boldsymbol{J}_c = 0$) media presenting respective electric permittivities ε_1 and ε_2, and magnetic permeabilities μ_1 and μ_2. The interface is located at $z = 0$. We define a plane wave polarized in the x dimension and traveling at the $+\hat{\mathbf{a}}_z$ direction normally incident upon the interface. As such, both the electric and magnetic fields are transverse to the interface, giving the wave the name of transverse electromagnetic (TEM). Fig. 2.2 and the solution form (2.32) for the wave equation can also represent this incident wave if we make $A_1 = E_0$ and $B_1 = 0$, as follows:

$$\boldsymbol{E}^i = \hat{\mathbf{a}}_x E_0 e^{-jk_1 z} \tag{2.40}$$

where $k_1 = \omega\sqrt{\mu_1\varepsilon_1}$. With Faraday–Lenz's law (2.18) and constitutive relation (2.21), we can calculate the magnetic field related to this wave as:

$$\begin{aligned}\boldsymbol{H}^i &= -\frac{1}{j\omega\mu_1}\nabla\times\boldsymbol{E}^i \\ &= -\hat{\mathbf{a}}_y\frac{1}{j\omega\mu_1}\frac{\partial \boldsymbol{E}^i}{\partial z} = \hat{\mathbf{a}}_y\frac{k_1}{\omega\mu_1}E_0e^{-jk_1 z} \\ &= \hat{\mathbf{a}}_y\frac{E_0}{\sqrt{\mu_1/\varepsilon_1}}e^{-jk_1 z} = \hat{\mathbf{a}}_y\frac{E_0}{\eta_1}e^{-jk_1 z}\end{aligned} \tag{2.41}$$

where $\eta_1 = \sqrt{\mu_1/\varepsilon_1}$ is the intrinsic impedance of medium 1, in Ω (ohms), a quantity defined by the ratio of the electric and magnetic fields of waves supported by this medium. The free space, for example, has an intrinsic impedance $\eta_0 = \sqrt{\mu_0/\varepsilon_0} \cong 120\pi\ \Omega$.

As it will become clear in the next steps, in Fig. 2.3a, an amount of the incident energy is transmitted to medium 2, and the rest is reflected back to medium 1 due to the boundary conditions (2.13) and (2.14) [11]. We define then the reflection and transmission coefficients for TEM waves as:

$$\Gamma_{TEM} = E^r/E^i \tag{2.42}$$

$$\tau_{TEM} = E^t/E^i \tag{2.43}$$

For simplicity, we assume no polarization conversion here. Hence, the reflected, and transmitted electric fields can be expressed as:

$$\boldsymbol{E}^r = \hat{\mathbf{a}}_x \Gamma_{TEM} E_0 e^{+jk_1 z} \tag{2.44}$$

$$\boldsymbol{E}^t = \hat{\mathbf{a}}_x \tau_{TEM} E_0 e^{-jk_2 z} \tag{2.45}$$

The positive sign in the exponential of (2.44) accounts for the direction of propagation of the reflected wave at the $-\hat{\mathbf{a}}_z$ direction. Applying Faraday–Lenz's law (2.18) to (2.44) and (2.45) leads to the following expressions for the transmitted and reflected magnetic fields:

$$\boldsymbol{H}^r = -\hat{\mathbf{a}}_y \Gamma_{TEM} \frac{E_0}{\eta_1} e^{+jk_1 z} \tag{2.46}$$

$$\boldsymbol{H}^t = \hat{\mathbf{a}}_y \tau_{TEM} \frac{E_0}{\eta_2} e^{-jk_2 z} \tag{2.47}$$

The orientation of the vectors related to the incident, reflected, and transmitted waves are defined in Fig. 2.3b. Note that the reflected magnetic field $\boldsymbol{H}^r$ has an orientation opposite to incident and transmitted magnetic fields $\boldsymbol{H}^i$ and $\boldsymbol{H}^t$. This comes from the application of the curl of Faraday–Lenz's law (2.18) in the electric field $\boldsymbol{E}^r$, whose propagation occurs at the $-\hat{\mathbf{a}}_z$ direction.

Since $\boldsymbol{J}_i = \boldsymbol{J}_c = 0$ and $\boldsymbol{M}_i = 0$, from boundary conditions (2.13) and (2.14), the total tangential electric and magnetic fields should be continuous across the interface. For $z = 0$, using (2.40), (2.44) and (2.45), we have:

$$\begin{gathered} \boldsymbol{E}^i + \boldsymbol{E}^r = \boldsymbol{E}^t \\ \hat{\mathbf{a}}_x E_0 e^{-jk_1 z} + \hat{\mathbf{a}}_x \Gamma_{TEM} E_0 e^{+jk_1 z} = \hat{\mathbf{a}}_x \tau_{TEM} E_0 e^{-jk_2 z} \\ 1 + \Gamma_{TEM} = \tau_{TEM} \end{gathered} \tag{2.48}$$

and from (2.41), (2.46), and (2.47), we have:

$$\begin{gathered} \boldsymbol{H}^i + \boldsymbol{H}^r = \boldsymbol{H}^t \\ \hat{\mathbf{a}}_y \frac{E_0}{\eta_1} e^{-jk_1 z} - \hat{\mathbf{a}}_y \Gamma_{TEM} \frac{E_0}{\eta_1} e^{+jk_1 z} = \hat{\mathbf{a}}_y \tau_{TEM} \frac{E_0}{\eta_2} e^{-jk_2 z} \\ \frac{1}{\eta_1}(1 - \Gamma_{TEM}) = \frac{1}{\eta_2} \tau_{TEM} \end{gathered} \tag{2.49}$$

Solving the system composed of (2.48) and (2.49) leads to:

$$\Gamma_{TEM} = \frac{\eta_2 - \eta_1}{\eta_2 + \eta_1} = |\Gamma_{TEM}|e^{j\varphi_\Gamma} \tag{2.50}$$

$$\tau_{TEM} = \frac{2\eta_2}{\eta_2 + \eta_1} = 1 + \Gamma_{TEM} = |\tau_{TEM}|e^{j\varphi_\tau} \tag{2.51}$$

where φ_Γ and φ_τ are the reflection and transmission coefficient phases, respectively. From (2.50) and (2.51), we may conclude that the reflection and transmission of waves through the interface of two media depend on ε_1, ε_2, μ_1, and μ_2. Putting it differently, the reflection and transmission of waves depend on the impedance matching between the media. It is important to retain here that reflection is a phenomenon that adjusts the ratio between electric and magnetic fields in each of the media per their respective intrinsic impedances while meeting the boundary conditions at the interface.

Now, suppose that medium 1 is the free space, whose intrinsic impedance is $\eta_1 = \eta_0 = 120\pi\ \Omega$ while medium 2 is a conductor that can be considered a PEC, with intrinsic impedance $\eta_2 \cong 0$. In this case, we conclude from (2.50) and (2.51) that the reflection and transmission coefficients are respectively $\Gamma_{TEM} = -1$ and $\tau_{TEM} = 0$. Then, using (2.44) and (2.45), we obtain the following expressions for the reflected and transmitted electric fields:

$$\boldsymbol{E}^r = -\hat{\mathbf{a}}_x E_0 e^{+jk_1 z} \tag{2.52}$$

$$\boldsymbol{E}^t = 0 \tag{2.53}$$

The total electric field in medium 1 is therefore:

$$\begin{aligned} \boldsymbol{E}_1 &= \boldsymbol{E}^i + \boldsymbol{E}^r \\ &= -\hat{\mathbf{a}}_x E_0 \left(e^{+jk_1 z} - e^{-jk_1 z}\right) \\ &= j2E_0 \sin(k_1 z)\hat{\mathbf{a}}_x \end{aligned} \tag{2.54}$$

which has the same form of the standing wave solution (2.33) with $C_1 = 0$ and $D_1 = j2E_0$. On the other hand, the total magnetic field in medium 1 is:

$$\begin{aligned} \boldsymbol{H}_1 &= \boldsymbol{H}^i + \boldsymbol{H}^r \\ &= \hat{\mathbf{a}}_y \frac{E_0}{\eta_1}\left(e^{-jk_1 z} + e^{+jk_1 z}\right) \\ &= 2\frac{E_0}{\eta_1}\cos(k_1 z)\hat{\mathbf{a}}_y \end{aligned} \tag{2.55}$$

which also has the same form (2.33) but with $C_1 = 2E_0\eta_1^{-1}$ and $D_1 = 0$. Note that the total electric field $\boldsymbol{E}_1$ is zero at $z = 0$, that is, over the interface, but the total magnetic field $\boldsymbol{H}_1$ is twice the incident field $\boldsymbol{H}^i$ thereon.

If medium 2 presents a high intrinsic impedance $\eta_2 \cong \infty$ instead, the reflection coefficient can be calculated as $\Gamma_{TEM} = +1$ and the situation is the opposite. We can calculate the total electric field in medium 1 as $\boldsymbol{E}_1 = 2E_0 \cos(k_1 z)\hat{\mathbf{a}}_x$, being twice $\boldsymbol{E}^i$ over the interface while the magnetic field is $\boldsymbol{H}_1 = j2E_0\eta_1^{-1} \sin(k_1 z)\hat{\mathbf{a}}_y$, being null over the interface.

The above development shows that the values of A_1, B_1, C_1, and D_1 in (2.32) and (2.33) depend on the nature of the problem. Unbounded problems, with no impedance discontinuities, will not present reflections and the presence of standing waves. In this case, (2.32) can represent traveling waves with A_1 and B_1 chosen according to the direction of propagation. Bounded problems where the intrinsic impedances of the media do not match will lead to severe reflections. As such, (2.33) can represent standing waves with the C_1, and D_1 also chosen accordingly. For bounded problems where the impedances do not really match but are close to each other, both traveling and standing waves will be present, and the solution will be a mixture between (2.32) and (2.33) with proper values for A_1, B_1, C_1, and D_1.

The concept of impedance matching has a special role in Electromagnetics, ruling not only the transfer of energy between media but also through impedance discontinuities in transmission lines. In fact, an analogy between both cases allows one to map the scenario of Fig. 2.3b into an equivalent transmission line schematic, as represented in Fig. 2.4. Here, the impedance Z_S of the source plays the role of the intrinsic impedance of medium 1, while the load impedance Z_L substitutes the intrinsic impedance of medium 2. As such, we can calculate a reflection coefficient Γ using the equation for transmission lines, which has the same form of (2.50).

The adoption of equivalent schematics simplifies a series of questions in Electromagnetics, including many problems related to metasurfaces. Their importance will become more obvious to the reader as analyses of specific metasurfaces are considered in later sections.

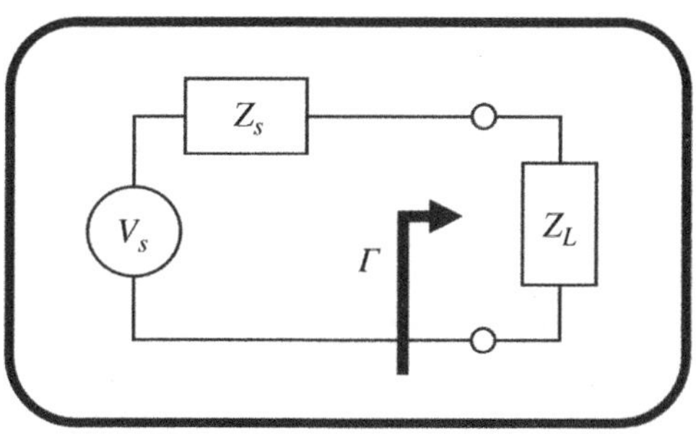

Fig. 2.4 Equivalent schematic for scenario of Fig. 2.3b

2.1.5 Oblique Incidence and the Refraction of Waves

Consider next that the plane wave coming from medium 1 impinges upon the interface with medium 2 with an oblique angle of incidence θ_i in relation to the normal of the interface, as in Fig. 2.5a. Both media are isotropic, source-free ($\boldsymbol{J}_i = q_{ev} = 0$, and $\boldsymbol{M}_i = q_{mv} = 0$), and lossless ($\boldsymbol{J}_c = 0$). The polarization of this wave, with only the magnetic field transverse to the interface, is called transverse magnetic (TM). Now, the vector of the electric field is not aligned with the x dimension. Using the vector orientation of Fig. 2.5a, we can represent the incident, reflected and transmitted electric fields as [12]:

$$\begin{aligned}\boldsymbol{E}^i &= (\hat{\mathbf{a}}_\mathbf{x}\cos\theta_i - \hat{\mathbf{a}}_\mathbf{z}\sin\theta_i)E_0 e^{-j\boldsymbol{k}^i\cdot\boldsymbol{r}}\\ &= (\hat{\mathbf{a}}_\mathbf{x}\cos\theta_i - \hat{\mathbf{a}}_\mathbf{z}\sin\theta_i)E_0 e^{-jk_1(x\sin\theta_i + z\cos\theta_i)}\end{aligned} \tag{2.56}$$

$$\begin{aligned}\boldsymbol{E}^r &= (\hat{\mathbf{a}}_\mathbf{x}\cos\theta_r + \hat{\mathbf{a}}_\mathbf{z}\sin\theta_r)\Gamma_{TM}E_0 e^{-j\boldsymbol{k}^r\cdot\boldsymbol{r}}\\ &= (\hat{\mathbf{a}}_\mathbf{x}\cos\theta_r + \hat{\mathbf{a}}_\mathbf{z}\sin\theta_r)\Gamma_{TM}E_0 e^{-jk_1(x\sin\theta_r - z\cos\theta_r)}\end{aligned} \tag{2.57}$$

$$\begin{aligned}\boldsymbol{E}^t &= (\hat{\mathbf{a}}_\mathbf{x}\cos\theta_t - \hat{\mathbf{a}}_\mathbf{z}\sin\theta_t)\tau_{TM}E_0 e^{-j\boldsymbol{k}^t\cdot\boldsymbol{r}}\\ &= (\hat{\mathbf{a}}_\mathbf{x}\cos\theta_t - \hat{\mathbf{a}}_\mathbf{z}\sin\theta_t)\tau_{TM}E_0 e^{-jk_2(x\sin\theta_t + z\cos\theta_t)}\end{aligned} \tag{2.58}$$

We recall once more boundary condition (2.14) to enforce continuity of the total tangential electric field across the interface. However, in the current situation, the tangential fields depend on the angle of incidence θ_i. Since the interface is located at $z = 0$ and the wave has TM polarization, picking the tangential components of (2.56)–(2.58) leads to:

$$\begin{aligned}\left(\boldsymbol{E}^i + \boldsymbol{E}^r\right)\cdot\hat{\mathbf{a}}_\mathbf{x} &= \boldsymbol{E}^t\cdot\hat{\mathbf{a}}_\mathbf{x}\\ \cos\theta_i e^{-jk_1 x\sin\theta_i} + \Gamma_{TM}\cos\theta_r e^{-jk_1 x\sin\theta_r} &= \tau_{TM}\cos\theta_t e^{-jk_2 x\sin\theta_t}\end{aligned} \tag{2.59}$$

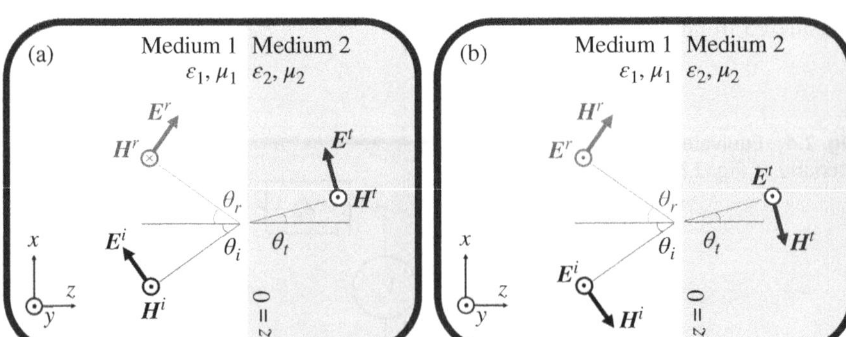

Fig. 2.5 Transmission and reflection at oblique incidence: (a) TM polarization; (b) TE polarization

In (2.59), $\cos\theta_i$, $\cos\theta_r$, $\cos\theta_t$, Γ_{TM}, and τ_{TM} are constants. Moreover, such equation must hold throughout the whole interface, that is, for all values of x. This can happen only if all phase terms in (2.59) are equal [13]:

$$k_1 x \sin\theta_i = k_1 x \sin\theta_r = k_2 x \sin\theta_t \tag{2.60}$$

This relation leads to Snell's laws:

$$\theta_r = \theta_i \text{ (Snell's law of reflection)} \tag{2.61}$$

$$k_1 \sin\theta_i = k_2 \sin\theta_t \text{ (Snell's law of refraction)} \tag{2.62}$$

It is evident that both sides of (2.62) match in regard to (2.38) on the interface. A practical interpretation for (2.62) is therefore that the crossing from medium 1 to 2 deflects waves to match the projection of k_2 to the projection of k_1 on the interface. The media support different wavelengths according to their respective intrinsic impedances and wavenumbers. Since continuity of tangential fields is required, these wavelengths should match at the interface, where they are projected according to (2.39). Refraction is the phenomenon that adjusts such matching by deflecting the angle of propagation of the transmitted wave.

We sketch in Fig. 2.6a the phenomena related to refraction with a plane wave that impinges upon the interface of media 1 and 2 with angle of incidence $\theta_i = 35°$. The electric permittivity of medium 2 is $\varepsilon_2 = 2.2\varepsilon_1$ whereas the magnetic permeability is $\mu_2 = \mu_1$. For the sake of clarity, we show only the incident and transmitted electric fields. The arrows indicate the angles of propagation with respect to the normal of the interface. Note the matching of the wavelengths of both fields projected over the interface. The angle of propagation of the transmitted wave is $\theta_t = 22.7°$ according to (2.62).

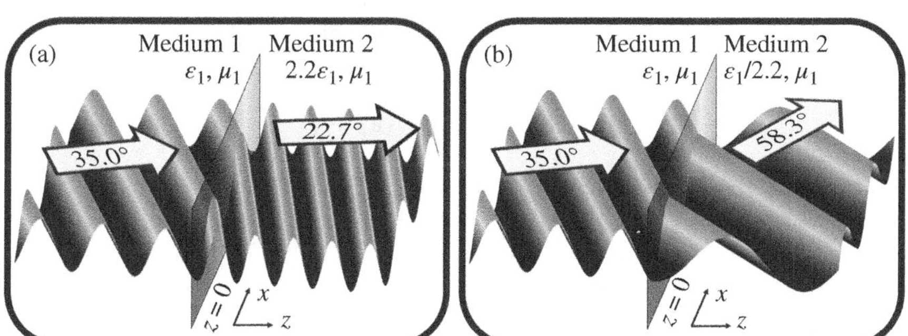

Fig. 2.6 Refraction of waves for different cases: (a) $\mu_2\varepsilon_2 = 2.2\mu_1\varepsilon_1$; (b) $\mu_2\varepsilon_2 = \mu_1\varepsilon_1/2.2$

If the electric permittivity of medium 2 was $\varepsilon_2 = \varepsilon_1/2.2$ instead, the angle of propagation of the transmitted wave would be $\theta_t = 58.3°$. Figure 2.6b illustrates this case. We can see again the matching of the wavelengths projected over the interface.

As we will see in the next sections, (2.61) and (2.62) are crucial relations for the study of metamaterials and metasurfaces. First, because it is (2.62) that confirms that metamaterials can indeed provide a negative phase velocity. And second, because metasurfaces can break both (2.61) and (2.62), providing arbitrary angles in the reflection and transmission of waves, which shows the power of the technology.

We can do a similar development for the transverse electric (TE) polarization, where the electric field is transverse to the interface instead of the magnetic field, as in Fig. 2.5b. The same relations (2.61) and (2.62) hold for this case.

2.1.6 Reflection and Transmission at Oblique Incidence

Now that we have the relation between the angles of the incident, reflected, and transmitted waves, we can use boundary conditions (2.13) and (2.14) to require continuity of total tangential fields across the interface and calculate the reflection and transmission coefficients. Using the vector orientation indicated in Fig. 2.5a, we can express the magnetic fields as:

$$\boldsymbol{H}^i = \hat{\mathbf{a}}_y \frac{E_0}{\eta_1} e^{-jk_1(x\sin\theta_i + z\cos\theta_i)} \tag{2.63}$$

$$\boldsymbol{H}^r = -\hat{\mathbf{a}}_y \frac{\Gamma_{TM} E_0}{\eta_1} e^{-jk_1(x\sin\theta_r - z\cos\theta_r)} \tag{2.64}$$

$$\boldsymbol{H}^t = \hat{\mathbf{a}}_y \frac{\tau_{TM} E_0}{\eta_2} e^{-jk_2(x\sin\theta_t + z\cos\theta_t)} \tag{2.65}$$

The interface is located at $z = 0$, and now we know from (2.61) that $\theta_i = \theta_r$ and $k_1 \sin\theta_i = k_2 \sin\theta_t$. Therefore, from boundary condition (2.13) and (2.63)–(2.65), we have:

$$\begin{aligned} \left(\boldsymbol{H}^{\mathrm{i}} + H^{\mathrm{r}}\right)\cdot\hat{\mathbf{a}}_{\mathrm{y}} &= \boldsymbol{H}^{\mathrm{t}}\cdot\hat{\mathbf{a}}_{\mathrm{y}} \\ \frac{E_0}{\eta_1}\left[e^{-jk_1 x\sin\theta_i} - \Gamma_{TM} e^{-jk_1 x\sin\theta_r}\right] &= \frac{E_0}{\eta_2}\tau_{TM} e^{-jk_2 x\sin\theta_t} \\ \frac{1}{\eta_1}(1 - \Gamma_{TM}) &= \frac{1}{\eta_2}\tau_{TM} \end{aligned} \tag{2.66}$$

Also, we can rewrite (2.59) as:

$$\cos\theta_i + \Gamma_{TM}\cos\theta_i = \tau_{TM}\cos\theta_t \tag{2.67}$$

Manipulating (2.66) and (2.67), we obtain:

$$\Gamma_{TM} = \frac{\eta_2 \cos\theta_t - \eta_1 \cos\theta_i}{\eta_2 \cos\theta_t + \eta_1 \cos\theta_i} = \frac{Z_w^t - Z_w^i}{Z_w^t + Z_w^i} \tag{2.68}$$

$$\tau_{TM} = \frac{2\eta_2 \cos\theta_i}{\eta_2 \cos\theta_t + \eta_1 \cos\theta_i} \tag{2.69}$$

We can see in (2.68) that $\cos\theta_i$ multiplies η_1, and $\cos\theta_t$ multiplies η_2. These multiplications account for the projection of the fields over the interface, indicating that, for reflection purposes, this projection is what matters. Consequently, we define wave impedances Z_w^i and Z_w^t [Ω] accordingly. As such, the form of (2.68) keeps similar to (2.50) despite the oblique incidence.

To express (2.68) and (2.69) exclusively in terms of the angle of incidence θ_i, we use relation (2.62) as:

$$\cos\theta_t = \sqrt{1 - \sin^2\theta_t} = \sqrt{1 - \frac{\mu_1\varepsilon_1}{\mu_2\varepsilon_2}\sin^2\theta_i} \tag{2.70}$$

Then, we rewrite (2.68) and (2.69) as:

$$\Gamma_{TM} = \frac{\eta_2\sqrt{1 - \frac{\mu_1\varepsilon_1}{\mu_2\varepsilon_2}\sin^2\theta_i} - \eta_1\cos\theta_i}{\eta_2\sqrt{1 - \frac{\mu_1\varepsilon_1}{\mu_2\varepsilon_2}\sin^2\theta_i} + \eta_1\cos\theta_i} \tag{2.71}$$

$$\tau_{TM} = \frac{2\eta_2\cos\theta_i}{\eta_2\sqrt{1 - \frac{\mu_1\varepsilon_1}{\mu_2\varepsilon_2}\sin^2\theta_i} + \eta_1\cos\theta_i} \tag{2.72}$$

It is evident from (2.71) and (2.72) that the reflection and transmission of waves through the interface of two media depend not only on the constitutive parameters ε_1, ε_2, μ_1, and μ_2, but also on the angle of incidence θ_i. One may also wish to make explicit in the reflection coefficient the features of a medium 2 characterized for different angles of incidence. To this end, we divide every term of (2.71) by $\cos\theta_i$, leading to:

$$\Gamma_{TM} = \frac{\eta_2\frac{\sqrt{1 - \frac{\mu_1\varepsilon_1}{\mu_2\varepsilon_2}\sin^2\theta_i}}{\cos\theta_i} - \eta_1}{\eta_2\frac{\sqrt{1 - \frac{\mu_1\varepsilon_1}{\mu_2\varepsilon_2}\sin^2\theta_i}}{\cos\theta_i} + \eta_1} = \frac{\eta_2^{TM} - \eta_1}{\eta_2^{TM} + \eta_1} \tag{2.73}$$

where η_2^{TM} is the effective impedance of medium 2, in Ω, for TM polarization at a specific angle of incidence θ_i.

We can also do a similar development for an incident wave with TE polarization, as in Fig. 2.5b. In this case, we calculate the reflection and transmission coefficients as [14]:

$$\Gamma_{TE} = \frac{\eta_2 \cos\theta_i - \eta_1\sqrt{1 - \dfrac{\mu_1\varepsilon_1}{\mu_2\varepsilon_2}\sin^2\theta_i}}{\eta_2 \cos\theta_i + \eta_1\sqrt{1 - \dfrac{\mu_1\varepsilon_1}{\mu_2\varepsilon_2}\sin^2\theta_i}} \tag{2.74}$$

$$\tau_{TE} = \frac{2\eta_2 \cos\theta_i}{\eta_2 \cos\theta_i + \eta_1\sqrt{1 - \dfrac{\mu_1\varepsilon_1}{\mu_2\varepsilon_2}\sin^2\theta_i}} \tag{2.75}$$

Analogous to (2.73), one can make explicit the features of medium 2. From (2.74), we have:

$$\Gamma_{TE} = \frac{\eta_2 \dfrac{\cos\theta_i}{\sqrt{1 - \dfrac{\mu_1\varepsilon_1}{\mu_2\varepsilon_2}\sin^2\theta_i}} - \eta_1}{\eta_2 \dfrac{\cos\theta_i}{\sqrt{1 - \dfrac{\mu_1\varepsilon_1}{\mu_2\varepsilon_2}\sin^2\theta_i}} + \eta_1} = \frac{\eta_2^{TE} - \eta_1}{\eta_2^{TE} + \eta_1} \tag{2.76}$$

where η_2^{TE} is the effective impedance of medium 2, in Ω, for TE polarization at a specific angle of incidence θ_i.

From (2.73) and (2.76), it is apparent that the effective impedance of medium 2 makes the reflection and transmission of waves dependent not only on the angle of incidence θ_i, but also on the polarization of the impinging wave. As one can imagine, the above conclusion will present an impact in the functioning of metasurfaces, requiring in some applications a number of techniques to enhance their performance in these matters.

Lastly, the fact that (2.73) and (2.76) present the same form of (2.50) allows the use of the schematic analogy (see Fig. 2.4) also for the case of oblique incidence of waves.

2.1.7 Surface and Evanescent Waves

Suppose that a TM-polarized wave impinges over an interface between two source-free ($\boldsymbol{J}_i = q_{ev} = 0$, and $\boldsymbol{M}_i = q_{mv} = 0$) and lossless ($\boldsymbol{J}_c = 0$) media with $\mu_1\varepsilon_1 > \mu_2\varepsilon_2$. According to (2.62), in this case $k_1 > k_2$, which leads to $\theta_t > \theta_i$. One can see in (2.70) that the condition $\mu_1\varepsilon_1 > \mu_2\varepsilon_2$ can lead to a complex value depending on the angle of incidence θ_i, specifically if:

$$\theta_i > \arcsin\sqrt{\frac{\mu_2\varepsilon_2}{\mu_1\varepsilon_1}} = \theta_c \tag{2.77}$$

where θ_c is known as the critical angle. Hence, from (2.58), we can rewrite the transmitted electric field as:

$$\begin{aligned}
\boldsymbol{E}^t &= (\hat{\mathbf{a}}_{\mathbf{x}}\cos\theta_t - \hat{\mathbf{a}}_{\mathbf{z}}\sin\theta_t)\tau_{TM}E_0 e^{-jk_2 z\cos\theta_t} e^{-jk_2 x\sin\theta_t} \\
&= (\hat{\mathbf{a}}_{\mathbf{x}}\cos\theta_t - \hat{\mathbf{a}}_{\mathbf{z}}\sin\theta_t)\tau_{TM}E_0 e^{-jk_2 z\sqrt{1-\frac{\mu_1\varepsilon_1}{\mu_2\varepsilon_2}\sin^2\theta_i}} e^{-jk_2 x\sqrt{\frac{\mu_1\varepsilon_1}{\mu_2\varepsilon_2}}\sin\theta_i} \\
&= (\hat{\mathbf{a}}_{\mathbf{x}}\cos\theta_t - \hat{\mathbf{a}}_{\mathbf{z}}\sin\theta_t)\tau_{TM}E_0 e^{-jk_2 z\left(-j\sqrt{\frac{\mu_1\varepsilon_1}{\mu_2\varepsilon_2}\sin^2\theta_i - 1}\right)} e^{-jk_2 x\sqrt{\frac{\mu_1\varepsilon_1}{\mu_2\varepsilon_2}}\sin\theta_i} \\
&= (\hat{\mathbf{a}}_{\mathbf{x}}\cos\theta_t - \hat{\mathbf{a}}_{\mathbf{z}}\sin\theta_t)\tau_{TM}E_0 e^{-k_2 z\sqrt{\frac{\mu_1\varepsilon_1}{\mu_2\varepsilon_2}\sin^2\theta_i - 1}} e^{-jk_2 x\sqrt{\frac{\mu_1\varepsilon_1}{\mu_2\varepsilon_2}}\sin\theta_i}
\end{aligned} \tag{2.78}$$

The phase of $\boldsymbol{E}^t$ varies and propagation occurs only in the x dimension, for what this wave is known as a surface wave. In the z dimension, the amplitude of $\boldsymbol{E}^t$ decays. For this reason, we can also call this wave an evanescent wave.

We illustrate the phenomena related to (2.78) first with the example of a plane wave impinging upon the interface of media 1 and 2 ($\varepsilon_1 = 2.2\varepsilon_2$, and $\mu_1 = \mu_2$) with an angle of incidence $\theta_i = \theta_c = 42.4°$. According to (2.70), this condition leads to an extreme angle of propagation of $\theta_t = 90°$ for the transmitted wave, which is a surface wave. Figure 2.7a illustrates only the incident and transmitted electric field. The arrows indicate their angles of propagation with respect to the normal of the interface.

Here, refraction deflects the angle of propagation of the transmitted wave until the limit ($\theta_t = 90°$) in order to match the wavelengths of both fields projected on

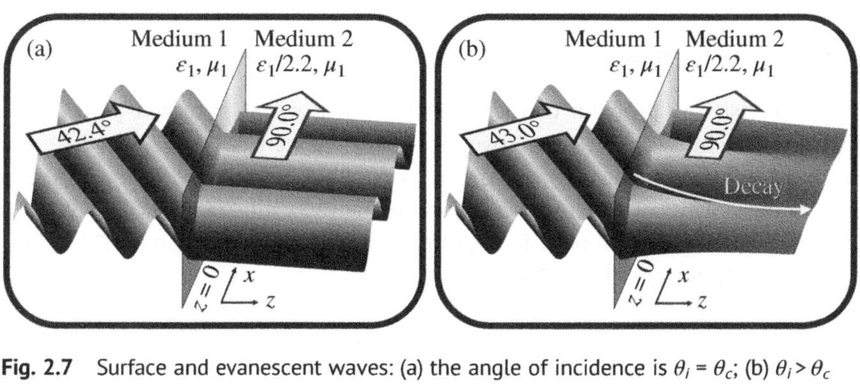

Fig. 2.7 Surface and evanescent waves: (a) the angle of incidence is $\theta_i = \theta_c$; (b) $\theta_i > \theta_c$

the interface. Substituting $\theta_i = \arcsin\sqrt{\mu_2\varepsilon_2/\mu_1\varepsilon_1}$ in (2.78) makes the first exponential equal 1, leading to no variation of fields in the z dimension.

Now, consider in Fig. 2.7b an angle of incidence $\theta_i = 43.0° > \theta_c$, meeting condition (2.77). In this case, refraction deflects again the angle of propagation of the transmitted wave until the limit ($\theta_t = 90°$). However, as (2.78) indicates, the transmitted waves suffer a decay in the $+\hat{\mathbf{a}}_z$ direction. The practical interpretation for this phenomenon is that the wavelength of the incident wave projected on the interface is shorter than the one that medium 2 can support. Therefore, the transmitted wave vanishes as the observation point moves away from the interface z dimension.

2.1.8 Takeaways

The concepts presented hitherto provide readers with everything they need to understand the physics behind unusual properties for the propagation of waves in media with both relative permittivity ε_r and permeability μ_r equal to −1. Further, if the math is important to understand how we arrive at these concepts, the interpretation of some phenomena is paramount to realize how metasurfaces interact electromagnetic waves.

Table 2.1 summarizes the main phenomena studied so far, as well as valuable interpretations thereof.

2.2 Unusual Properties for the Propagation of Waves

In the hypothetical medium proposed by Veselago [3], both relative electric permittivity ε_r and magnetic permeability μ_r are equal to −1. In this section, we investigate the consequences of such features for the propagation of electromagnetic waves.

2.2.1 Negative Refraction

Let us revisit the problem of the transmission of a TM-polarized plane wave through the interface between two isotropic, source-free ($\boldsymbol{J}_i = q_{ev} = 0$, and $\boldsymbol{M}_i = q_{mv} = 0$), and lossless ($\boldsymbol{J}_c = 0$) media. An oblique angle of incidence θ_i in relation to the normal of the interface exists, as Fig. 2.5a illustrated earlier. We can calculate the angle of propagation θ_t of the transmitted wave using (2.70). However, two solutions exist for (2.70):

$$\theta_t = \pm \arccos\sqrt{1 - \frac{\mu_1\varepsilon_1}{\mu_2\varepsilon_2}\sin^2\theta_i} \tag{2.79}$$

Table 2.1 Summary of important fundamentals in electromagnetics

Fundamental	Takeaway	Reference
Maxwell's equations	Electric and magnetic fields and flux densities are all related at any point in space and time.	Eqs. (2.1)–(2.4)
Boundary conditions	Fields of electromagnetic waves are continuous functions of position and time and exhibit continuous derivatives, except when impinging over surfaces or interfaces between media, when they can meet charges and currents and present discontinuities.	Eqs. (2.11)–(2.14)
Complex representation	Calculations can be simplified at a single frequency f.	Eqs. (2.17)–(2.22)
Wave equations	The features of electromagnetic waves, such as the wavelength λ, the wavenumber k, and the phase velocity v_{ph}, are defined by the frequency f and the medium's properties.	Eqs. (2.29), (2.31), Figs. 2.1, and 2.2
Impedance matching	The reflection and transmission of waves through the interface of two media are ruled by the matching of their intrinsic impedances.	Eqs. (2.50), (2.51), and Fig. 2.3
Equivalent circuit	Intrinsic impedances can be mapped onto lumped impedances in a schematic, which can simplify problems related to metasurfaces.	Fig. 2.4
Refraction	Waves transmitted through interfaces between media are deflected in order to respect the continuity of tangential fields required by boundary conditions.	Eq. (2.62) and Fig. 2.6
Oblique incidence	The effective transmission and reflection coefficients depend on the angle of incidence θ_i and the polarization of the impinging wave.	Eq. (2.73) and (2.76)
Evanescent waves	When the wavelength of the incident wave projected over the interface is shorter than the wavelength supported by medium 2, the wave vanishes in medium 2.	Eq. (2.78) and Fig. 2.7b

At a first glance, one may think that (2.79) provides us an ambiguity that only experimentation can solve. However, the boundary conditions play again an important role on the resolution of the problem, this time by imposing continuity of the electric flux density normal to the interface, as defined by (2.11).

In Fig. 2.8, we show only the incident and transmitted fields for this case. Further, we decompose the electric fields in the z and x dimensions. Observe that for the case $\theta_t > 0$, E_z^i and E_z^t are in the same direction. On the other hand, such vectors

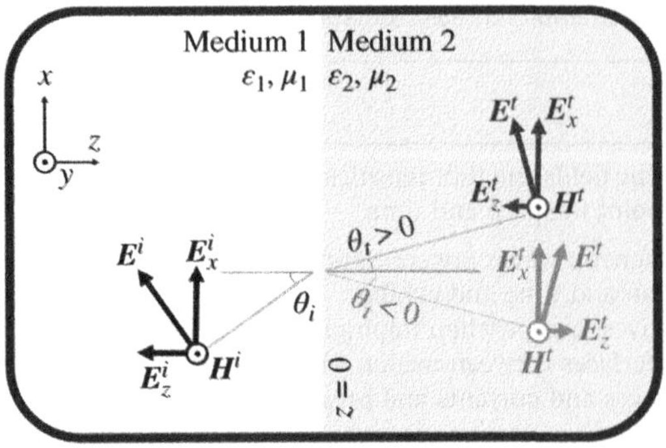

Fig. 2.8 Disposition of fields for the two possible angles of propagation θ_t of the transmitted wave

present opposite directions when $\theta_t < 0$. Since the media are source-free, the condition $\varepsilon_1 E^i_z = \varepsilon_2 E^t_z$ must hold according to boundary condition (2.11) and constitutive relation (2.22). Therefore, we can obtain an angle of propagation for the transmitted wave $\theta_t > 0$ only if ε_1 and ε_2 present the same sign. On the other hand, $\theta_t < 0$ only if ε_1 and ε_2 present opposite signs, a case called negative refraction.

The latter is exactly the case when medium 1 is the free space and medium 2 is Veselago's medium ($\varepsilon_2 = -\varepsilon_1$, and $\mu_2 = -\mu_1$). Figure 2.9 depicts a negative refraction when the angle of incidence is $\theta_i = 35°$ (we represent only incident and transmitted electric fields). Note that the negative refraction is not the only unusual phenomenon in Fig. 2.9. The arrow over medium 2 indicates that the wavefronts of the transmitted wave propagate not from the interface, but toward it.

2.2.2 Negative Phase Velocity and Backward Waves

The propagation of wavefronts toward the interface is another consequence of boundary conditions. Note in Fig. 2.9 that the incident electric field, in the x dimension, propagates in the positive direction ($+\hat{\mathbf{a}}_x$). The only way of having

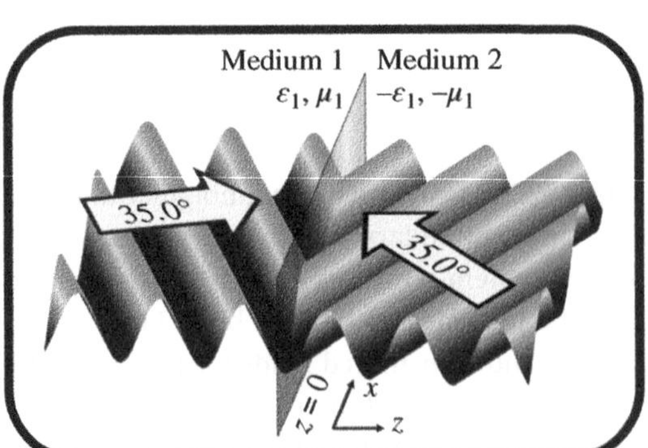

Fig. 2.9 Negative refraction and backward waves

continuity of tangential electric field over the interface is that the projection of the transmitted electric field also propagates at the $+\hat{\mathbf{a}}_x$ direction.

According to (2.58), a propagation for the transmitted electric field in the $+\hat{\mathbf{a}}_x$ direction with $\theta_t < 0$ is possible only if the wavenumber in medium 2 is $k_2 < 0$. Actually, (2.30) allows for a negative wavenumber when both the permittivity and magnetic permeability are negative, as follows:

$$k = \omega\sqrt{(-|\mu|)(-|\varepsilon|)} = \omega\sqrt{(-1)^2|\mu||\varepsilon|} = -\omega\sqrt{|\mu||\varepsilon|} \tag{2.80}$$

Next, using (2.36), we calculate the phase velocity as:

$$v_{ph} = \frac{\omega}{k} = -\frac{1}{\sqrt{|\mu||\varepsilon|}} \tag{2.81}$$

We call waves with a negative wavenumber k and phase velocity v_{ph} backward waves.

It is worthwhile to note that simultaneously negative permittivity and permeability do not reverse the direction of the Poynting vector $\mathcal{P}$, which indicates the flux of the electromagnetic energy. We can check that by using negative values for μ and ε in (2.6) and (2.7) and realizing that reverse directions for both electric and magnetic fields occur in (2.15) which in turn do not change the direction of $\mathcal{P}$. That is why the phase velocity $\mathbf{v}_{ph}$ and Poynting vector $\mathcal{P}$ of backward waves present opposite directions as Fig. 1.7b sketched earlier. With backward waves, energy can propagate from a Tx antenna to an Rx antenna whereas the wavefronts within the energy envelope propagate from the Rx to the Tx antenna. However, thinking that the wavefronts depart from the Rx antenna before its source energy arrives goes against the principle of causality. This question deserves an explanation and will be clarified in Section 2.3.

2.2.3 Amplification of Evanescent Waves

Another unusual property of a medium with simultaneous negative permittivity and permeability is the amplification of evanescent waves. We can verify this phenomenon by directly substituting negative values for these quantities in (2.78):

$$\begin{aligned}\boldsymbol{E}^t &= (\hat{\mathbf{a}}_\mathbf{x}\cos\theta_t - \hat{\mathbf{a}}_\mathbf{z}\sin\theta_t)\tau_{TM}E_0 e^{-(-|k_2|)z\sqrt{\frac{\mu_1\varepsilon_1}{(-|\mu|)(-|\varepsilon|)}\sin^2\theta_i - 1}}\, e^{-j(-|k_2|)x\sqrt{\frac{\mu_1\varepsilon_1}{(-1)^2|\mu_2||\varepsilon_2|}}\sin\theta_i} \\ &= (\hat{\mathbf{a}}_\mathbf{x}\cos\theta_t - \hat{\mathbf{a}}_\mathbf{z}\sin\theta_t)\tau_{TM}E_0 e^{+|k_2|z\sqrt{\frac{\mu_1\varepsilon_1}{|\mu_2||\varepsilon_2|}\sin^2\theta_i - 1}}\, e^{-j|k_2|x\sqrt{\frac{\mu_1\varepsilon_1}{|\mu_2||\varepsilon_2|}}\sin\theta_i}\end{aligned} \tag{2.82}$$

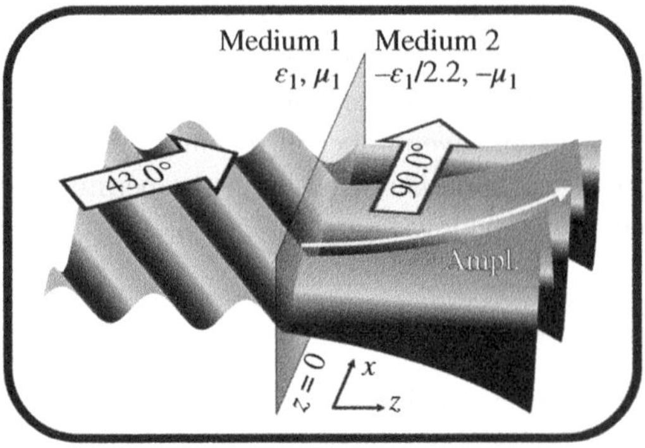

Fig. 2.10 Amplification of evanescent waves

In (2.82), the first exponential has a positive argument, leading to an amplification of the transmitted electric field in the $+\hat{\mathbf{a}}_z$ direction. The second exponential keeps like in (2.78) in spite of the negative wavenumber k_2, meaning that propagation still occurs strictly at the x dimension, as well as phase variations.

We illustrate the amplification of evanescent waves with an example similar to the one of Fig. 2.7b, but now the plane wave impinges upon the interface of a conventional medium and a Veselago's one. In this case, we define that $\varepsilon_2 = -\varepsilon_1/2.2$, $\mu_2 = -\mu_1$, and the angle of incidence is $\theta_i = 43.0°$. Fig. 2.10 depicts only the incident and transmitted electric fields. Now, the propagation in Veselago's medium occurs tangential to the interface, exactly like in Fig. 2.7b. However, now, we have an amplification of the transmitted electric field in the $+\hat{\mathbf{a}}_z$ direction instead of having a decay. Considering that both media are passive, one can call into question the energy conservation principle. This subject also requires an explanation, which will be given in Section 2.3.

2.3 Demystifying the Break of Fundamental Laws

The unusual properties that a medium with simultaneous negative permittivity and permeability ($\varepsilon < 0$, and $\mu < 0$) would provide for the propagation of electromagnetic waves, like negative refraction, backward waves, and the amplification of evanescent, raise questions upon the causality and energy conservation principles.

Indeed, the derivation of such phenomena can suggest misleading concepts if one does not pay attention to the fact that the equations are in the complex representation, valid for a single frequency. In this scenario, thinking that a switch will be turned on and then these unusual phenomena will readily take place is not precise, because a single-frequency operation means that continuous waves impregnate both media. In fact, the time-domain solution for this problem should consider the convolution of fields with the response of the media, as in (2.6) and (2.7), while respecting Maxwell's equations in the instantaneous form (2.1)–(2.4). The complexity of this scenario is way above the one made explicit in the complex representation. There is where the catch is. The time-domain solution to a medium with both $\varepsilon_r = -1$ and $\mu_r = -1$ driven from the instant $t = 0$ by a time-harmonic source approaches continuous wave fields only as $t \rightarrow \infty$ [15].

In practice, we can consider that the unusual phenomena will take place when fields achieve a steady state, that is, when transient signals are insignificant in comparison to the energy driving the media. Therefore, wavefronts propagating from the Rx antenna to the Tx antenna do not necessarily mean a break in causality. Something happened before the establishment of this unusual mode of propagation. Similarly, an amplification of evanescent waves in a passive medium does not necessarily mean that energy conservation has been broken, considering that the medium could store energy before the achievement of the steady state.

These concepts are tricky, but the related mechanisms become clear when one understands how electromagnetic waves and matter interact.

2.3.1 Mechanisms Behind Constitutive Parameters of Media

The intrinsic impedance $\eta_0 = \sqrt{\mu_0/\varepsilon_0} \cong 120\pi\ \Omega$ defines the relation between the electric and magnetic fields of electromagnetic waves when they propagate through the free space, as we discussed in Section 2.1.4. However, when we apply electric and magnetic fields within materials, they can interact with the bound charges and lead to an intrinsic impedance different from η_0.

Consider a volume v of finite extension in the x dimension and infinite extension in both y and z dimensions embedded in the free space. Volume v is made of an isotropic dielectric material with a number N_e [units/m^3] of positive bound charges $+q_e$ [C] (protons) and negative bound charges $-q_e$ (electrons) per unit volume Δv in a lattice disposition. Fig. 2.11a depicts such volume in the absence of external fields. For the sake of simplicity, consider that the material's molecules are non-polar, and the negative charges are randomly moving around the positive charges they are bound to.

In Fig. 2.11a, the electric and magnetic fields originated from the bound charges cancel each other and can be considered null on a macroscopic scale. Next, consider in Fig. 2.11b an applied electric field $\boldsymbol{E}_0$ in the region the volume v is embedded in. For now, consider that $\boldsymbol{E}_0$ is static. Such field displaces the bound charges within volume v through a distance $\boldsymbol{l}_n$ [m], causing the alignment and polarization thereof. As a result, an electric field $\boldsymbol{E}_n$ is established from each polarized positive bound charge to its respective negative bound charge. Due to its infinite extension in both y and z dimensions, volume v has an also infinite number of

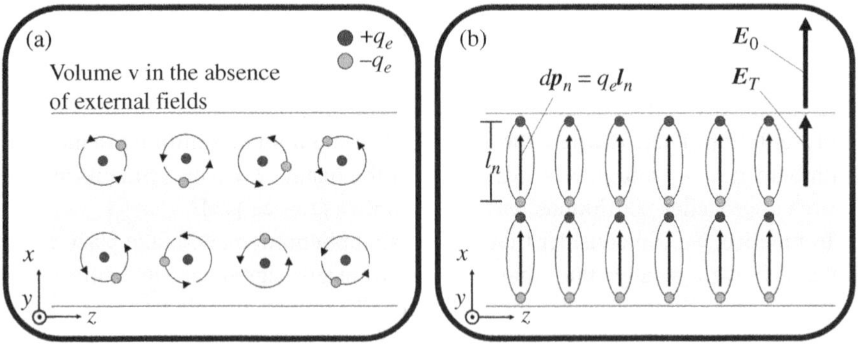

Fig. 2.11 Material under the influence of an external electric field: (a) a random disposition of particles exists without the field; (b) the polarization of charges changes the resulting field

polarized charge pairs. In the microscopic scale, we can calculate the total electric field $\boldsymbol{E}_T$ at an observation point as the superposition of the original field $\boldsymbol{E}_0$ with the fields originated by each of these pairs in the lattice:

$$\boldsymbol{E}_T = \boldsymbol{E}_0 + \sum_{n=1}^{\infty} \boldsymbol{E}_n \tag{2.83}$$

To avoid using a complicated formulation for the microscopic lattice structure, which would require the account for a field distribution and the distance of each of the infinity charges to an observation point, it is easier to consider in a macroscopic scale that an electric polarization vector $\boldsymbol{P}$ exists. This is possible because the size of the particles is much smaller than the wavelength, allowing us to consider average fields around each particle. When volume v is polarized, N_e electric dipole moments $\boldsymbol{dp}_n = q_e \boldsymbol{l}_n$ $[C \times m]$ are created per unit volume Δv, as in Fig. 2.11b. The electric polarization vector $\boldsymbol{P}$ is defined by the density of dipole moments $\boldsymbol{dp}_n$ per unit volume Δv [16]:

$$\boldsymbol{P} = \lim_{\Delta v \to 0} \left(\frac{1}{\Delta v} \sum_{n=1}^{N_e \Delta v} \boldsymbol{dp}_n \right) \tag{2.84}$$

The unit of $\boldsymbol{P}$ is $C \times m/m^3 = C/m^2$. In the illustration of Fig. 2.11b, the polarization of the particles makes positive charges move next to the top interface between the volume v and free space. Also, negative charges approach the bottom interface. In the middle, positive and negative charges cancel each other. Therefore, the electric polarization vector $\boldsymbol{P}$ works as a surface electric charge density q_{es}. Using boundary condition (2.11), we have:

$$\varepsilon_0 \boldsymbol{E}_0 - \varepsilon_0 \boldsymbol{E}_T = \boldsymbol{P} \tag{2.85}$$

which leads to:

$$\varepsilon_0 \boldsymbol{E}_0 = \varepsilon_0 \boldsymbol{E}_T + \boldsymbol{P} = \varepsilon \boldsymbol{E}_T \tag{2.86}$$

Observe that, considering the geometrical alignment of $\boldsymbol{E}_0$, $\boldsymbol{E}_T$ and $\boldsymbol{P}$ in Fig. 2.11b, (2.86) indicates that $|E_T| < |E_0|$ for the static case. This is reasonable since the charges are aligned in opposition to the external field $\boldsymbol{E}_0$.

Similar to the exemplified case, an external magnetic field $\boldsymbol{H}_0$ can impose a torque that aligns molecules according to currents resulted from the orbiting and spinning of electrons in magnetic materials. Then, we can define a magnetic permeability μ to represent the material's magnetic response in the macroscopic scale, simplifying the calculation of the total magnetic field $\boldsymbol{H}_T$ therein [17].

2.3.2 Typical Resonant Behaviors of Material

Now consider that the applied electric field $\boldsymbol{E}_0$ in Fig. 2.11b is a time-harmonic field of frequency f. The oscillations of the applied field displace the bound charges back and forth making the electric polarization vector $\boldsymbol{P}$ oscillate accordingly.

In an ideal material, where all internal forces are neglected, nothing opposes the movement of the bound charges, which are perfectly synchronized with the total electric field $\boldsymbol{E}_T$. For this reason, we can expect a purely real electric permittivity, meaning that the displacement $\boldsymbol{l}_n$ of the particles will be in phase with $\boldsymbol{E}_T$. However, in practice, the particles present a mass whose inertia we cannot neglect. Also, their displacement is damped through friction. Moreover, other complex phenomena of atomic, inter, and intramolecular nature compel the material to return to its original arrangement. As a result, the displacement $\boldsymbol{l}_n$ of bound charges can be out of phase with the total field $\boldsymbol{E}_T$, and we should consider a complex relative permittivity [18].

Due to interactions between atomic, inter, and intramolecular forces, the particles of a material may oscillate at a natural frequency f_0 even in the absence of any external force. Then, when an external electric field $\boldsymbol{E}_0$ is applied, we can find different resonant behaviors depending on its frequency f and the material natural frequency f_0 [18].

It is easier to understand these behaviors by making an analogy with a situation that we can actually see. Thus, consider the example of a swing set made of a kid suspended by a rope of specific length and elasticity, and also by bolts, chains, hangers, and connectors, all with specific masses and friction features. The kid can swing on her own. In this case, she will do so at the natural frequency f_0 of the system, which depends on the properties of the swing set parts, air resistance, and gravity. In Fig. 2.12, an external force, let us say the kid's older brother, conducts her back and forth at different frequencies f. The older brother should make a huge effort in Fig. 2.12a to slowly displace the kid through a considerable range.

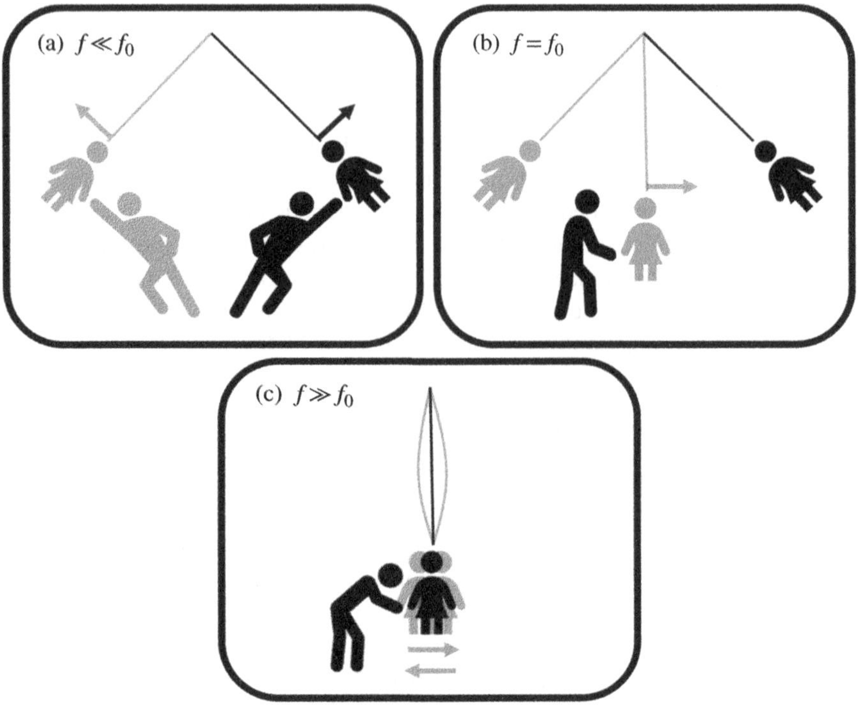

Fig. 2.12 Resonant behaviors: (a) at low frequencies, force and displacement are in phase; (b) at resonance, force and displacement are 90° out of phase and the system stores energy; (c) at high frequencies, force and displacement are 180° out of phase

The effort is maximum when the rope is at the extremities. Therefore, when $f \ll f_0$, force and displacement are in phase.

Consider now in Fig. 2.12b that, after some training, the older brother finds out that there is a rhythm he can be more effective with. He realizes that if he applies an impulse to the kid whenever she is in the middle of the displacement range, he can add more and more energy to the system after each cycle, making the play funnier (and riskier). Hence, when $f = f_0$, force and displacement are 90° out of phase, the system is at resonance and stores energy.

Lastly, in Fig. 2.12c, the older brother starts to shake the kid at a frequency f much higher than the natural frequency f_0. He makes a huge effort again. When the kid is in the left, he applies force to the right and vice-versa. Thus, when $f \gg f_0$, force and displacement are 180° out of phase.

Resonance behaviors similar to those exemplified in Fig. 2.12 can also occur when a time-harmonic electric field $\boldsymbol{E}_0$ of frequency f is applied over a dielectric with a natural frequency f_0. We can use the Lorentz oscillator model to quantitatively represent such behaviors in the complex permittivity of the material. In this approach, we model the instantaneous displacement ℓ_n of the dipole moments in the material as an equivalent mass-spring-damper system, as follows [19]:

$$\underbrace{m_e \frac{\partial^2 \boldsymbol{\ell}_n}{\partial t^2}}_{\text{inertia}} + \underbrace{m_e d_e \frac{\partial \boldsymbol{\ell}_n}{\partial t}}_{\text{friction}} + \underbrace{m_e \omega_0^2 \boldsymbol{\ell}_n}_{\substack{\text{restoring}\\ \text{force}}} = \underbrace{q_e \boldsymbol{\mathcal{E}}_T}_{\substack{\text{driving}\\ \text{force}}} \tag{2.87}$$

where

- m_e is the mass of an electron, in kilograms [kg];
- d_e is the damping coefficient for the friction of electrons [s^{-1}];
- $\omega_0 = 2\pi f_0$ is the natural angular frequency of the dielectric [rad/s].

Using the complex representation, (2.87) becomes:

$$(-m_e\omega^2 - j\omega m_e d_e + m_e\omega_0^2)\boldsymbol{l}_n = q_e \boldsymbol{E}_T \tag{2.88}$$

Solving (2.88) for $\boldsymbol{l}_n$, we have:

$$\boldsymbol{l}_n = \frac{q_e}{m_e} \frac{\boldsymbol{E}_T}{(\omega_0^2 - \omega^2 - j\omega d_e)} \tag{2.89}$$

Since the electric dipole moment is defined as $d\boldsymbol{p}_n = q_e \boldsymbol{l}_n$, we have:

$$d\boldsymbol{p}_n = \frac{q_e^{\,2}}{m_e} \frac{\boldsymbol{E}_T}{(\omega_0^2 - \omega^2 - j\omega d_e)} \tag{2.90}$$

From (2.90), we can define an electric polarizability p_ε $[F \times m^2]$ for a single particle as:

$$p_\varepsilon = \frac{d\boldsymbol{p}_n}{\boldsymbol{E}_T} = \frac{q_e^{\,2}}{m_e} \frac{1}{(\omega_0^2 - \omega^2 - j\omega d_e)} \tag{2.91}$$

It is important to retain here that the electric polarizability p_ε is a quantity representing how sensitive a particle is to create dipole moments when under the influence of an electric field. This concept is exhaustively exploited both in metamaterials and metasurfaces and is of central importance to the reader. In anisotropic media, $\hat{p}_\varepsilon$ is a 3×3 tensor representing the relationship between each component of $d\boldsymbol{p}_n$ and each component of $\boldsymbol{E}_T$ denoted as:

$$\begin{bmatrix} dp_{n_x} \\ dp_{n_y} \\ dp_{n_z} \end{bmatrix} = \begin{bmatrix} p_{\varepsilon_{xx}} & p_{\varepsilon_{xy}} & p_{\varepsilon_{xz}} \\ p_{\varepsilon_{yx}} & p_{\varepsilon_{yy}} & p_{\varepsilon_{yz}} \\ p_{\varepsilon_{zx}} & p_{\varepsilon_{zy}} & p_{\varepsilon_{zz}} \end{bmatrix} \begin{bmatrix} E_x \\ E_y \\ E_z \end{bmatrix} \tag{2.92}$$

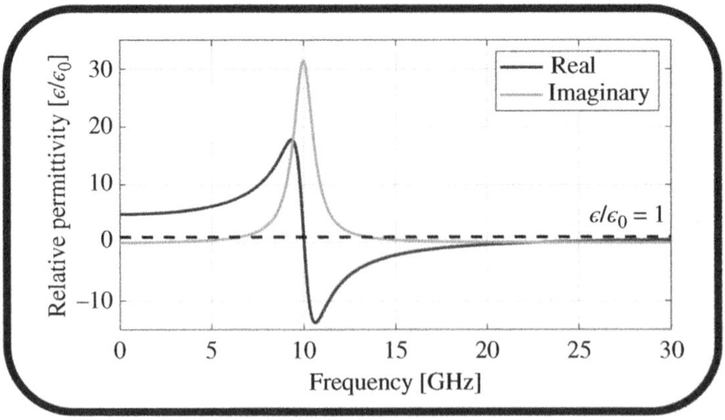

Fig. 2.13 Relative permittivity $\varepsilon_r = \varepsilon/\varepsilon_0$ of a generic dielectric with a natural frequency at $f_0 = 10$ GHz, $N_e = 4.962 \times 10^{18}$ units/m^3, and $d_e = 5.0 \times 10^9$ s^{-1}

Replacing (2.90) and (2.84) in (2.86), and then using (2.91), we can represent the electric permittivity of the material as:

$$\hat{\varepsilon} = \varepsilon_0 + \frac{N_e q_e{}^2}{m_e} \frac{1}{(\omega_0^2 - \omega^2) - j\omega d_e} = \varepsilon_0 + \hat{p}_\varepsilon N_e \tag{2.93}$$

Next, we exemplify the behavior of an isotropic, generic dielectric with a natural frequency at $f_0 = 10$ GHz, considering $N_e = 4.962 \times 10^{18}$ positive bound charges per m^3, a value $q_e = 1.602 \times 10^{-19}$ C for the positive charges, a mass $m_e = 9.109 \times 10^{-31}$ kg for the electrons, and a damping coefficient $d_e = 5.0 \times 10^9$ s^{-1}. This set of features does not represent any one particular material. Figure 2.13 depicts the relative electric permittivity using (2.93).

Note in Fig. 2.13 that at low frequencies the relative electric permittivity ε_r is purely real and greater than one. This results from the fact that, according to (2.89), a frequency f much lower than the material natural frequency f_0 leads to a displacement $\boldsymbol{l}_n$ in phase with the total field $\boldsymbol{E}_T$. This situation is analogous to the example of Fig. 2.12a where the force made by the older brother is in phase with the displacement of the kid. Using (2.86), we can see that $|E_T| < |E_0|$ in this case, which approaches the static case of Fig. 2.11b.

On the other hand, at high frequencies the relative electric permittivity is purely real and below the $\varepsilon_r = 1$ level. In (2.89), a frequency f much higher than f_0 results in negative values, meaning that the displacement $\boldsymbol{l}_n$ is 180° out of phase with the total field $\boldsymbol{E}_T$. This is similar to what happens in Fig. 2.12c, where the movements of the older brother are too fast, and the force he makes is 180° out of phase with the kid's displacement. In this case, a negative electric polarizability p_ε leads to a relative permittivity $\varepsilon_r < 1$ in Fig. 2.13. As such, the magnitude of the total field $|E_T|$ is higher than the one of the applied field $|E_0|$ in (2.86).

We can see in Fig. 2.13 that when $f = f_0 = 10$ GHz, the imaginary part of ε_r is maximum. An $\boldsymbol{l}_n$ and $\boldsymbol{E}_T$ 90° out of phase characterizes this situation. A similarity with the swing set is again seen, now in Fig. 2.12b, where the older brother realizes that the right rhythm can maximize the kid's displacement. Here, the electric polarizability p_ε is purely imaginary and, analogous to the example of Fig. 2.12b, can achieve high magnitudes due to the storage of energy if a small damping d_e is present.

Note in Fig. 2.13 that it is possible to obtain at some frequencies, like $f = 15$ GHz, permittivities with a negative real part and neglectable imaginary part. Applying such a permittivity in (2.86), we obtain that $\varepsilon_0 \boldsymbol{E}_0 \cong -|\varepsilon|\boldsymbol{E}_T$. This is an indication that the polarizability p_ε, besides being negative, is so strong that the total field $\boldsymbol{E}_T$ is out of phase not only with the displacement $\boldsymbol{l}_n$ of the particles, but also with the applied field $\boldsymbol{E}_0$. Plasmas can provide permittivities with a negative real part and neglectable imaginary part at microwave frequencies [20].

We may also draw the above conclusions concerning resonant phenomena for the magnetic permeability μ of a material [21]. We emphasize that among these conclusions is the fact that strong, resonating magnetic polarizabilities p_μ $[H \times m^2]$ can also provide permeabilities with a negative real part and neglectable imaginary part. Ferrites can provide this kind of permeability at microwave frequencies [20].

Although the macroscopic approach with polarization vectors facilitates calculations, understanding what is happening within the material structure is essential to visualize how the interaction between electromagnetic waves and matter can provide unusual properties. The electric and magnetic polarizabilities p_ε and p_μ play a central role in this matter because they make explicit the displacement of charges according to a total electric or magnetic field. In other words, they make clear how the fields of the polarized particles are superimposed with the original fields, leading to a total field that may be atypic.

Since resonant processes take place before the establishment of negative permittivity and permeability, a medium exhibiting both simultaneously would break neither causality nor energy conservation principles with effects resulting from these properties. It is fair to say that unusual phenomena like negative refraction, backward waves, and amplification of evanescent waves, studied in Section 2.2, are the macro manifestation of local oscillations of both bound charges and currents. Electric and magnetic fields applied at a specific frequency close to the natural frequencies of the electric and magnetic responses of the material would make bound charges move, let us say, in a weird rhythm. The establishment of this condition takes place when $t \rightarrow \infty$ or at least in a steady state, after energy is stored. In the case of the amplification of evanescent waves, it is important to add that, after reaching the steady state, there is no transmission of energy in the direction the fields grow. Propagation is strictly parallel to the interface, as (2.82) indicates. That said, a medium with such properties would not go against fundamental laws.

2.4 Final Remarks of the Chapter

The unusual properties of metamaterials and metasurfaces caused spirited discussions among researchers in a first moment. The physics associated to them may be tricky depending on the form of Maxwell's equations we use in the analysis. This chapter made a review on the most pertinent points we should understand to visualize how metamaterials and metasurfaces work.

It was shown that resonant phenomena can lead to a negative electric permittivity and magnetic permeability. Further, a negative refraction, the existence of backward waves and the amplification of evanescent waves can take place without the break of any fundamental law if these features occur at the same frequency.

However, one cannot find in nature any material whose electric and magnetic responses present respective natural frequencies so as to achieve a negative electric permittivity and magnetic permeability at the same frequency.

Such impossibility becomes reasonable when we realize that, in natural materials, both electric and magnetic responses depend on the same particles, that is, on bound charges and currents resulted from the movement thereof. Additionally, the lack of a simultaneous negative permittivity and permeability in nature may be a suggestion that they are conflicting phenomena in conventional material.

Finally, if this suggestion proceeds, it is also reasonable that independent sources must provide the electric and magnetic responses of a medium to allow simultaneous negative properties. Here is where the concept of metamaterials comes into play.

References

1 Houck, A.A., Brock, J.B., Chuang, I.L. (2003). Experimental observations of a left-handed material that obeys snell's law. *Physics Review Letters* 90 (137401): 1–4. https://doi.org/10.1103/PhysRevLett.90.137401

2 Yu, N., Genevet, P., Kats, M.A. et al. (2011). Light propagation with phase discontinuities: generalized laws of reflection and refraction. *Science* 334 (6054): 333–337. https://doi.org/10.1126/science.1210713

3 Veselago, V.G. (1968). The electrodynamics of substances with simultaneously negative values of ε and μ. *Physics-Uspekhi* 10 (4): 509–514.

4 Balanis, C.A. (2012). *Advanced engineering electromagnetics*, 2nd ed. New York, NY: Wiley, Section 1.2.1, pp. 2–3.
5 Balanis, C.A. (2012). *Advanced engineering electromagnetics*, 2nd ed. New York, NY: Wiley, Section 1.3, pp. 5–7.
6 Balanis, C.A. (2012). *Advanced engineering electromagnetics*, 2nd ed. New York, NY: Wiley, Section 1.5.3, pp. 17–18.
7 Balanis, C.A. (2012). *Advanced engineering electromagnetics*, 2nd ed. New York, NY: Wiley, Section 1.6, pp. 18–21.
8 Balanis, C.A. (2012). *Advanced engineering electromagnetics*, 2nd ed. New York, NY: Wiley, Section 1.7.1, pp. 22.
9 Balanis, C.A. (2012). *Advanced engineering electromagnetics*, 2nd ed. New York, NY: Wiley, Section 1.7.3, pp. 25–29.
10 Balanis, C.A. (2012). *Advanced engineering electromagnetics*, 2nd ed. New York, NY: Wiley, Section 3.4.1A, pp. 102–107.
11 Balanis, C.A. (2012). *Advanced engineering electromagnetics*, 2nd ed. New York, NY: Wiley, Section 5.2, pp. 173–175.
12 Balanis, C.A. (2012). *Advanced engineering electromagnetics*, 2nd ed. New York, NY: Wiley, Section 5.3.2, pp. 182–184.
13 Hayt, W.H., Buck, J.A. (2012). *Engineering electromagnetics*, 8th ed. New York, NY: McGraw-Hill, section 12.5, Eq. 61, pp. 431.
14 Balanis, C.A. (2012). *Advanced engineering electromagnetics*, 2nd ed. New York, NY: Wiley, Section 5.3.1, pp. 178–180.
15 Yaghjian, A.D., Hansen, T.B. (2006). Plane-wave solutions to frequency-domain and time-domain scattering from magnetodielectric slabs. *Physical Review E* 73: 046608. https://doi.org/10.1103/PhysRevE.73.046608
16 Balanis, C.A. (2012). *Advanced engineering electromagnetics*, 2nd ed. New York, NY: Wiley, Section 2.2, pp. 41–46.
17 Balanis, C.A. (2012). *Advanced engineering electromagnetics*, 2nd ed. New York, NY: Wiley, Section 2.2, Section 2.3, pp. 48–51.
18 Balanis, C.A. (2012). *Advanced engineering electromagnetics*, 2nd ed. New York, NY: Wiley, Section 2.2, Section 2.9.1, pp. 68–79.
19 Lorentz, H.A. (2013). *Versuch einer Theorie der electrischen und optischen Erscheinungen in bewegten Körpern Leiden*. Cambridge, UK: Cambridge University Press (1895). https://doi.org/10.1017/CBO9781139381406
20 Balanis, C.A. (2012). *Advanced engineering electromagnetics*, 2nd ed. New York, NY: Wiley, Section 2.9.2, Section 5.7.1, pp. 229.
21 Balanis, C.A. (2012). *Advanced engineering electromagnetics*, 2nd ed. New York, NY: Wiley, Section 2.9.2, pp. 79–89.

3

Mechanisms Behind the Operation of Metasurfaces

Resonant phenomena can lead to a negative real part for both electric permittivity and magnetic permeability in materials excited by electromagnetic waves. Ultimately, negative real parts at the same frequency for both can yield a negative refraction, the existence of backward waves, and the amplification of evanescent waves.

We saw in Section 2.3.1 that a single quantity (the polarization vector $\boldsymbol{P}$) can represent the electric polarization of an infinity set of bound charges in a volume v, avoiding the use of a complicated formulation for the microscopic structure. Such representation is possible because the subwavelength size of the particles allows the adoption of average electric fields within each unit volume Δv. After, we saw in Section 2.3.2 that the electric polarizability p_e of local bound charges may present resonant behaviors due to the properties of the material. For this reason, oscillating bound charges may provide extra, out-of-phase fields stronger than the original one when under the influence of an external electric field.

Analogous phenomena are possible for the magnetic response of materials, originated from the movement of bound charges and an applied magnetic field. However, both electric and magnetic responses depend on the same particles in natural materials, and we do not know any one of them that has strong electric and magnetic resonances at the same frequency.

One can imagine that artificial resonant metal elements disposed in a lattice with a subwavelength periodicity could perform a role analog to that of bound charge pairs in natural materials. This is exactly the idea behind metamaterials, which we can also apply to metasurfaces.

This chapter first extends the concept of polarizabilities to artificial materials, explaining how we can obtain a negative electric permittivity and magnetic permeability therefrom. Then, we apply these concepts in the two-dimensional case, i.e., metasurfaces and present a series of approaches to make their analyses. Finally, we discuss challenges common to most of metasurfaces.

Metasurface-Driven Electronic Warfare, First Edition. Rafael Gonçalves Licursi de Mello.

Published 2025 by John Wiley & Sons, Inc.
Companion website: www.wiley.com/go/licursidemello/metasurfacedrivenelectronicwarfare1

3.1 Obtaining Arbitrary Medium Properties

From an electromagnetic point of view, it does not matter if pairs of bound charges generate extra fields in natural materials or metal structures do so in artificial materials. To exemplify such an idea, we refer to the seminal work of J. B. Pendry et al., where a negative electric permittivity was obtained from a structure composed of thin wires arranged in a cubic lattice [1]. Pendry et al. assumed that electrons in a medium completely made of a good conductor are not bound to any nucleus. As such, a restoring force in the material is meaningless and we disregard the third term of (2.87), making (2.93) become:

$$\varepsilon_{metal} = \varepsilon_0 - \frac{N_e {q_e}^2}{m_e} \frac{1}{\omega^2 + j\omega d_e} = \varepsilon_0 \left(1 - \frac{\omega_p^2}{\omega^2 + j\omega d_e} \right) \tag{3.1}$$

where $\omega_p^2 = N_e {q_e}^2/\varepsilon_0 m_e$ is known as the plasma frequency, that is, the frequency at which the permittivity passes from negative to positive in conductors whose d_e is small. Here, we interpret N_e as the electron density. We know this particular use of Lorentz model today as the Drude model for the electric response of metals [2]. At frequencies much lower than ω_p^2, the permittivity ε_{metal} presents a large imaginary part. At such conditions, a good impedance matching with the free space is impossible and the small amount of accepted energy quickly vanishes within the metal. To obtain a permittivity ε_{metal} with a negative real part and neglectable imaginary part, we need an operation below but close to the plasma frequency ω_p.

The goal of Pendry et al. [1] was to shift ω_p to the microwave frequency range. Their perspective was that constraining the transit of electrons within thin wires would provide an effective electron density much smaller than N_e. Moreover, an applied electric field would induce conductive currents on each wire, which in turn would generate a magnetic field over its neighbors. Consequently, such maneuver would considerably reduce the electron mobility, which would represent an effective mass much larger than m_e.

To highlight a general mechanism that we can apply to other types of metamaterials, we simplify the approach of Pendry et al. [1] representing in Fig. 3.1a an infinite wire over which we apply an external electric field $\boldsymbol{E}_0$.

The applied electric field $\boldsymbol{E}_0$ in Fig. 3.1a induces a conduction electric current $\boldsymbol{I}_w$ over the infinite wire due to constitutive relation (2.23). Such current establishes a curl distribution of magnetic field $\boldsymbol{H}_w$ due to Ampère's law (2.17). The magnetic field $\boldsymbol{H}_w$ in turn creates an electric field $\boldsymbol{E}_w$, also with a curl nature, due to Faraday–Lenz's law (2.18).

Now, consider in Fig. 3.1b a thin-wire medium periodically arranged in the x, y, and z dimensions. The medium has infinite extension in all directions. We can calculate the total electric field $\boldsymbol{E}_T$ within this medium at an observation point as the

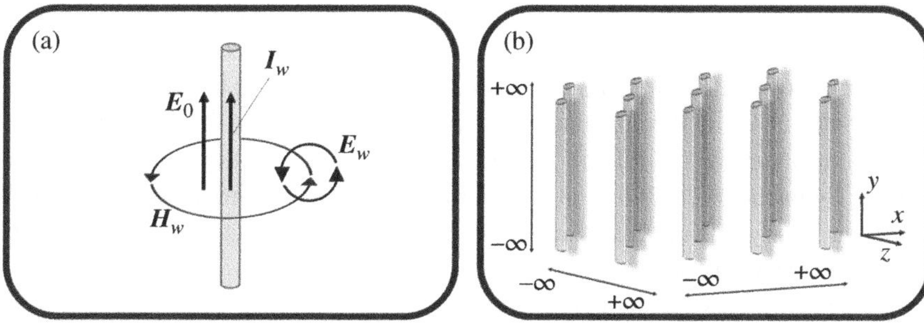

Fig. 3.1 Thin-wire metamaterial: (a) unit cell; (b) infinite medium

superposition of the applied electric field $\boldsymbol{E}_0$ with the fields $\boldsymbol{E}_w$ originated by each unit cell, as:

$$\boldsymbol{E}_T = \boldsymbol{E}_0 + \sum_{w=1}^{\infty} \boldsymbol{E}_w \tag{3.2}$$

We can obtain a final, rigorous expression for the fields $\boldsymbol{E}_w$ and $\boldsymbol{E}_T$ in the thin-wire medium after a lengthy mathematical derivation [3]. However, here we are mostly interested in observing two specific points:

- first, (3.2) is identical in form to (2.83) which means that in this matter the unit cells of the metamaterial work like pairs of charged particles in natural materials;
- second, the electric field $\boldsymbol{E}_w$ acts over the wire with a direction opposite to that of the applied field $\boldsymbol{E}_0$, as indicated in Fig. 3.1a, meaning that such field plays the role of the restoring force in natural materials and opening the possibility for resonant behaviors.

Further analogies ultimately lead to the replacement of (3.2) in (2.85) and (2.86), allowing us to assign an effective electric polarizability $\hat{p}_{\varepsilon_{eff}}$ and effective electric permittivity $\hat{\varepsilon}_{eff}$ to the artificial structure. We can also derive rigorous expressions for these quantities, in which we indeed verify resonant behaviors [3]. Similar to the natural case, the polarizability $\hat{p}_{\varepsilon_{eff}}$ represents how sensitive a unit cell is to create currents that will lead to extra fields when under the influence of an applied field. Like the effective permittivity $\hat{\varepsilon}_{eff}$, $\hat{p}_{\varepsilon_{eff}}$ is a 3×3 tensor representing the relationship between each component of the quantities it associates.

We can compare the operation principle of metamaterials with that of natural materials because the periodicity in the lattice is of the subwavelength order for both. As such, we can associate an average field with each small element in the lattice for both cases. The fundamental difference is that in natural materials it is the displacement of bound charges that generates the extra electric field whereas in metamaterials such field arises from induced currents over the metal geometry. In other words, while it is the material's properties that define the electromagnetic features in natural materials, in metamaterials it is mostly the metal shape that does so. By properly designing the geometry of the metamaterial unit cell, as well as its periodicity in the three axes, we can obtain arbitrary electric polarizability and permittivity tensors.

Another seminal work by Pendry et al. presents a unit cell called split-ring resonator to obtain arbitrary magnetic permeabilities [4]. When we apply a magnetic field $\boldsymbol{H}_0$ upon the metal geometry as in Fig. 3.2a, an electric field $\boldsymbol{E}_w$ results from Faraday–Lenz's law (2.18).

The electric field $\boldsymbol{E}_w$ induces in turn a conduction electric current $\boldsymbol{I}_w$ over the split rings due to constitutive relation (2.23). Then, this current creates an extra magnetic field $\boldsymbol{H}_w$ due to Ampère's law (2.17). Note that the $\boldsymbol{H}_w$ has direction opposite to $\boldsymbol{H}_0$ in the center of the split-ring resonator. Therefore, conclusions similar to the electric case lead to an effective magnetic polarizability $\hat{p}_{\mu_{eff}}$ and permeability $\hat{\mu}_{eff}$ for an infinite medium composed of split-ring resonators such as the one that Fig. 3.2b sketches. Again, $\hat{p}_{\mu_{eff}}$ and $\hat{\mu}_{eff}$ are 3×3 tensors that the unit cell's shape and periodicity can define. A thorough derivation for $\boldsymbol{H}_w$, $\boldsymbol{H}_T$, and μ_{eff} in the split-resonator medium was provided by Pendry et al. [4].

After these two works [1, 4], the task to emulate a medium like the one proposed by Veselago was to put the thin wire and split-ring resonators interleaved in a same lattice. The difficult was, of course, to adjust the structures to avoid a mutual coupling and present resonances at the desired frequencies. D. R. Smith et al. [5] accomplished this milestone, unlocking the possibility to validate some fundamental laws in their most atypical manifestations.

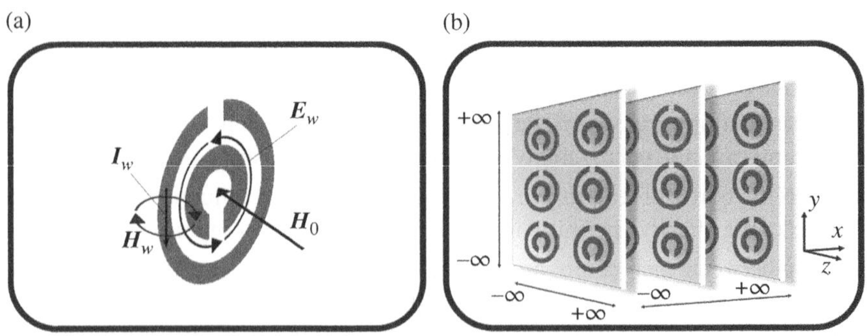

Fig. 3.2 Split-ring resonator metamaterial: (a) unit cell; (b) infinite medium

S. Hrabar et al. confirmed the existence of backward waves experimentally [6]. At this point, we have enough background to visualize what happens inside the metamaterial so that such waves take place without breaking any physics law. Before the establishment of backward waves within the metamaterial, we must somehow apply time-harmonic fields therein to stimulate the unit cells at a frequency near resonance. These fields induce currents over the unit cell, generating strong, extra electric and magnetic fields that are superimposed to the original fields. A relative phase between the total fields of neighbor cells exists due to the distribution of the applied fields over the structure. In the macroscale, the weird rhythm with which both electric and magnetic total fields of neighbor cells oscillate causes the final effect of an electromagnetic wave whose phase velocity has direction opposite to the group velocity.

Putting it differently, we can see each unit cell as a swing set where one stores energy, similar to the older brother in Fig. 2.12b. After charge, a specific synchronization (even if weird) will exist in the three-dimensional grid of swing sets according to their features and the specific delay with which "the older brothers" drive them.

The fact that each unit cell is first charged, and then backward waves appear is what ensures that the causality principle is not broken here.

Another approach to describe the behavior of metamaterials is to regard the unit cells as simple metal inclusions separated by free space (and/or dielectric material). As we saw in Section 2.1.4, the metal parts separated by free space constitute a series of impedance discontinuities. This configuration gives rise within the cells to multiple reflections and standing waves, akin to those in Fig. 2.1b. A phase difference exists between standing waves of neighbor cells due to their geometrical separation and the consequent phase delay of the applied fields. The standing waves, oscillating with this relative phase, are superimposed with traveling waves, causing the final, unusual effect.

We can quantitatively predict the behavior of a periodic structure excited by an electromagnetic wave as a function of an arbitrary phase delay between neighbor cells with Floquet's theorem [7]. We can use such theorem to obtain a rigorous derivation for the response of periodic structures considering impedance discontinuities and standing waves within each unit cell [8].

B.-I. Popa et al. experimentally demonstrated the amplification of evanescent waves [9]. Similar to the case of backward waves, first we charge each unit cell, and then they give a local contribution to the total field with a relative phase in relation to their neighbors. The final effect of the local fields interacting with the applied fields is the amplification of the evanescent wave when the observation point moves away from the interface. Propagation occurs orthogonally to the interface, which means that there is no power transfer in the direction the fields grow. As such, the amplification is just the effect of a previously charged medium, which in no way breaks the principle of energy conservation.

D. Schurig et al. experimentally demonstrated the so-called invisibility cloak [10]. The idea was to cover a scatterer with various layers of metamaterials providing a spatial variation of the permittivity and permeability. We can obtain a phase velocity larger than that of the free space inside the metamaterial by using some layers with relative permittivity or permeability less than one, as evident in (2.30) and (2.36). In this manner, wavefronts traveling a longer path around the scatterer (but within the metamaterial) match wavefronts outside the structure. Again, the unit cells resonate with a specific phase delay in relation to their neighbors, and the final effect is seen in the phase velocity without breaking any physics principle.

3.2 Obtaining Arbitrary Boundary Conditions

After the scientific community acknowledged metamaterials' unusual properties, they envisioned a series of applications such as the construction of high-resolution lenses [11, 12], the improvement of microwave devices [13, 14], or the covering of objects with the invisibility cloak [10, 15]. The latter was particularly interesting from an EW point of view, because it would allow the RCS reduction of aircraft and war vessels. At that point, however, the community raised a question on the real advantages related to the use of metamaterials: how can we apply the unusual capabilities of these artificial structures on real products and systems?

The idea of working with media with arbitrary constitutive parameters had a clear barrier: we would need to build these media. Metamaterials are bulky, heavy, complex, and expensive to fabricate. Furthermore, their principle of operation, based on resonances, makes them frequency dependent. As an example, we recall the narrow BW of only 2.4% of the invisibility cloak [15]. In practical terms, supposing that it would be possible to cover a platform with such a cloak, we would reduce the RCS only for narrowband radars tuned at the very same frequency, with no agility on this parameter. We should admit that one such capability would not be that useful.

The narrowband operation problem could be at least alleviated if a frequency reconfigurability could be employed. However, the three-dimensional lattice considerably limits the tuning methods, reinforcing the limitations of the metamaterial solution.

Metasurfaces are the alternative for the bulky arrangement of metamaterials, not only due to their physical features (compactness, lightweight, low cost) but also because the two-dimensional disposition allows the individual biasing of the unit cells for reconfigurability purposes.

3.2.1 The Mushroom-Like Metasurface

The seminal work on metasurfaces [16] was about a resonating structure to emulate the behavior of a perfect magnetic conductor (PMC). A medium made of PEC provides a reflection coefficient $\Gamma_{TEM} = -1$ for waves normally incident from the free space, that is, it reflects these waves with a phase shift $\varphi_\Gamma = 180°$. We can calculate that using (2.50) considering the intrinsic impedance $\eta_1 = \eta_0 = 120\pi\ \Omega$ for the free space and $\eta_2 = 0$ for the PEC medium, as we did in (2.52). The PMC is a material with $\Gamma_{TEM} = +1$, which means that it reflects electromagnetic waves with a phase shift $\varphi_\Gamma = 0°$. Such property finds many applications in microwave devices and is key to the design of unidirectional low-profile antennas using reflectors, as we will see in Chapter 6.

Actually, PMCs cannot be found in nature. Originally, such material was a theoretical glimpse of a dual element for the PEC. However, we can see in (2.50) that an intrinsic impedance $\eta_2 = \infty$ provides a $\Gamma_{TEM} = +1$. D.F. Sievenpiper et al. obtained such reflection coefficient with a two-dimensional lattice of periodic metal patches connected through metal vias to a ground plane [16]. For this disposition, the structure, depicted in Fig. 3.3a, is known as mushroom-like metasurface.

Calculating the exact field solution for the mushroom structure would be a hard task, but this is not necessary. Sievenpiper et al. [16] simplified this process by representing the unit cells by equivalent lumped elements in parallel. Specifically, the authors assigned an equivalent capacitance C_{eq} related to the charges concentrated over the edges of the patches and an equivalent inductance L_{eq} associated to the flow of currents through the vias and ground plane, as indicated in Fig. 3.3b. This assumption is reasonable when the unit cells are much smaller than the wavelength. The lumped-element approximation allows us to regard the

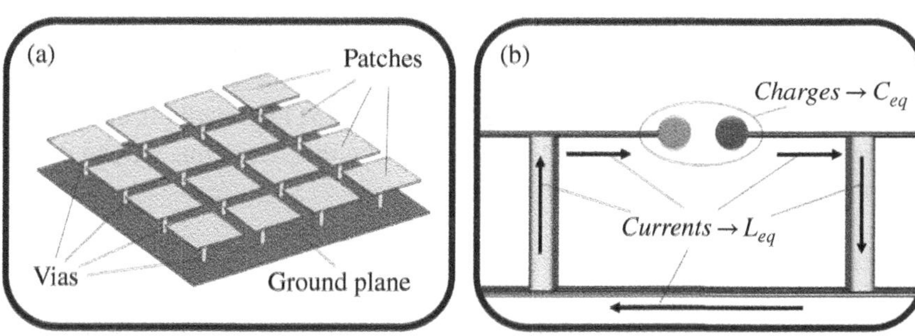

Fig. 3.3 Mushroom-like metasurface: (a) panorama view; (b) simplification of unit cell by lumped elements

metasurface as a simple parallel LC-circuit presenting a resonance at a frequency f_0, which we can calculate using the usual formula:

$$f_0 = \frac{1}{2\pi\sqrt{L_{eq}C_{eq}}} \tag{3.3}$$

At this frequency, the parallel circuit has infinite impedance, which, using (2.50), provides the desired reflection coefficient phase $\varphi_{\Gamma} = 0°$. Thus, one can consider that the mushroom-like metasurface emulates the behavior of a PMC at a single frequency f_0. For this reason, this structure is also known as artificial magnetic conductor (AMC).

Sievenpiper et al. [16] used this approach to obtain an AMC resonating at 15 GHz, approximately, which they associated with a horizontal wire in a low-profile antenna design. Such approach is useful but not general. Other types of AMC exist [17–19] with different shapes for the metal patches, which are isolated from the ground plane by a dielectric. The presence of vias is not mandatory, and currents can noticeably differ from those of the mushroom-like structure. Thus, we need a method that can also model the behavior of these structures.

3.2.2 Surface Impedances

After we derived the reflection coefficient at the interface of the free space and a PEC medium in (2.52), we saw in (2.54) and (2.55) that, in this case, the total electric field in medium 1 and over the interface is $\boldsymbol{E}_1 = 0$, but the total magnetic field is twice the incident magnetic field, that is, $\boldsymbol{H}_1 = 2\boldsymbol{H}^i$. Since in this case the transmitted fields are zero, we can rewrite boundary conditions (2.13) and (2.14) in the complex representation as:

$$\hat{\mathbf{a}}_{\mathbf{n}} \times \boldsymbol{H}_1 = \hat{\mathbf{a}}_{\mathbf{n}} \times 2\boldsymbol{H}^i = \boldsymbol{J}_s \tag{3.4}$$

$$-\hat{\mathbf{a}}_{\mathbf{n}} \times \boldsymbol{E}_1 = 0 = \boldsymbol{M}_s \tag{3.5}$$

meaning that a surface electric current density $\boldsymbol{J}_s \neq 0$ exists over the interface, but not a magnetic one ($\boldsymbol{M}_s = 0$). Since the transmitted fields are zero, from the point of view of medium 1, it does not matter the extension of the PEC medium, and we can even replace such medium by a PEC surface. From this reasoning, we can define the concept of surface impedance Z_s as:

$$Z_s = \frac{E_1}{H_1} \tag{3.6}$$

While the intrinsic impedance represents the ratio of the electric and magnetic fields of waves supported by a medium, the surface impedance Z_s represents the ratio of the electric and magnetic fields over a surface. Using (3.4) and (3.5) in (3.6) leads to $Z_s = 0$ for the PEC surface, which is coherent with an intrinsic impedance $\eta_2 = 0$ for the PEC medium.

We also saw after the derivation of (2.54) and (2.55) that the reflection coefficient at the interface of the free space with a medium with intrinsic impedance $\eta_2 = \infty$ (like one made of PMC) would result in a total magnetic field $\boldsymbol{H}_1 = 0$ over the interface and a total electric field twice the incident electric field ($\boldsymbol{E}_1 = 2\boldsymbol{E}^i$). In this case, from (2.13) and (2.14), we have:

$$\hat{\mathbf{a}}_n \times \boldsymbol{H}_1 = 0 = \boldsymbol{J}_s \tag{3.7}$$

$$-\hat{\mathbf{a}}_\mathbf{n} \times \boldsymbol{E}_1 = -\hat{\mathbf{a}}_\mathbf{n} \times 2\boldsymbol{E}^i = \boldsymbol{M}_s \tag{3.8}$$

indicating that now it is the surface magnetic current density that is $\boldsymbol{M}_s \neq 0$, whereas $\boldsymbol{J}_s = 0$. Using (3.7) and (3.8) in (3.6) results in a surface impedance $Z_s = \infty$ for the PMC surface, again in coherence with the intrinsic impedance of medium 2.

Now, a more generalized understanding about the principle of operation of the mushroom-like metasurface is possible. Fig 3.4a depicts the electric field $\boldsymbol{E}_0$ of an incident wave.

The electric field $\boldsymbol{E}_0$ induces a current $\boldsymbol{I}_w$ over the patches due to constitutive relation (2.23). The displacement of charges, however, also makes this current flow through the vias and the ground plane, forming a loop that leads to a magnetic field $\boldsymbol{H}_w$ due to Ampère's law (2.17). Due to Faraday–Lenz's law (2.18), the latter creates an electric field $\boldsymbol{E}_w$ which has a direction opposite to the original electric field $\boldsymbol{E}_0$ in the plane of the metal patches. All these vectors are shown in Fig. 3.4a with numbers indicating which one causes the other.

Next, Fig. 3.4b illustrates the magnetic field $\boldsymbol{H}_0$ of an incident wave. Due to Faraday–Lenz's law (2.18), $\boldsymbol{H}_0$ induces an electric field $\boldsymbol{E}_w$, which in turn induces

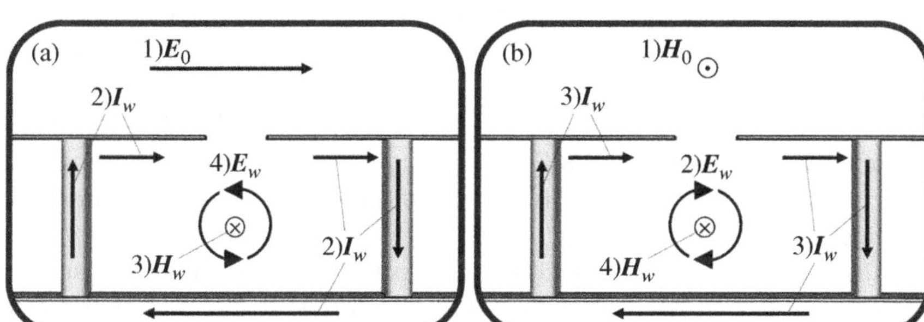

Fig. 3.4 Electromagnetic response of the mushroom-like unit cell: (a) electric response; (b) magnetic response; the numbers indicate the causality order between the quantities

a current $\boldsymbol{I}_w$ over the patches, the vias and the ground plane. Due to Ampère's law (2.17), this current $\boldsymbol{I}_w$ generates a magnetic field $\boldsymbol{H}_w$ which is in a direction opposite to $\boldsymbol{H}_0$ in the plane of the metal patches.

After having studied the concepts of Section 3.1, perhaps the reader has already realized where we are going to get. In the mushroom-like metasurface, the original fields ($\boldsymbol{E}_0$ and $\boldsymbol{H}_0$) lead to extra fields ($\boldsymbol{E}_w$ and $\boldsymbol{H}_w$, respectively) exactly as in the metamaterial case. The latter are in directions opposite to the former, indicating that they act as restoring forces and can lead to resonances. This is nothing but the concept of polarizability again, now applied to two-dimensional structures. Due to their geometries, the unit cells have a certain sensitiveness to create currents that will generate extra fields $\boldsymbol{E}_w$ and $\boldsymbol{H}_w$ when under the influence of $\boldsymbol{E}_0$ and $\boldsymbol{H}_0$. Ultimately, at a specific frequency, $\boldsymbol{E}_w$ and $\boldsymbol{H}_w$ can become such that they make the total fields $\boldsymbol{E}_1 = 2\boldsymbol{E}^i$ and $\boldsymbol{H}_1 = 0$ in the reference plane of the patches. In other words, the resonances can lead to the amount of the fields that would take place over a surface made of PMC, emulating a surface impedance $Z_s = \infty$ in the plane of the patches.

The main difference in relation to metamaterials is that now the polarizabilities do not concern changing fields within media with a subsequent adoption of effective constitutive parameters. This time, the polarizabilities change fields over surfaces, allowing us to consider arbitrary boundary conditions.

3.2.3 Transmission-Line Equivalent Schematic

The concept of surface impedance opens the possibility to use the transmission-line equivalent schematic, introduced in Fig. 2.4, to model different kinds of AMC, with different patch shapes, without vias, and even with multiple layers. To this end, we need first to know the surface impedance of each layer separately.

Consider the example of a single-layer AMC without vias over which TEM waves are normally incident from the free space. We can see such AMC as an FSS of impedance Z_{FSS} (the patches) over a grounded slab [20]. The grounded slab is made of a dielectric with permeability μ_{diel}, permittivity ε_{diel}, and thickness l_{diel} over a PEC ground plane. Hence, we can see the grounded slab as a transmission line of characteristic impedance $Z_{diel} = \sqrt{\mu_{diel}/\varepsilon_{diel}}$ and length l_{diel} leading to a short circuit, whose impedance is $Z_{short} = 0$. Fig. 3.5a sketches the transmission-line equivalent of such AMC.

In the case of an FSS composed of square patches, we can obtain the surface impedance Z_{FSS} analytically [21]. For more complicated shapes, we can estimate Z_{FSS} by simulating or measuring its reflection coefficient when illuminated by a normally incident plane wave. The metal-patch FSS is a transmissive surface when it does not resonate alone. Therefore, we should consider that in this simulation/measurement its surface impedance Z_{FSS} will be in parallel with an equivalent

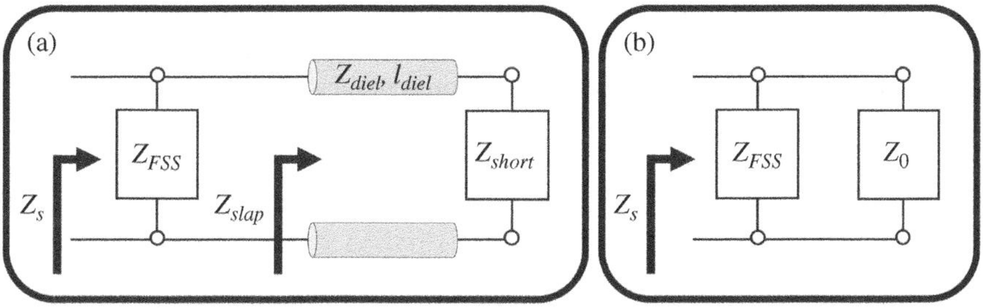

Fig. 3.5 Transmission-line equivalents: (a) AMC without vias; "Gonçalves Licursi de Mello [20]/MDPI/CC BY 4.0"; (b) FSS measured alone

impedance Z_0 of the free space, as in Fig. 3.5b. Then, we calculate the reflection coefficient Γ_S in the reference plane of the patches using (2.50) as:

$$\Gamma_S = \frac{Z_s - Z_0}{Z_s + Z_0} \tag{3.9}$$

From which we can deduce that:

$$Z_s = Z_0 \frac{1 + \Gamma_S}{1 - \Gamma_S} \tag{3.10}$$

$$\frac{1}{Z_s} = \frac{1}{Z_0}\frac{1 - \Gamma_S}{1 + \Gamma_S} = \frac{1}{Z_{FSS}} + \frac{1}{Z_0} \tag{3.11}$$

$$Z_{FSS} = -Z_0 \frac{(1 + \Gamma_S)}{2\Gamma_S} \tag{3.12}$$

Now that we have the impedance of the FSS, we can use the conventional formulation for transmission lines [22] to calculate the input impedance Z_{slab} of the grounded slab indicated in Fig. 3.5a as:

$$\begin{aligned} Z_{slab} &= Z_{diel} \frac{Z_{short} + jZ_{diel}\tan(k_{slab} l_{diel})}{Z_{diel} + jZ_{short}\tan(k_{slab} l_{diel})} \\ &= jZ_{diel}\tan(k_{slab} l_{diel}) \end{aligned} \tag{3.13}$$

where $k_{slab} = \omega\sqrt{\mu_{diel}\varepsilon_{diel}}$ is the wavenumber within the dielectric. Note that Z_{slab} is inductive if the thickness of the dielectric is $l_{diel} \leq \lambda/4$. Moreover, if $l_{diel} \ll \lambda$, then:

$$\begin{aligned} &\lim_{l \to 0} (Z_{slab}) = jZ_{diel} k_{slab} l_{diel} \\ &Z_{slab} \cong j\omega\mu_{diel} l_{diel} = j\omega L_{slab} \end{aligned} \tag{3.14}$$

where $L_{slab} = \mu_{diel} l_{diel}$ is an inductance that we can assign to the grounded slab. L_{slab} is analogous to L_{eq} in the case with vias.

The surface impedance Z_s of the AMC at the reference plane of the FSS is the parallel of Z_{slab} and Z_{FSS}. For the specific case where the FSS impedance of (3.12) is purely capacitive, that is, we may represent it by $Z_{FSS} = 1/j\omega C_{FSS}$, we have:

$$\begin{aligned} \frac{1}{Z_s} &= \frac{1}{Z_{FSS}} + \frac{1}{Z_{slab}} \\ Z_s &= j\frac{\omega L_{slab}}{1-\omega^2 L_{slab} C_{FSS}} \end{aligned} \tag{3.15}$$

The capacitance C_{FSS} is also analogous to C_{eq} in the case without vias. When the angular frequency is $\omega_0 = 1/\sqrt{L_{slab}C_{FSS}}$, the surface impedance of the AMC is $Z_s = j\infty$, which leads to a reflection coefficient phase $\varphi_\Gamma = 0°$.

The transmission-line equivalence is useful not only for the design of AMCs. As it will become clearer to the reader in later chapters, this approach can substantially simplify the analysis of the interaction of fields within other metasurfaces. Yet, in the end, it also reveals that the incident fields induce currents, which in turn create extra fields that can lead to resonant processes. The concept of polarizabilities applied to two-dimensional structures is reinforced here. Actually, such concept is general for metasurfaces. We can use it together with a centuries-old theorem to create metasurface arbitrary responses.

3.2.4 Arbitrary Reflection Phase

AMCs receive this name because they emulate the behavior of a PMC, providing a reflection coefficient phase $\varphi_\Gamma = 0°$. We obtain this phase when the structure resonates at an angular frequency $\omega_0 = 1/\sqrt{L_{slab}C_{FSS}}$ leading to a surface impedance $Z_s = j\infty$. However, (3.15) actually indicates that AMCs provide arbitrary phases.

As an example, Fig. 3.6 depicts the phase response of an AMC with $C_{FSS} = 0.25$ pF and $L_{slab} = 1$ nH based on (3.15)The resonant frequency is $f_0 = 10$ GHz, where $\varphi_\Gamma = 0°$. We can see that, as soon as f diverges from 10 GHz, $\varphi_\Gamma(f)$ differs from 0°. Far from the resonant frequency, $\varphi_\Gamma(f)$ asymptotically goes to ±180°, where the AMC actually behaves like a PEC.

We can shift the resonant frequency by using other capacitance values in (3.15), obtaining thereby other reflection coefficient phases at a same frequency. For example, Fig. 3.6 also shows that $\varphi_\Gamma = +100°$ at 10 GHz when $C_{FSS} = 0.20$ pF. Besides $\varphi_\Gamma = -100°$ when $C_{FSS} = 0.30$ pF. Therefore, we can say that AMCs may provide not only a reflection coefficient phase $\varphi_\Gamma = 0°$ at the resonant frequency f_0, but also arbitrary phases at a chosen frequency.

AMCs may also provide more than one resonant frequency. For example, structures with nested shapes can do that [23]. In this configuration, we can see the structure as a multiband AMC because it provides the desired phases in specific bands around the resonant frequencies. Two nested metal squares without vias

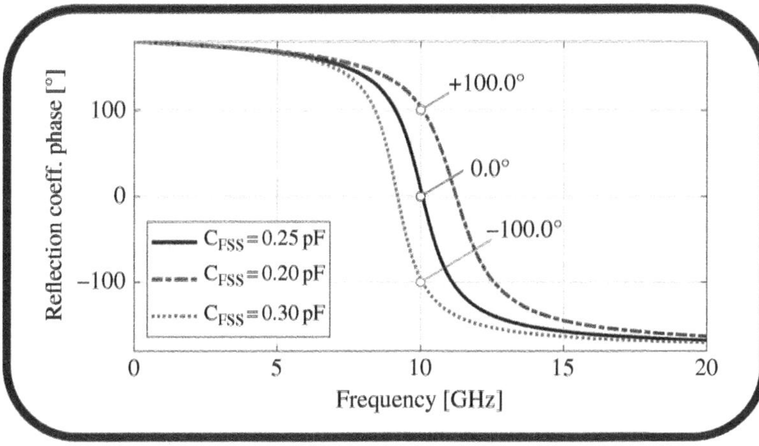

Fig. 3.6 Reflection coefficient phase of an AMC, based on (3.15)

over a ground plane can handle this task [24]. Each square is responsible for a resonance. Fig. 1.11 depicted earlier one such a metasurface designed for dual-band operation.

We can extend this concept to three nested metal squares [25, 26]. Chapter 6 will provide a more detailed explanation on these metasurfaces in the context of multiband antennas for ES applications.

3.2.5 Huygens' Principle

A principle dating back to 1690 provides a different approach for modeling the interactions of electromagnetic waves and metasurfaces. The Huygens principle qualitatively states that a propagating wave actually is a series of subsequent wavefronts. Each point on a primary wavefront serves as a source for a spherical wave. A secondary wavefront results from the superposition of the spherical waves from all points in the primary wavefront [27]. Fig. 3.7a illustrates this principle, where original sources radiate spherical waves that interfere constructively in the direction of propagation and destructively elsewhere. We can replace the propagation of the wavefront in an infinitesimal distance by a line of secondary sources, leading each one to newly radiated spherical waves. This cycle repeats as long as the wave propagates.

From the perspective of a receiver, it does not matter where are the original sources. Therefore, one may imagine that an arbitrary distribution of fields $\boldsymbol{E}_0$ and $\boldsymbol{H}_0$ that could be generated by sources $\boldsymbol{J}_i$ and $\boldsymbol{M}_i$ can be instead generated

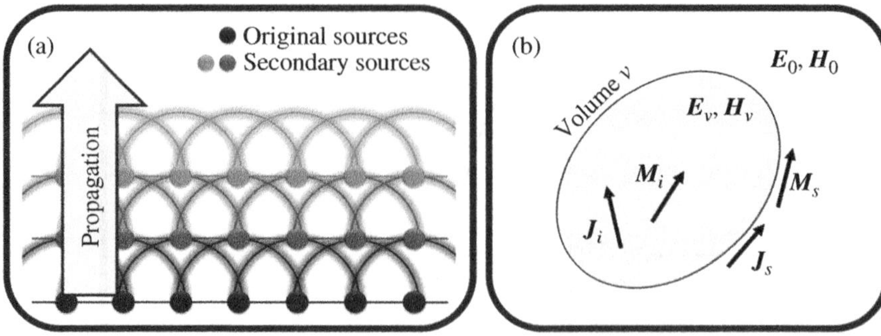

Fig. 3.7 Huygens' principle: (a) points in a wavefront serve as sources for spherical waves, which in turn form a new wavefront; (b) current densities $\boldsymbol{J}_s$ and $\boldsymbol{M}_s$ can emulate the original sources $\boldsymbol{J}_i$ and $\boldsymbol{M}_i$ if they generate equivalent fields over the surface they are located

by surface currents replacing the wavefronts. We can find this idea in Love's equivalence principle, which quantitatively formalizes Huygens' principle [28].

Figure 3.7b illustrates such principle. In an original scenario, impressed current densities $\boldsymbol{J}_i$ and $\boldsymbol{M}_i$ generate distributions of electric and magnetic fields everywhere ($\boldsymbol{E}_0 = \boldsymbol{E}_v$ and $\boldsymbol{H}_0 = \boldsymbol{H}_v$). In the original scenario, $\boldsymbol{J}_s$ and $\boldsymbol{M}_s$ are zero and volume v is a virtual volume encompassing $\boldsymbol{J}_i$ and $\boldsymbol{M}_i$. Love's equivalence principle states that we can substitute this configuration. The same fields $\boldsymbol{E}_0$ and $\boldsymbol{H}_0$ can be instead generated by current densities $\boldsymbol{J}_s$ and $\boldsymbol{M}_s$ impressed over the surface of volume v. In this new configuration, the fields inside volume v are forced to be $\boldsymbol{E}_v = 0$ and $\boldsymbol{H}_v = 0$, and then, from boundary conditions (2.13) and (2.14), we obtain the current densities required over the surface of volume v as:

$$\boldsymbol{J}_s = \hat{\mathbf{a}}_\mathbf{n} \times \boldsymbol{H}_0 \tag{3.16}$$

$$\boldsymbol{M}_s = -\hat{\mathbf{a}}_\mathbf{n} \times \boldsymbol{E}_0 \tag{3.17}$$

When Love's principle is used, $\boldsymbol{J}_s$ and $\boldsymbol{M}_s$ must be impressed somehow over the surface of the volume v.

In fact, arbitrary fields $\boldsymbol{E}_0$ and $\boldsymbol{H}_0$ can be generated passively from fields already existing within volume v, as stated by the surface equivalence principle of Schelkunoff [29]. Consider now in Fig. 3.7b that sources $\boldsymbol{J}_i$ and $\boldsymbol{M}_i$ lead to fields $\boldsymbol{E}_v$ and $\boldsymbol{H}_v$ inside volume v. Arbitrary fields $\boldsymbol{E}_0$ and $\boldsymbol{H}_0$ may exist outside volume v if surface current densities $\boldsymbol{J}_s$ and $\boldsymbol{M}_s$ make boundary conditions (2.13) and (2.14) hold over its closed surface:

$$\boldsymbol{J}_s = \hat{\mathbf{a}}_\mathbf{n} \times (\boldsymbol{H}_0 - \boldsymbol{H}_v) \tag{3.18}$$

$$\boldsymbol{M}_s = -\hat{\mathbf{a}}_\mathbf{n} \times (\boldsymbol{E}_0 - \boldsymbol{E}_v) \tag{3.19}$$

Expressing differently, Schelkunoff's surface equivalence principle states that, theoretically, we can obtain an arbitrary field distribution. All we have to do is to generate somehow the surface current densities that satisfy boundary conditions (3.18) and (3.19). Now, $\boldsymbol{J}_s$ and $\boldsymbol{M}_s$ are not necessarily impressed. Electric and magnetic polarizabilities of particles over the surface can induce them according to fields generated from sources $\boldsymbol{J}_i$ and $\boldsymbol{M}_i$ inside volume v.

3.2.6 Arbitrary Transmission Phase

At the limit of generalization, we can regard metasurface unit cells as polarizable particles that can generate the surface current densities $\boldsymbol{J}_s$ and $\boldsymbol{M}_s$ required to convert arbitrary fields $\boldsymbol{E}_1$ and $\boldsymbol{H}_1$ into arbitrary fields $\boldsymbol{E}_2$ and $\boldsymbol{H}_2$.

Like we did with metamaterials, we will consider electric and magnetic polarizabilities to this end [30]. True, these concepts are much harder to understand than reducing unit cells to lumped elements. However, they allow a generalized approach to model the interactions between electromagnetic waves and metasurfaces.

Next, we formalize the sensitiveness of a metasurface unit cell to create surface current densities $\boldsymbol{J}_s$ and $\boldsymbol{M}_s$ when excited by electric and magnetic fields tangent to the surface. These fields will present different intensities in each side of the metasurface depending on its reflection and transmission properties. Therefore, such formalization considers the average electric and magnetic tangential fields as $\boldsymbol{E}_{avg} = (\boldsymbol{E}_1 + \boldsymbol{E}_2)/2$ and $\boldsymbol{H}_{avg} = (\boldsymbol{H}_1 + \boldsymbol{H}_2)/2$ [31]. We can respectively define the surface current densities $\boldsymbol{J}_s$ and $\boldsymbol{M}_s$ as [32]:

$$\begin{bmatrix} \boldsymbol{J}_s \\ \boldsymbol{M}_s \end{bmatrix} = j\omega \begin{bmatrix} \hat{p}_{ee} & \hat{p}_{em} \\ \hat{p}_{me} & \hat{p}_{mm} \end{bmatrix} \begin{bmatrix} \boldsymbol{E}_{avg} \\ \boldsymbol{H}_{avg} \end{bmatrix} \tag{3.20}$$

where $\hat{p}_{ee}$, $\hat{p}_{em}$, $\hat{p}_{me}$, and $\hat{p}_{mm}$ are surface polarizabilities. In expanded form, (3.20) becomes:

$$\boldsymbol{J}_s = j\omega\hat{p}_{ee}\boldsymbol{E}_{avg} + j\omega\hat{p}_{em}\boldsymbol{H}_{avg} \tag{3.21}$$

$$\boldsymbol{M}_s = j\omega\hat{p}_{me}\boldsymbol{E}_{avg} + j\omega\hat{p}_{mm}\boldsymbol{H}_{avg} \tag{3.22}$$

The polarizability $\hat{p}_{ee}$ is related to the electric response of the metasurface and represents the ability of the average tangential electric field $\boldsymbol{E}_{avg}$ to create surface current density $\boldsymbol{J}_s$. Conversely, $\hat{p}_{em}$ is associated to a magneto-electric response of the artificial structure and relates the arise of $\boldsymbol{J}_s$ to the average tangential magnetic field $\boldsymbol{H}_{avg}$. The polarizabilities $\hat{p}_{me}$ and $\hat{p}_{mm}$ play analogous roles for $\boldsymbol{M}_s$, the former also associated to the metasurface magneto-electric response and the latter to the magnetic response. In anisotropic metasurfaces, each of the surface polarizabilities $\hat{p}_{ee}$, $\hat{p}_{em}$, $\hat{p}_{me}$, and $\hat{p}_{mm}$ is a 2×2 tensor representing the relationship between the two polarization components of the associated quantities in the plane of the metasurface. For example, if a metasurface lies in the plane $z = 0$, (3.21) can be expanded as:

$$\begin{bmatrix} J_x \\ J_y \end{bmatrix} = j\omega \begin{bmatrix} p_{ee_{xx}} & p_{ee_{xy}} \\ p_{ee_{yx}} & p_{ee_{yy}} \end{bmatrix} \begin{bmatrix} E_{avg_x} \\ E_{avg_y} \end{bmatrix} + j\omega \begin{bmatrix} p_{em_{xx}} & p_{em_{xy}} \\ p_{em_{yx}} & p_{em_{yy}} \end{bmatrix} \begin{bmatrix} H_{avg_x} \\ H_{avg_y} \end{bmatrix} \tag{3.23}$$

In Section 2.1.4, we calculated the reflection and transmission coefficients of a wave impinging over the interface between two media as a function of their intrinsic impedance. Now, we are interested in calculating the same coefficients as a function of the surface polarizabilities of a metasurface over which a wave impinges. To this end, consider in Fig. 3.8a a metasurface with infinite extension in the x and y dimensions, located at the plane $z = 0$, and embedded in the free space. A plane wave denoted by (2.40) polarized in the x dimension and traveling at the $+\hat{\mathbf{a}}_z$ direction is normally incident upon this metasurface. In order to follow the same notation of Section 2.1.4, we define that side 1 is the region where $z < 0$ whereas $z > 0$ is within side 2. Let us first consider an isotropic metasurface with electric response only, that is, $p_{ee} \neq 0$, and $p_{em} = p_{me} = p_{mm} = 0$. We define the normal of the surface as the direction $-\hat{\mathbf{a}}_z$, as Fig. 3.8b indicates together with the other vectors. From boundary condition (2.13) and then using (2.40), (2.41), and (2.44)–(2.47), we have:

$$\begin{aligned} -\hat{\mathbf{a}}_z \times (\boldsymbol{H}_1 - \boldsymbol{H}_2) &= j\omega p_{ee} \boldsymbol{E}_{avg} \\ -\hat{\mathbf{a}}_z \times \hat{\mathbf{a}}_y \frac{E_0}{\eta_0}(1 - \Gamma_{TEM} - \tau_{TEM}) &= j\omega p_{ee} \frac{E_0}{2}(1 + \Gamma_{TEM} + \tau_{TEM})\hat{\mathbf{a}}_x \\ 1 - \Gamma_{TEM} - \tau_{TEM} &= j\omega \frac{p_{ee}\eta_0}{2}(1 + \Gamma_{TEM} + \tau_{TEM}) \end{aligned} \tag{3.24}$$

Also, from boundary condition (2.14), we have:

$$\begin{aligned} \hat{\mathbf{a}}_z \times (\boldsymbol{E}_1 - \boldsymbol{E}_2) &= 0 \\ \tau_{TEM} &= 1 + \Gamma_{TEM} \end{aligned} \tag{3.25}$$

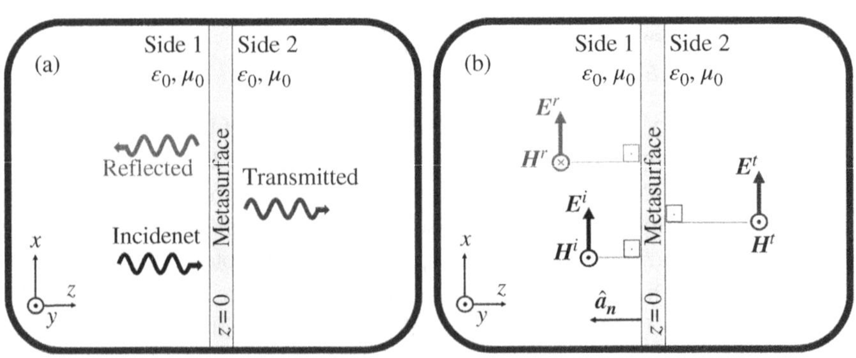

Fig. 3.8 Transmission and reflection of waves across metasurface: (a) disposition of waves; (b) associated vectors

Combining (3.24) and (3.25), the reflection and transmission coefficients for a metasurface with only electric response are:

$$\Gamma_{TEM} = \frac{-1}{1 - j\frac{2}{\omega p_{ee}\eta_0}} \tag{3.26}$$

$$\tau_{TEM} = \frac{1}{1 + j\frac{\omega p_{ee}\eta_0}{2}} \tag{3.27}$$

Equation (3.26) indicates that we cannot avoid reflections in metasurfaces whose response is strictly electric. We can draw a similar conclusion for the case with only magnetic response ($p_{mm} \neq 0$, and $p_{ee} = p_{em} = p_{me} = 0$).

Following, consider that the isotropic metasurface of Fig. 3.8 has electric and magnetic responses only, that is, $p_{ee} \neq 0$ and $p_{mm} \neq 0$, but $p_{em} = p_{me} = 0$. Boundary condition (2.13) leads again to (3.24) but (2.14) leads to:

$$\begin{aligned} \hat{\mathbf{a}}_{\mathbf{z}} \times (\boldsymbol{E}_1 - \boldsymbol{E}_2) &= j\omega p_{mm}\boldsymbol{H}_{avg} \\ 1 + \Gamma_{TEM} - \tau_{TEM} &= j\frac{\omega p_{mm}}{2\eta_0}(1 - \Gamma_{TEM} + \tau_{TEM}) \end{aligned} \tag{3.28}$$

Combining (3.24) and (3.28), the reflection and transmission coefficients for a metasurface with only electric and magnetic responses are:

$$\Gamma_{TEM} = j\omega\frac{1}{2\eta_0}\frac{p_{mm} - p_{ee}\eta_0^2}{\left(1 + j\frac{\omega p_{ee}\eta_0}{2}\right)\left(1 + j\frac{\omega p_{mm}}{2\eta_0}\right)} \tag{3.29}$$

$$\tau_{TEM} = \frac{1 + \frac{\omega^2 p_{ee} p_{mm}}{4}}{\left(1 + j\frac{\omega p_{ee}\eta_0}{2}\right)\left(1 + j\frac{\omega p_{mm}}{2\eta_0}\right)} \tag{3.30}$$

Equation (3.29) makes evident that a reflection coefficient $\Gamma_{TEM} = 0$ is possible. Specifically, if:

$$\frac{p_{mm}}{p_{ee}} = \eta_0^2 \tag{3.31}$$

which is known as the reflectionless condition. Using this relation in (3.30) leads to:

$$\begin{aligned} \tau_{TEM} &= \frac{1 + \frac{\omega^2 p_{ee}^2 \eta_0^2}{4}}{\left(1 + j\frac{\omega p_{ee}\eta_0}{2}\right)\left(1 + j\frac{\omega p_{ee}\eta_0}{2}\right)} \\ \tau_{TEM} &= \frac{1 - j\frac{\omega p_{ee}\eta_0}{2}}{1 + j\frac{\omega p_{ee}\eta_0}{2}} = e^{-j2\arctan(\omega p_{ee}\eta_0/2)} \end{aligned} \tag{3.32}$$

The meaning of (3.32) is that a full transmission with arbitrary phase is possible if the polarizabilities are purely real. Associated with (3.29), this result shows that metasurfaces allow for a reflectionless transmission with full control of the phase as long as (3.31) holds.

The above conclusions show only a part of the capabilities of metasurfaces. Indeed, we could also have considered in our development polarizability tensors, magneto-electric responses, and so forth. Such a formulation predicts the asymmetric transmission, and the generation of linear, circular, or elliptical polarization of waves from an arbitrary excitation without reflections. We can find this derivation in [32].

3.2.7 Arbitrary Reflection and Refraction Directions

One way of implementing the polarizabilities to obtain surface current densities $\boldsymbol{J}_s$ and $\boldsymbol{M}_s$ required to carry out a specific task is to employ electric and magnetic dipoles in an orthogonal disposition. Metasurfaces that explicitly do so are known as Huygens metasurfaces [33]. As an example, C. Pfeiffer et al. [34] proposed a Huygens metasurface for the reflectionless steering of electromagnetic beams at 10 GHz. The beam steering in this work exploits the generalized law of refraction [35].

We derived in Section 2.1.5 Snell's reflection and refraction laws (2.61) and (2.62) to have a continuity of tangential fields through the interface between two media. The angle of propagation of both reflected and transmitted waves are such that these fields are continuous. Continuity of fields was necessary in that case because no surface current densities $\boldsymbol{J}_s$ and $\boldsymbol{M}_s$ existed over the interface. However, in the presence of such currents, the situation changes, and the fields jump. The community had to review Snell's laws with the arrival of metasurfaces and the consequent possibility of implementing $\boldsymbol{J}_s$ and $\boldsymbol{M}_s$ by means of polarizabilities. In this context, N. Yu et al. [35] derived new expressions accounting for phase discontinuities in the anomalous reflection and transmission of waves.

Consider in Fig. 3.9 an electromagnetic wave impinging over a metasurface embedded in the free space with angle of incidence θ_i in relation to the normal of the surface.

We defined in Section 2.1.4 respective expressions for the reflection and transmission coefficients as $\Gamma_{TEM} = |\Gamma_{TEM}|e^{j\varphi_\Gamma}$ and $\tau_{TEM} = |\tau_{TEM}|e^{j\varphi_\tau}$. This time, however, we introduce linear gradients in the x dimension in the phases of these coefficients:

$$\varphi_\Gamma = \varphi_{\Gamma 0} + \frac{x d\varphi_\Gamma}{dx} \tag{3.33}$$

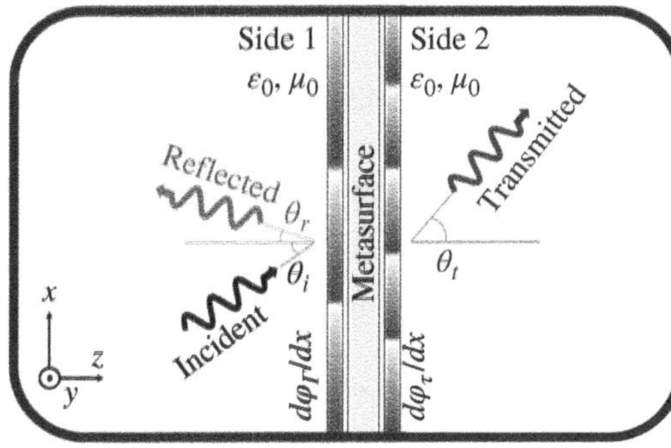

Fig. 3.9 Anomalous reflection and refraction of waves

$$\varphi_\tau = \varphi_{\tau 0} + \frac{x d\varphi_\tau}{dx} \tag{3.34}$$

where $\varphi_{\Gamma 0}$ and $\varphi_{\tau 0}$ are constants. The linear gradients are indicated in Fig. 3.9.

The generalized laws of reflection and refraction [35] state that the linear phase gradient in the x dimension causes a deflection in the direction of propagation in relation to the yz plane. Therefore, we should adjust the relations that (2.61) and (2.62) represent for the case of Fig. 3.9. In this case (where sides 1 and 2 are media with the same properties), the angle of propagation after reflection/transmission is:

$$\theta_r = \arcsin\left(\frac{\lambda}{2\pi}\frac{d\varphi_\Gamma}{dx} + \sin\theta_i\right) \tag{3.35}$$

$$\theta_t = \arcsin\left(\frac{\lambda}{2\pi}\frac{d\varphi_\tau}{dx} + \sin\theta_i\right) \tag{3.36}$$

Reflective and transmissive metasurfaces whose parameters vary from unit cell to unit cell can provide phase gradients to exploit the generalized laws of reflection and refraction (3.35) and (3.36). Metasurfaces purposely designed to this end are known as phase-gradient metasurfaces.

For instance, we can exploit the generalized law of reflection (3.35) with AMCs whose neighbor unit cells present different patch dimensions. In this way, each unit cell has a different capacitance value C_{FSS} and provides a different reflection phase at a specific frequency, as Fig. 3.6 previously depicted. This configuration can implement the linear phase gradient $xd\varphi_\Gamma/dx$. Another possibility is to use lumped elements in parallel to the intrinsic capacitances C_{FSS}. Sievenpiper et al. [36] used varactors connecting neighbor unit cells of a mushroom-like metasurface to locally control the reflection phases. They achieved a dynamic steering of $\theta_r = \pm 45°$ for the reflected beams.

On the other hand, we can exploit the generalized law of refraction (3.36) with Huygens metasurfaces. C. Pfeiffer et al. [34] used unit cells of the type of Fig. 3.10a to obtain the electric and magnetic responses required for a reflectionless transmission with arbitrary phase. They obtained a linear phase gradient $xd\varphi_\tau/dx$ with a transmission coefficient of magnitude $|\tau_{TEM}| \cong 1$ by implementing a variation in

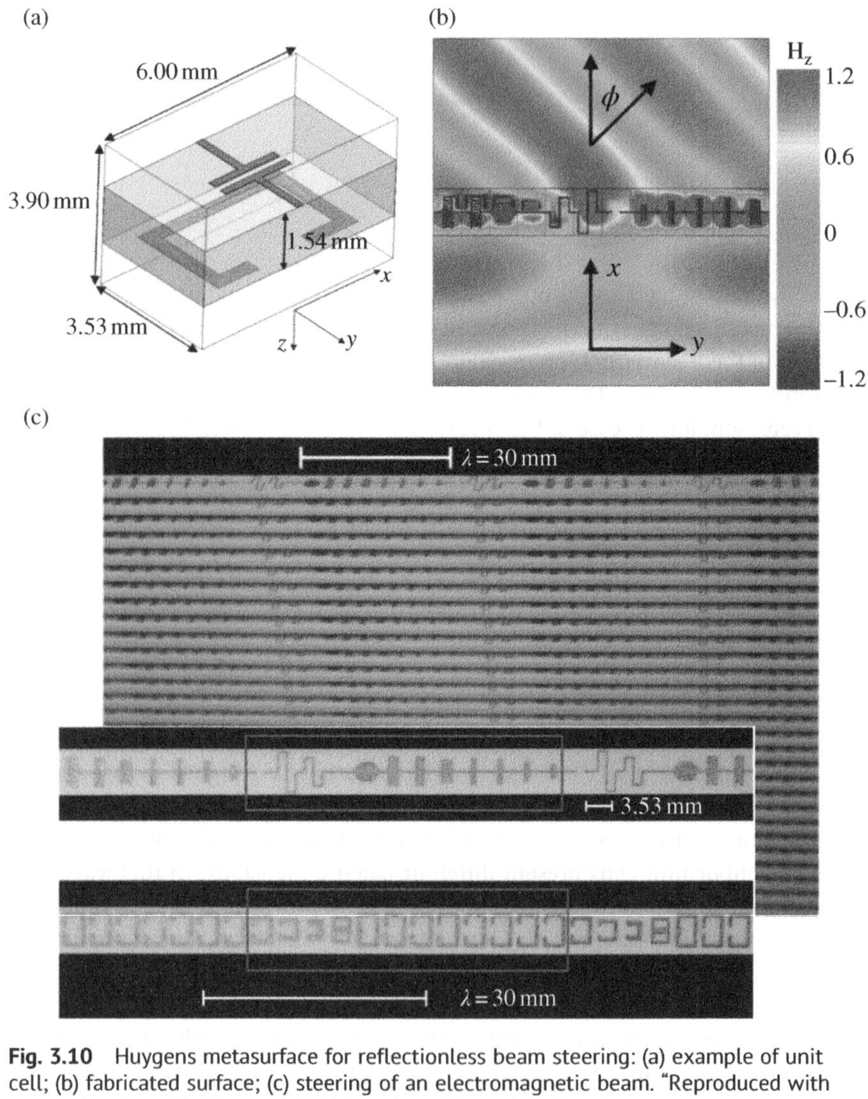

Fig. 3.10 Huygens metasurface for reflectionless beam steering: (a) example of unit cell; (b) fabricated surface; (c) steering of an electromagnetic beam. "Reproduced with permission from Pfeiffer et al. [34]/American Physical Society."

the geometry of neighbor unit cells to provide different electric and magnetic polarizabilities. The required surface polarizabilities p_{ee} and p_{mm} were calculated using (3.32) ensuring that relation (3.31) holds. To obtain them, they simulated a series of unit cells with commercial software and built a lookup table. The unit cell variations were grouped in supercells of 30 mm, which corresponds to the free-space wavelength at 10 GHz. Figure 3.10b shows one component of the magnetic field, making explicit the beam steering toward the direction denominated $\Phi = 45°$ in that work [34] ($\theta_t = 45°$ in our notation). The angle of incidence was $\theta_i = 0°$. The transmission coefficient magnitude was measured as $|\tau_{TEM}| = 0.86$. Fig. 3.10c shows the fabricated structure.

The disposition of the metasurface of [34] is inconvenient. We require such disposition to make the fields interact with the unit cells in a way that they provide the desired electric and magnetic responses. An alternative to such disposition is to implement Huygens metasurfaces with three layers. J.P.S Wong et al. obtained a beam steering of 80° with $|\tau_{TEM}| \cong 1$ in simulations [37].

The examples of [34, 37] show one of the many capabilities of metasurfaces, namely the arbitrary refraction or steering of electromagnetic beams. In the metamaterial golden years, they pursued a negative refraction, something much less useful. The bottom line here is that nowadays the discussion about "magical" properties is already closed. We orient the design of metasurfaces to provide game-changing capabilities in electromagnetics.

3.3 Diversity of Approaches

Many approaches exist to model metasurfaces and their interactions with electromagnetic waves. Throughout this book, we will explain them as they appear. We cannot consider any of these approaches the best for all the cases. Simple procedures can lead to sufficiently accurate results in complex problems where detailed methods are unpractical. On the other hand, general approaches, besides offering a good precision, also provide invaluable physical insight.

One thing is certain: expecting to find a detailed analysis, for example, about polarizabilities in every metasurface paper is a good way to get disappointed, and often. Most works will show one or two indications of the physical phenomenon they are investigating. Next, they will validate them with measurements of the reflection and transmission coefficients of a structure purposedly fabricated to this end. And that is not a problem from a scientific point of view.

Electromagnetics is not simple, and the fact that we cannot see waves with our own eyes makes things harder, often requiring analogies. Since the beginning, great scientists, like Faraday and Ampère, struggled to dissolve the veils of electromagnetics, making analogies with the Newtonian theory of fields in that case. If we need to make analogies with swing sets, why not? If evidence for an interesting mechanism cannot be shown through diverse means, we can still retain one learning or another and look other contexts.

But, in the end, we should not forget: we must go from theory to practice, and, ultimately, to the theater of operation – and as fast as we can.

3.4 Metasurfaces' Limitations

Some limitations are common to most types of metasurfaces. It is worth understanding them right away, as they will walk alongside the presentation of most of the applications in this book.

3.4.1 Metasurfaces' Frequency Dependence

The study of Chapter 2 and Sections 3.1 and 3.2 shows that the operation of metasurfaces is based on the resonance of unit cells at a specific frequency f_0. In other words, the performance of metasurfaces inherently depends on frequency. The properties metasurfaces provide change by operating at frequencies far apart in the spectrum. We can associate one or more operating frequency ranges, each one with a specific BW, to each metasurface depending on the desired functionalities and its activating mechanisms.

We can derive a rigorous formulation for the limitation on the BW of AMCs [38]. The reflection coefficient phase $\varphi_\Gamma(f_0) = 0°$ we expect from AMCs, seen in Sections 3.2.1–3.2.3, makes them ideal to work as antenna reflectors in low-profile designs. A particular antenna design uses an AMC and a parallel radiating element separated by a spacing much smaller than a wavelength [16]. In this design, constructive interferences between the direct electric field of the radiating element and the fields reflected on the AMC occur at the broadside direction when the AMC reflection coefficient phase is $\varphi_\Gamma = 0°$. Even though AMCs provide $\varphi_\Gamma = 0°$ only at the resonant frequency f_0, they provide phases close to 0° at frequencies f in the vicinity of f_0. In this way, such design can still lead to constructive interferences at the broadside direction at these frequencies. Because of that, Sievenpiper et al. [16] defined the BW of their AMC as the frequency range where $-90° < \varphi_\Gamma(f) < +90°$.

Another example is the case of metasurface absorbers, which dissipate the energy of impinging electromagnetic waves in lossy materials or components. For instance, N. Engheta [39] proposed a resistive sheet placed over the metal patches of an AMC. The sheet matches at resonance the surface impedance Z_s and the free-space impedance η_0, consequently dissipating the energy of impinging waves. In this case, we expect a BW of same order of the one of the employed AMC.

However, usually BW will be defined differently given the functionality of the absorber. For example, it will be defined as the frequency range where the reflection coefficient magnitude is $|\Gamma_{TEM}| < -10$ dB.

Two main alternatives exist to the problem of frequency dependence in metasurfaces. The first one is to work on the mechanism that activates the desired functionality itself. To this end, some works investigate different AMC unit cells to obtain a reflection coefficient phase respecting the condition $-90° < \varphi_\Gamma(f) < +90°$ in wider BWs [19, 40]. Other works research on multilayer metasurface absorbers to match the free-space impedance in wide BWs whereby obtaining a wideband absorption [41].

The second alternative is to exploit reconfigurability. One can change the electromagnetic characteristics of the unit cells and whereby shift their natural frequencies in the spectrum by using tunable components such as PIN diodes, and varactors [42]. As such, the metasurface can operate in a wide frequency range, but not instantaneously. Good examples with varactors embedded in the unit cells could tune the resonance of an AMC from 3.5 to 4.5 GHz [36] and the absorption frequency from 4.45 to 5.64 GHz [43].

3.4.2 Metasurfaces' Angle/Polarization Dependence

The reflection and transmission of waves through impedance discontinuities (either caused by a change between two media or by encountering a surface) depend on the matching between the impedances of both zones. We saw in Section 2.1.4 that the surface transmits and reflects specific amounts of energy so that the relation between electric and magnetic fields in both zones respects its respective impedance. The problem is that this phenomenon takes into account the fields tangent to the surface, as we studied in Section 2.1.6.

It is evident from (2.71) and (2.72) that the reflection and transmission of waves depend on the angle of incidence θ_i. Moreover, (2.74) and (2.75) make explicit that the polarization of the impinging wave (TE or TM) also plays an important role in this matter. The polarizabilities presented in Section 3.2.6 allow an alternative explanation [44].

One way of circumventing the angle/polarization problem in metasurfaces is to use a structure that compensates these variations in the relation between electric and magnetic fields projected over the surface. A high-permittivity dielectric, called wide-angle impedance matcher (WAIM) [45], could perform this task when placed over an absorber [41]. The WAIM kept the absorber performance stable for angles of incidence ranging from 0° to 60° for both TE and TM polarizations. The correct choice of the unit cell shape (a Jerusalem cross) could also make this compensation. Such unit cell stabilized the reflection coefficient phase φ_Γ of an AMC that could be reconfigured in two working states for incidence angles from 0° to 45° for both TE and TM polarizations [46].

The fact that metasurfaces represent impedance discontinuities also allows one to regard them as media whose effective parameters can be extracted with special methods [47]. An analysis based on this approach developed an AMC whose resonance does not shift in frequency when the angle or the polarization changes [48]. Similar approaches stabilized the angle/polarization response of an antenna array with one [49] or two layers [50] of a special type of WAIM, called metasurface WAIM.

Another possibility is to reduce the contribution of the geometrical disposition between the unit cell shape and the incident waves for the metasurface response. The introduction of lumped elements (inductors) between the unit cells of an AMC stabilized its reflection coefficient phase φ_r for incidence angles from 0° to 60° for both polarizations [51]. A similar approach made an energy harvesting absorber stable for angles of incidence from 0° to 45° for both polarizations again [52].

3.4.3 Low Profile and Conformal Shape in Metasurfaces

Metasurfaces represent a huge advancement in electromagnetics in terms of thickness. As previously stated, we can dispose AMCs and parallel radiating elements with a spacing much smaller than a wavelength λ in antenna designs [16]. Traditional metal plates, whose behavior is close to a PEC, need a spacing of $\lambda/4$, as we will detail in Chapter 6. Another low-profile example is the metasurface absorber proposed in [39], which presents a thickness much smaller than a wavelength, in opposition to traditional Salisbury lenses that are about a $\lambda/4$ thick [53].

Metasurfaces provide a significant advantage in the matter of compactness in relation to metamaterials. Nevertheless, some applications, like Aeronautics and Space to name a few, demand extremely low-profile devices that fit the fuselage without causing extra drag.

The limitations of metasurfaces in this matter is that the solution to a problem often involves the addition of extra layers. For example, the angle/polarization stability and the wideband impedance matching of the WAIM-based absorber we commented in Section 3.4.2 was achieved with the stack of six layers [41]. As a result, the metasurface presents a total thickness of 11.5 mm, which corresponds to $\lambda/7$ at the lowest operating frequency. In terms of wavelength λ, this is a remarkable result. However, not every aircraft can handle a 11.5-mm thick coating. We can make similar observations over AMCs that achieve low resonant frequencies through the use of extra layers [54].

Concerning the shapes, we usually design metasurfaces in a planar disposition. The problem regards again some applications requiring structures conformal to the fuselage. Extensive research exists today both to develop conformal metasurfaces to specific applications [55–62] and to map arbitrary metasurfaces into desired conformal shapes [63, 64]. In the end, it is fair to say that most of the open literature provides derivations and analyses about metasurfaces taking into account planar surfaces. That said, we should expect that not always there will be a ready-to-go metasurface reported in the literature for specific applications requiring conformal shapes.

The fabrication of metasurfaces in arbitrary, conformal shapes can be challenging. However, different groups around the world have recently shown significant advancements in 3D printing methods [65–67]. As such, designs that traditional methods eventually do not handle will soon be possible to fabricate.

3.5 Final Remarks of the Chapter

Resonant phenomena within metasurfaces' constituent elements, that is, the unit cells, provide arbitrary properties for the interaction with electromagnetic waves. Understanding them is not trivial and we introduced in this chapter the principle of operation of metamaterials, making evident that their atypical properties do not break any laws of physics.

Following, we approached the operation principle of metasurfaces, which is fundamentally the same of metamaterials, first with a very specific case: the mushroom-like metasurface, whose analysis simplifies the unit cells to lumped elements. Then, we presented a less specific method, namely the transmission-line approach. Lastly, we arrived at a general method that considers metasurface unit cells as polarizable particles that can provide the currents required for the conversion of arbitrary field distributions.

A large number of methods exist today to tackle the plurality of functionalities that these artificial structures can carry out. We will present them as needed in the next chapters. We also studied, making use of formerly derived concepts, some limitations that we will meet throughout this book. For some applications, they may be a challenge to resolve before metasurfaces hit the theater of operation.

Following this path was not easy. However, as it will become more evident to the reader as they make analyses of metasurface applications in later chapters, this effort will surely be rewarding. Now, it is time to bring theory to practice, adding to this scenario an operational point of view:

> "Talent and genius operate outside the rules, and theory conflicts with practice."
>
> —Carl P. G. von Clausewitz, military theorist.

References

1 Pendry, J.B., Holden, A.J., Robbins, D.J. et al. (1998). Low frequency plasmons in thin-wire structures. *Journal of Physics: Condensed Matter* 10 (4785): 23–25. https://doi.org/10.1088/0953-8984/10/22/007

2 Drude, P. (1900). Zur Elektronentheorie der Metalle. *Annalen der Physik* 306 (3): 566–613. https://doi.org/10.1002/andp.19003060312

3 Tretyakov, S. (2003). *Analytical modeling in applied electromagnetics*. London, UK: Artech House, Section 5.5, pp. 164–179.

4 Pendry, J.B., Holden, A.J., Robbins, D.J. (1999). Magnetism from conductors and enhanced nonlinear phenomena. *IEEE Transactions on Microwave Theory and Techniques* 47 (11): 2075–2084. https://doi.org/10.1109/22.798002

5 Smith, D.R., Padilla, W.J., Vier, D.C. et al. (2003). Composite medium with simultaneously negative permeability and permittivity. *Physics Review Letters* 84 (4184): 1–4. https://doi.org/10.1103/PhysRevLett.84.4184

6 Hrabar, S., Bartolic, J. (2003). Backward wave propagation in waveguide filled with negative permeability meta material. *Proceedings of the IEEE Antennas and Propagation Society International Symposium Digest Held in Conjunction with: USNC/CNC/URSI North American Radio Science Meeting (Cat. No.03CH37450)*, Columbus, OH, USA (22–27 June 2003), New York, US: IEEE. https://doi.org/10.1109/APS.2003.1217413

7 Floquet, G. (1883). Sur les équations différentielles linéaires à coefficients périodiques. *The Annales Scientifiques de l'École Normale Supérieure* 12: 47–88. https://doi.org/10.24033/asens.220

8 Collin, R.E. (1991). *Field theory of guided waves*, 2nd ed. New York, NY: IEEE Press, Chapter 9, pp. 605–608.

9 Popa, B.-I., Cummer, S.A. (2006). Direct measurement of evanescent wave enhancement inside passive metamaterials. *Physics Review E* 73 (016617): 1–5. https://doi.org/10.1103/PhysRevE.73.016617

10 Schurig, D., Mock, J.J., Justice, B.J. et al. (2006). Metamaterial electromagnetic cloak at microwave frequencies. *Science* 314 (5801): 977–980. https://doi.org/10.1126/science.1133628

11 Pendry, J.B. (2000). Negative refraction makes a perfect lens. *Physical Review Letter* 85 (18): 3966–3969. https://doi.org/10.1103/PhysRevLett.85.3966

12 Fang, N., Lee, H., Sun, C. et al. (2005). Sub-diffraction-limited optical imaging with a silver Superlens. *Science* 308 (5721): 534–537. https://doi.org/10.1126/science.1108759

13 Patel, C.A., Patel, S.K. (2015). Pyramidal horn antenna design loaded by metamaterial for performance enhancement. *Proceedings of the IEEE MTT-S International Conference on Numerical Electromagnetic and Multiphysics Modeling*

and Optimization (NEMO), Ottawa, ON, Canada (11–14 August 2015), New York, US: IEEE. https://doi.org/10.1109/NEMO.2015.7414989

14 Kumar, P., Ali, T., Pai, M.M.M. (2021). Electromagnetic metamaterials: a new paradigm of antenna design. *IEEE Access* 9: 18722–18751. https://doi.org/10.1109/ACCESS.2021.3053100

15 Ivsic, B., Sipus, Z., Hrabar, S. (2009). Analysis of uniaxial multilayer cylinders used for invisible cloak realization. *IEEE Transactions on Antennas and Propagation* 57 (5): 1521–1527. https://doi.org/10.1109/TAP.2009.2016695

16 Sievenpiper, D.F., Zhang, L., Broas, R.F.J., et al (1999). High-impedance electromagnetic surfaces with a forbidden frequency band. *IEEE Transactions on Microwave Theory and Techniques* 47 (11): 2059–2074. https://doi.org/10.1109/22.798001

17 Feresidis, A.P., Goussetis, G., Wang, S. et al. (2005). Artificial magnetic conductor surfaces and their application to low-profile high-gain planar antennas. *IEEE Transactions on Antennas and Propagation* 53 (1): 209–215. https://doi.org/10.1109/TAP.2004.840528

18 Goussetis, G., Feresidis, A.P., Vardaxoglou, J.C. (2006). Tailoring the AMC and EBG characteristics of periodic metallic arrays printed on grounded dielectric substrate. *IEEE Transactions on Antennas and Propagation* 54 (1): 82–89. https://doi.org/10.1109/TAP.2005.861575

19 Bellion, A., Cable, M. (2012). A new wideband and compact high impedance surface. *Proceedings of the 15th International Symposium on Antenna Technology and Applied Electromagnetics*, Toulouse, France (25–28 June 2012), New York, US: IEEE. https://doi.org/10.1109/ANTEM.2012.6262322

20 Gonçalves Licursi de Mello, R. (2023). Metasurfaces and blinking jamming: convergent study, comparative analysis, and challenges. *MDPI Micromachines* 14 (7): 1405. https://doi.org/10.3390/mi14071405

21 Luukkonen, O., Simovski, C., Granet, G. et al. (2008). Simple and accurate analytical model of planar grids and high-impedance surfaces comprising metal strips or patches. *IEEE Transactions on Antennas and Propagation* 56 (6): 1624–1632. https://doi.org/10.1109/TAP.2008.923327

22 Hayt, W.H., Buck, J.A. (2012). *Engineering electromagnetics*, 8th ed. New York, NY: McGraw-Hill, Section 10.11, Eq. 98, pp. 329.

23 Mantash, M., Tarot, A. (2016). On the bandwidth and geometry of dual-band AMC structures. *10th European Conference on Antennas and Propagation (EuCAP)*, Davos, Switzerland (10–15 April 2016), New York, US: IEEE. https://doi.org/10.1109/EuCAP.2016.7481589

24 Gonçalves Licursi de Mello, R., Lepage, A.C., Begaud X. (2023). A low-profile, triple-band, and wideband antenna using dual-band AMC. *MDPI Sensors* 23 (4): 1920. https://doi.org/10.3390/s23041920

25 Gonçalves Licursi de Mello, R., Lepage, A.C., Begaud X. (2022). Réflecteur multibande à métamatériaux. *Speech presented at the Journées Scientifiques URSI France 2022* in Palaiseau, France (22–23 March 2022).

26 Gonçalves Licursi de Mello, R., Lepage, A.C., Begaud X. (2023). A Compact, quad-band, and wideband antenna using triple-band AMC. *Proceedings of 13th International Conference on Metamaterials, Photonic Crystals, and Plasmonics (Meta2023)*, Paris, France (18–21 July 2023), Paris, France: Metaconferences. https://metaconferences.org/META23/files/meta23_proceedings.pdf

27 Baker, B.B., Copson, E.T. (1950). *The mathematical theory of Huygens' principle*, 2nd ed. New York, NY: Oxford University Press.

28 Love, A.E.H. (1901). The integration of the equations of propagation of electric waves. *Philosophical Transactions of the Royal Society, Series A* 197: 1–45.

29 Schelkunoff, S.A. (1936). Some equivalence theorems of electromagnetics and their application to radiation problems. *Bell Labs Technical Journal* 15: 92–112, 1936.

30 Holloway, C.L., Kuester, E.F., Gordon, J.A. et al. (2012). An overview of the theory and applications of metasurfaces: the two-dimensional equivalents of metamaterials. *IEEE Antennas and Propagation Magazine* 54 (2): 10–35. https://doi.org/10.1109/MAP.2012.6230714

31 Kuester, E.F., Mohamed, M.A., Piket-May, M. et al. (2003). Averaged transition conditions for electromagnetic fields at a metafilm. *IEEE Transactions on Antennas and Propagation* 51 (10): 2641–2651. https://doi.org/10.1109/TAP.2003.817560

32 Pfeiffer, C., Grbic, A. (2014). Bianisotropic metasurfaces for optimal polarization control: analysis and synthesis. *Physical Review Applied* 2 (044011): 1–11. https://doi.org/10.1103/PhysRevApplied.2.044011

33 Chen, M., Kim, M., Wong, A.M.H. et al. (2018). Huygens' metasurfaces from microwaves to optics: a review. *Nanophotonics* 7 (6): 1207–1231. https://doi.org/10.1515/nanoph-2017-0117

34 Pfeiffer, C., Grbic, A. (2013). Metamaterial Huygens' surfaces: tailoring wave fronts with reflectionless sheets. *Physics Review Letters* 110 (197401): 1–5. https://doi.org/10.1103/PhysRevLett.110.197401

35 Yu, N., Genevet, P., Kats, M.A. et al. (2011). Light propagation with phase discontinuities: generalized laws of reflection and refraction. *Science* 334 (6054): 333–337. https://doi.org/10.1126/science.1210713

36 Sievenpiper, D.F., Schaffner, J.H., Song, H.J. (2003). Two-dimensional beam steering using an electrically tunable impedance surface. *IEEE Transactions on Antennas and Propagation* 51 (10): 2713–2722. https://doi.org/10.1109/TAP.2003.817558

37 Wong, J.P.S., Epstein, A., Eleftheriades, G.V. (2015). Reflectionless wide-angle refracting metasurfaces. *IEEE Antennas and Wireless Propagation Letters* 15: 1293–1296. https://doi.org/10.1109/LAWP.2015.2505629

38 Brewitt-Taylor, C.R. (2007). Limitation on the bandwidth of artificial perfect magnetic conductor surfaces. *IET Microwaves, Antennas & Propagation* 1 (1): 255–260. https://doi.org/10.1049/iet-map:20050309

39 Engheta, N. (2002). Thin absorbing screens using metamaterial surfaces. *Proceedings of the IEEE Antennas and Propagation Society Internationl Symposium (IEEE Cat. No.02CH37313)*, San Antonio, TX, USA (16–21 June 2002), New York, US: IEEE. https://doi.org/10.1109/APS.2002.1016106

40 Yang, W., Hua, G., Hong, W. (2009). Wideband artificial magnetic conductor structure for Ku-band antenna applications. *Proceedings of the 3rd IEEE International Symposium on Microwave, Antenna, Propagation and EMC Technologies for Wireless Communications*, Beijing, China (27–29 October 2009), New York, US: IEEE. https://doi.org/10.1109/MAPE.2009.5355800

41 Begaud, X., Lepage, A.C., Varault, S. et al. (2018). Ultra-wideband and wide-angle microwave metamaterial absorber. *MDPI Materials* 11 (10): 1–9. https://doi.org/10.3390/ma11102045

42 Shao, L., Zhu, W. (2021). Electrically reconfigurable microwave metasurfaces with active lumped elements: a mini review. *Frontiers in Materials* 8 (689665): 1–7. https://doi.org/10.3389/fmats.2021.689665

43 Zhao, J., Cheng, Q., Chen, J. et al. (2013). A tunable metamaterial absorber using varactor diodes. *New Journal of Physics* 15 (043049): 2–11. https://doi.org/10.1088/1367-2630/15/4/043049

44 Gordon, J.A., Holloway, C.L., Dienstfrey, A. (2009). A physical explanation of angle-independent reflection and transmission properties of metafilms/metasurfaces. *IEEE Antennas and Wireless Propagation Letters* 8: 1127–1130. https://doi.org/10.1109/LAWP.2009.2033216

45 Magill, E., Wheeler, H. (1966). Wide-angle impedance matching of a planar array antenna by a dielectric sheet. *IEEE Transactions on Antennas and Propagation* 14 (1): 49–53. https://doi.org/10.1109/TAP.1966.1138622

46 Shabanpour, J., Simovski, C.R. (2022). Angular and polarization stability of broadband reconfigurable intelligent surfaces of binary type. *IEEE Access* 10: 126253–126268. https://doi.org/10.1109/ACCESS.2022.3226264

47 Szabó, Z., Park, G.-H., Hedge, R. et al. (2010). A unique extraction of metamaterial parameters based on Kramers–Kronig relationship. *IEEE Transactions on Microwave Theory and Techniques* 58 (10): 2646–2653. https://doi.org/10.1109/TMTT.2010.2065310

48 Hashemi, S.M., Tretyakov, S.A., Soleimani, M. et al. (2013). Dual-polarized angularly stable high-impedance surface. *IEEE Transactions on Antennas and Propagation* 61 (8): 4101–4108. https://doi.org/10.1109/TAP.2013.2263216

49 Bah, A.O., Qin, P.-Y., Ziolkowski, R.W. et al. (2018). Realization of an ultra-thin metasurface to facilitate wide bandwidth, wide angle beam scanning. *Scientific Reports* 8 (4761): 1–11. https://doi.org/10.1038/s41598-018-23288-4

50 Bah, A.O., Qin, P.-Y., Ziolkowski, R.W. (2019). A wideband low-profile tightly coupled antenna array with a very high figure of merit. *IEEE Transactions on Antennas and Propagation* 67 (4): 2332–2343. https://doi.org/10.1109/TAP.2019.2891460

51 Álvarez, H.F., de Cos, M.E., Las-Heras, F. (2018). AMC's angular stability improvement through the introduction of lumped components. *IEEE Antennas and Wireless Propagation Letters* 17 (5): 813–816. https://doi.org/10.1109/LAWP.2018.2817198

52 Amer, A.A.G., Sapuan, S.Z., Nasimuddin, N. et al. (2021). A broadband wide-angle metasurface absorber for energy harvesting applications. *Proceedings of the International Conference of Technology, Science and Administration (ICTSA)*, Taiz, Yemen, (22–24 March 2021), New York, US: IEEE. https://doi.org/10.1109/ICTSA52017.2021.9406540

53 Fante, R.L., McCormack, M.T. (1988). Reflection properties of the Salisbury screen. *IEEE Transactions on Antennas and Propagation* 36 (10): 1443–1454. https://doi.org/10.1109/8.8632

54 Monorchio, A., Manara, G., Lanuzza, L. (2002). Synthesis of artificial magnetic conductors by using multilayered frequency selective surfaces. *IEEE Antennas and Wireless Propagation Letters* 1: 196–199. https://doi.org/10.1109/LAWP.2002.807956

55 Chang, Q., Ji, J., Chen, K. et al. (2024). Transparent and ultra-thin flexible checkerboard metasurface for radar–infrared bi-stealth. *MDPI Sensors* 24 (1531): 1–13. https://doi.org/10.3390/s24051531

56 Mizmizi, M., Ayoubi, R.A., Tagliaferri, D. (2023). Conformal metasurfaces: a novel solution for vehicular communications. *IEEE Transactions on Wireless Communications* 22 (4): 2804–2817. https://doi.org/10.1109/TWC.2022.3213901

57 Lee, H., Kwon, D.-H. (2022). Conformal metasurfaces on a grounded dielectric substrate for leaky-wave antennas. *Proceedings of the IEEE International Symposium on Antennas and Propagation and USNC-URSI Radio Science Meeting (AP-S/URSI)*, Denver, CO, USA (10–15 July 2022), New York, US: IEEE. https://doi.org/10.1109/AP-S/USNC-URSI47032.2022.9886883

58 Soares, I.V., Vadher, P., Skrivervik, A.K. et al. (2023). Analysis of non-canonical body-conformal arrays with polarization decomposition. *Proceedings of the 17th European Conference on Antennas and Propagation (EuCAP 2023)*, Florence, Italy (26–31 March 2023), New York, US: IEEE. https://doi.org/10.23919/EuCAP57121.2023.10133000

59 Khuyen, B.X., Tung, B.S., Kim, Y.J. et al. (2018). Ultra-subwavelength thickness for dual/triple-band metamaterial absorber at very low frequency. *Scientific Reports* 8 (11632): 1–9. https://doi.org/10.1038/s41598-018-29896-4

60 Yoo, Y.J., Zheng, H.Y., Kim, Y.J. et al. (2014). Flexible and elastic metamaterial absorber for low frequency, based on small-size unit cell. *Applied Physics Letters* 105 (041902): 1–4. https://doi.org/10.1063/1.4885095

61 Kim, Y.J., Hwang, J.S., Khuyen, B.X. et al. (2018). Flexible ultrathin metamaterial absorber for wide frequency band, based on conductive fibers. *Science and Technology of Advanced Materials* 19 (1): 711–717. https://doi.org/10.1080/14686996.2018.1527170

62 Fan, S., Song, Y. (2020). Ultra-wideband flexible absorber in microwave frequency band. *MDPI Materials* 13 (4883): 1–11. https://doi.org/10.3390/ma13214883

63 Wu, K., Coquet, P., Wang, Q.J. et al. (2018). Modelling of free-form conformal metasurfaces. *Nature Communications* 9 (3494): 1–8. https://doi.org/10.1038/s41467-018-05579-6

64 Valle, C.L., Carranza, G.T., Rumpf, R.C. (2023). Conformal frequency selective surfaces for arbitrary curvature. *IEEE Transactions on Antennas and Propagation* 71 (1): 612–620. https://doi.org/10.1109/TAP.2022.3216960

65 Whittaker, T., Zhang, S., Powell, A. et al. (2023). 3D printing materials and techniques for antennas and metamaterials: a survey of the latest advances. *IEEE Antennas and Propagation Magazine* 65 (3): 10–20. https://doi.org/10.1109/MAP.2022.3229298

66 Xu, H.X., Hu, G., Wang, Y. et al. (2021). Polarization-insensitive 3D conformal-skin metasurface cloak. *Light: Science & Applications* 10 (75): 1–13. https://doi.org/10.1038/s41377-021-00507-8

67 Ge, R., Striker, R., Braaten, B.D. (2021). On changing the phase of the radiated field from a microstrip patch antenna using a 3D-printed conformal metasurface. *IEEE International Symposium on Antennas and Propagation and USNC-URSI Radio Science Meeting (APS/URSI)*, Singapore, Singapore (04–10 December 2021). https://doi.org/10.1109/APS/URSI47566.2021.9704576

4

Passive Stealth with Metasurfaces

Metasurfaces have found many applications since they became a hot topic in electromagnetics two decades ago. However, it is not too much to say that the vast majority of professionals who are not specialized in metasurfaces at a first glance think only about stealth applications whenever they listen to the subject. The reason is that one of the first metamaterial applications, which the main popular science media extensively reported back in the day, was the cloak of invisibility [1] which we commented in Section 3.1. Indeed, the appeal of a device that can coat others whereby making them invisible looks irresistible. Further, when the first advances in metasurfaces took place, the community saw their absorptive properties [2] as a means of achieving a low RCS in aircraft and war vessels. For these reasons, and also for stealth is one of the easiest applications to understand, we will pick it up as our starting point for the insertion of metasurface technology in the EW context.

In the historical cat-and-mouse EW game, stealth was a response to the advent of radar itself. In brief, the technology has traditionally relied on the shaping of platforms and the use of coatings that can dissipate electromagnetic energy at frequencies that most of the dangerous radars would work. Nonetheless, the shooting of the Lockheed-Martin F-117 Nighthawk down at Kosovo after a couple of decades of program [3] showed that the game would not finish any soon. If the use of VHF and UHF radars has shown to be key against stealth since then, the possibility of designing metasurfaces at arbitrary frequencies indicates that the game is still unfolding at full steam.

In this chapter, we discuss RCS reduction with passive metasurfaces. First, we explain classical concepts, like what is RCS, the radar equation, and traditional stealth methods. Next, we present FSSs, which work as bandpass or band-stop spatial filters and many consider as the precursor of metasurfaces. Then, we see metasurfaces that spread the energy throughout diverse directions, polarizations, or frequencies. After that, we discuss metasurfaces that reduce the RCS by absorbing

Metasurface-Driven Electronic Warfare, First Edition. Rafael Gonçalves Licursi de Mello.

Published 2025 by John Wiley & Sons, Inc.
Companion website: www.wiley.com/go/licursidemello/metasurfacedrivenelectronicwarfare1

energy. Finally, we present bandpass FSSs that actually absorb the energy out of band, called rasorbers.

For comparison, we will consider the performance of the metasurfaces presented in this chapter in the simulations conducted in Chapter 5.

4.1 The Radar Equation and the Concept of Radar Cross-Section

When a wave impinges over a scatterer, it spreads energy throughout all the directions around the object. Figure 4.1 illustrates this phenomenon.

RCS is a figure of merit representing the echo features of a scatterer of electromagnetic waves. The scattering depends on the frequency, the direction θ_i, Φ_i and polarization of the impinging wave, and the direction θ_s, Φ_s and polarization of the scattered waves. For ease of calculation, we can see the RCS as an equivalent area that collects the power of impinging waves for subsequent reradiation, considering all these factors.

Formally, we can express the RCS as a quantity $\widehat{RCS}\,[\mathrm{m}^2]$ whose square root is a complex number with magnitude $\left|\sqrt{\widehat{RCS}}\right|$ and phase φ_{RCS}, as follows [4]:

$$\sqrt{\widehat{RCS}(\theta_s, \Phi_s; \theta_i, \Phi_i)} = 2\sqrt{\pi} r_{sct} \frac{\boldsymbol{E}^s(\theta_s, \Phi_s)}{\boldsymbol{E}^i(\theta_i, \Phi_i)} = \left|\sqrt{\widehat{RCS}}\right| e^{j\varphi_{RCS}} \tag{4.1}$$

where

- r_{sct} is the distance between the scatterer and the observation point [m];
- $\boldsymbol{E}^s$ is the scattered electric field arriving at this point [V/m];
- $\boldsymbol{E}^i$ is the electric field impinging upon the scatterer [V/m].

Note that $\widehat{RCS}$ does not vary with r_{sct} because $\boldsymbol{E}^s$ is inversely proportional to this parameter. Generally, $\widehat{RCS}$ is a 2×2 tensor associating the polarizations of the

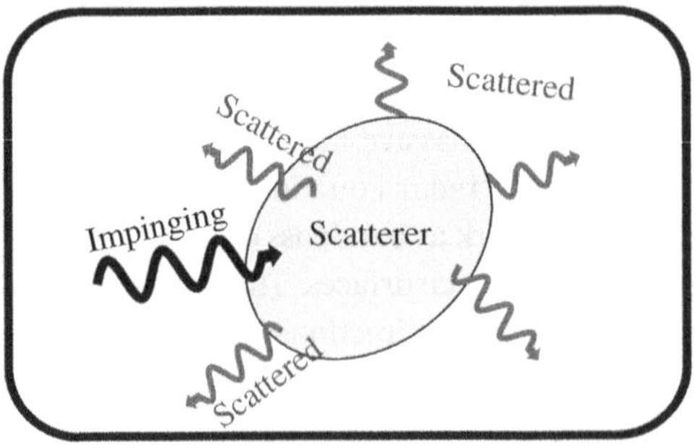

Fig. 4.1 Scattering of electromagnetic waves

incident and scattered waves. In the simulations of this book, we will always decompose the polarization of waves horizontally and vertically. Hence, in matrix form, we can conveniently express (4.1) as:

$$\sqrt{\widehat{RCS}} = 2\sqrt{\pi} r_{sct} \begin{bmatrix} E_h^s/E_h^i & E_h^s/E_v^i \\ E_v^s/E_h^i & E_v^s/E_v^i \end{bmatrix} = \begin{bmatrix} \sqrt{RCS_{hh}} & \sqrt{RCS_{hv}} \\ \sqrt{RCS_{vh}} & \sqrt{RCS_{vv}} \end{bmatrix} \tag{4.2}$$

The subscripts h/v indicate the orientation of the incident and scattered electric fields. We adopt this notation throughout the rest of this book.

Discriminating the polarization in the RCS formulation is important because real antennas do not present perfect polarization purity. Therefore, a radar intended to be vertically polarized will also transmit some power in the horizontal polarization and vice versa. The same is valid for the receiving of signals. The relationship between the two polarization components is crucial for a series of metasurface applications in EW.

Now, consider a radar transmitting electromagnetic waves with vertical and horizontal polarization, that is, for $m = h$ and $m = v$. The waves impinge over a scattering platform with a density of power [W/m^2]:

$$W_{rdr,m} = \frac{P_t G_{t,m}(\theta_{plt}, \Phi_{plt})}{4\pi r_{rdrTx}^2} \tag{4.3}$$

where

- P_t is the power transmitted by the radar [W];
- $G_{t,m}$ is the realized gain of the Tx antenna in the horizontal or vertical polarization ($m = h$ or v) in relation to an ideal isotropic antenna [dBi];
- θ_{plt}, Φ_{plt} is the angular position of the platform referred to the orientation axes of the radar Tx antenna [rad];
- r_{rdrTx} is the distance separating platform and radar transmitter [m].

We consider that the equivalent RCS area of the platform collects some power coming from $W_{rdr,m}$. Next, the platform reradiate this power equally throughout the area of a sphere with vertical and horizontal polarization, that is, for $n = h$ and $n = v$. Remember that each element of $\widehat{RCS}$ already computes the dependence on the directions of incidence and scattering. As such, the power density coming from the scatterer and arriving at the radar receiver is:

$$W_{sct,nm} = W_{rdr,m} \frac{RCS_{nm}(\theta'_{rdrRx}, \Phi'_{rdrRx}; \theta'_{rdrTx}, \Phi'_{rdrTx})}{4\pi r_{rdrRx}^2} \tag{4.4}$$

where

- RCS_{nm} is an element of the $\widehat{RCS}$ tensor [m^2];
- r_{rdrRx} is the distance between the platform and the radar receiver [m];

- θ'_{rdrRx}, Φ'_{rdrRx} and θ'_{rdrTx}, Φ'_{rdrTx} are the angular positions of the radar receiver and transmitter, respectively, both referred to the platform orientation axes [rad].

Inside the RCS_{nm} term of (4.4), $r_{sct} = r_{rdrRx}$ because the radar receiver is located at the observation point. In this book, we use the apostrophes to specify different reference coordinate systems in the same context. The bistatic skin return of a platform illuminated by this radar is the total power collected by the radar receiver antenna [W]:

$$S = \sum\sum S_{nm} \tag{4.5}$$

where

$$S_{nm} = W_{sct,nm} \frac{\lambda^2 G_{r,n}\left(\theta''_{plt}, \Phi''_{plt}\right)}{4\pi} \tag{4.6}$$

and

- $G_{r,n}$ is the realized gain of the Rx antenna in the horizontal or vertical polarization ($n = h$ or v) [dBi];
- θ''_{plt}, Φ''_{plt} is the angular position of the platform referred to the orientation axes of the radar Rx antenna [rad].

Equation (4.6) is known as the bistatic radar equation.

When the radar transmitter and receiver are located at the same position, $\theta_{plt} = \theta''_{plt}$, $\Phi_{plt} = \Phi''_{plt}$, $r_{rdrTx} = r_{rdrRx}$, $\theta'_{rdrRx} = \theta'_{rdrTx}$, and $\Phi'_{rdrRx} = \Phi'_{rdrTx}$. In this case, (4.6) simplifies to the monostatic radar equation, which we denote as:

$$S_{nm} = \frac{P_t\lambda^2}{(4\pi)^3 r^4_{rdr}} G_{t,m}(\theta_{plt}, \Phi_{plt}) G_{r,n}(\theta_{plt}, \Phi_{plt}) RCS_{nm}(\theta'_{rdr}, \Phi'_{rdr}) \tag{4.7}$$

where

- r_{rdr} is the distance between the platform and the radar [m];
- θ'_{rdr}, Φ'_{rdr} is the angular position of the radar referred to the platform orientation axes [rad].

The RCS definition stated above is valid only for values of r_{sct} much larger than the scatterer dimensions. The reason is that the scattering of waves on complex scatterers, such as aircraft, drones, satellites, and shipment, occurs at multiple parts of the platform and due to diverse mechanisms. The impedance mismatch between the free space and surfaces throughout the platform leads to specular reflections according to Snell's law (2.61) as discussed in Section 2.1.5. Besides, discontinuities between moving parts, tips, cavities, and corners of the fuselage and embedded weapons raise diffraction components [5]. Moreover, conventional antennas in the platform reradiate half of the power intercepted from impinging waves when they matched to their receivers. This is the best of the case. When they are not matched, they reradiate more than half of the intercepted power [6]. Hence, the RCS is the result of reflections, diffractions, and reradiation on diverse parts of the platform, each one contributing more or less to the scattering.

To simplify the analysis, we can interpret the RCS of complex scatterers as the sum of the RCSs of n simpler scattering centers as [4]:

$$\sqrt{\widehat{RCS}} = \sum_{i=1}^{n} \left| \sqrt{\widehat{RCS}_i} \right| e^{j\varphi_{RCS_i}} = \left| \sqrt{\widehat{RCS}} \right| e^{j\varphi_{RCS}} \tag{4.8}$$

where

- $\widehat{RCS}_i$ is the RCS of the ith scattering center [m^2];
- φ_{RCS_i} is the phase of its square root [rad].

In (4.8), φ_{RCS_i} accounts for the phases related to the scattering phenomenon and the position of each scattering center in relation to the observation point.

Reducing the RCS provides an invaluable operational advantage because such figure of merit has a direct impact on the maximum range $r_{\max}$ of radar detection [m]. We can calculate such range from the minimum signal-to-noise ratio $SNR_{\min} = S_{\min}/N_0$ [unitless] required for proper operation of the radar. We consider here that S_{min} is the minimum required signal level, $N_0 = BWk_bT$ is the noise level in an impedance-matched receiver [W], k_b is the Boltzmann constant, and T is the temperature, in kelvins [K]. For the sake of simplicity, consider a particular monostatic, vertically polarized radar with perfect polarization purity. Manipulating (4.7), we have:

$$r_{\max} = \sqrt[4]{\frac{P_t\lambda^2}{(4\pi)^3 SNR_{\min} BWk_bT} G_{t,v}^2(\theta_{plt}, \Phi_{plt}) RCS_{vv}(\theta'_{rdr}, \Phi'_{rdr})} \tag{4.9}$$

In (4.9), $G_{t,v}(\theta_{plt}, \Phi_{plt}) = G_{r,v}(\theta_{plt}, \Phi_{plt})$. The maximum range $r_{\max}$ decreases with the fourth root of the RCS reduction. In other words, reducing the RCS by 12 dB decreases $r_{\max}$ to half its original value, allowing a platform to penetrate much deeper within the radar zone. This advantage can mean the difference between an accomplished mission or a lethality in the friendly forces.

4.2 Traditional Stealth Technologies

Despite the simplicity in calculating the skin return with (4.6) and (4.7), the precise result depends on the knowledge of $\widehat{RCS}$. However, in stealth applications, the goal is simply to minimize the scattering for specific conditions, like frequency, polarization, or angular disposition of the radar transmitter and receiver referred to the platform. For the latter, the platform orientation axes in relation to the radar transmitter and receiver are relevant. That means that we should design the platform to carry out missions with a specific geometrical disposition. For instance, we can design destroyer vessels or attack aircraft to present a low RCS at the angle they offer the most when attacking.

We can exploit two main mechanisms in the traditional RCS reduction: the scattering of electromagnetic waves in directions other than the radar receiver or the absorption thereof [7].

The scattering method consists of adopting shapes with faceted flat plates and wedges whose flare angle is less than 180° because the backscatter from this kind of wedge is lower than that of convex surfaces. Also, the design of the platform should orient surfaces so that Snell's law (2.61) will direct most of the specular reflections away from the radar [5]. This method can reduce the monostatic RCS considerably. However, it cannot ensure that receivers of bistatic or multistatic radars will receive a low signal.

The absorption method, on the other hand, exploits the matching between the impedances of the free space and the platform material. According to (2.50), (2.73), and (2.76), reflections over a platform will tend to zero if its effective impedance is close to the free-space intrinsic impedance η_0. We can achieve the reduction in the RCS by coating strategic parts of the platform with special paints with dissipative particles or other artificial radar absorbing materials (RAM). As such, we can reduce the $\widehat{RCS}_i$ terms in the summation of (4.8). This approach is effective at narrow bands around resonant frequencies. We can achieve wide bands with multilayered, bulky structures. Besides, the coating may require regular maintenance depending on the operational conditions it is subjected to [8]. Among the latest technologies here, we should highlight multi-walled carbon nanotubes [9] and metal-organic framework [10] composites.

Combining both the scattering and absorption methods can provide a low platform RCS. In aeronautics, the Lockheed SR-71 Blackbird [11] and the F-117 Nighthawk [12] are good examples of this traditional combination.

4.3 Frequency Selective Surfaces

Long before the community coined the term "metasurface", they did many studies on the interaction of electromagnetic waves with grids of metal shapes or slots [13–18]. They used these grids mainly for the spatial filtering of electromagnetic waves, which is why today we call them FSSs.

FSSs have many similarities with metasurfaces in terms of working principle and construction. We can see them as a precursor of the metasurfaces. Small unit cells that can resonate when excited by electromagnetic waves is also

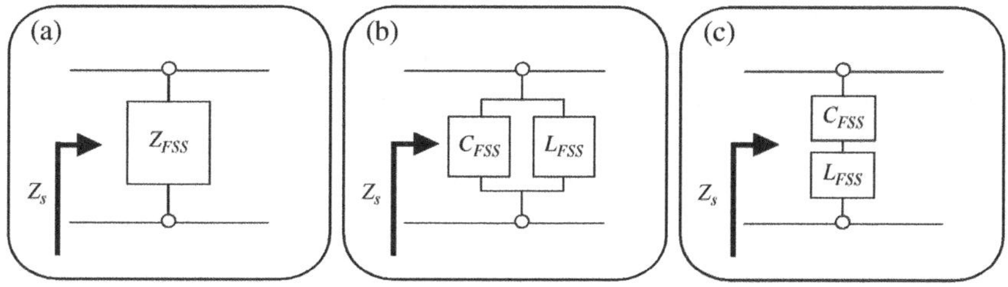

Fig. 4.2 Transmission-line equivalents: (a) generic FSS; (b) bandpass FSS; (c) band-stop FSS

what constitute FSSs. Moreover, the transmission-line equivalence also applies [16–18].

Consider in Fig. 4.2 the transmission-line equivalent of a generic, lossless FSS. We can represent the structure with a shunt impedance Z_{FSS}, as in Fig. 4.2a.

Depending on the construction of the FSS, we can obtain either a bandpass or a band-stop behavior. Metal plates where slots are etched will tend to reflect the impinging waves except when the slots resonate. Therefore, this configuration works as a bandpass filter. In this case, we can model the FSS impedance Z_{FSS} as a parallel circuit composed of a capacitance C_{FSS} and an inductance L_{FSS}, as in Fig. 4.2b. Conversely, an FSS composed of metal shapes deployed over a dielectric will tend to transmit the impinging waves except when the metal shapes resonate, working as a band-stop filter. Now, we can model Z_{FSS} as a circuit in series, as in Fig. 4.2c.

In the early 2000s, many restrictions over the work of B.A. Munk were lifted, and he could disclose a series of techniques for the aircraft RCS reduction he had developed for the United States government [19]. In aeronautics, the design of radomes using bandpass FSSs enabled transmission of electromagnetic waves strictly in the operational frequency band of an antenna, when it is properly matched with its receiver [20]. Out of band, the FSS would avoid impinging waves from stimulating the antenna. The radome shape was so that it would direct most of the specular reflections away from the source of radiation due to Snell's law (2.61). We can also use bandpass FSSs for EP purposes against directed-energy weapons, as it will become clearer to the reader in Chapter 8.

In Naval, besides the previous bandpass applications, we can also conveniently exploit band-stop FSSs. Many naval radars and communication systems work below 3.0 GHz. Hence, we can design band-stop FSSs to reflect waves from other frequency bands, like the X-band, as long as the FSS could present a considerable BW [21].

The features of the FSS, such as resonant frequency, BW, and stability with polarization and angle of incidence, depend mainly on the element shape and the associated periodicity [22], similar to what happens in metasurfaces.

Another approach Munk proposed was to act directly on the antenna RCS, reducing it through impedance matching, which could work well in a narrow BW, or through the choice of proper geometries for the antenna shape [23]. Many consider the work of Munk a starting point for what metasurfaces would bring to the RCS reduction technology in the upcoming years [24].

4.4 Scattering Metasurfaces

The mechanisms metasurfaces exploit are similar to those of traditional stealth methods. Metasurfaces can promote both the scattering and absorption of the electromagnetic energy. We can achieve the reduction in the RCS by coating the strategic parts of the platform, mainly those that contribute the most to the RCS. In this manner, a reduction in the $\widehat{RCS}_i$ terms in the summation of (4.8) takes place. This technique is an alternative to the use of dissipative paints and RAMs of traditional stealth methods, allowing for a different maintenance regime and the choice of specific frequency ranges.

4.4.1 Checkerboard Metasurfaces

Checkerboard metasurfaces are structures where we interchangeably employ two different kinds of materials, whether natural or artificial. We have used this structure for RCS reduction purposes at least since the 2010s [25]. To understand them, let us first consider two infinite, planar, PEC and PMC surfaces with surface impedances $Z_s = 0$ and $Z_s = \infty$, respectively. According to (2.50), the PEC surface provides a reflection coefficient $\Gamma_{TEM} = -1$, whereas the PMC surface $\Gamma_{TEM} = +1$.

Now consider a surface where PEC and PMC squares are disposed in a checkerboard configuration. Fig. 4.3a illustrates the two materials. When impinging normally upon this metasurface, waves reflected by the PEC and PMC squares will interfere destructively, leading to a mitigation of the RCS.

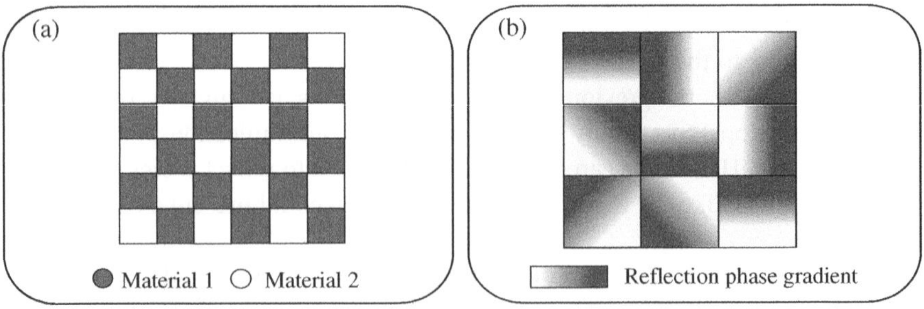

Fig. 4.3 Scattering metasurfaces for RCS reduction: (a) checkerboard; (b) phase-gradient

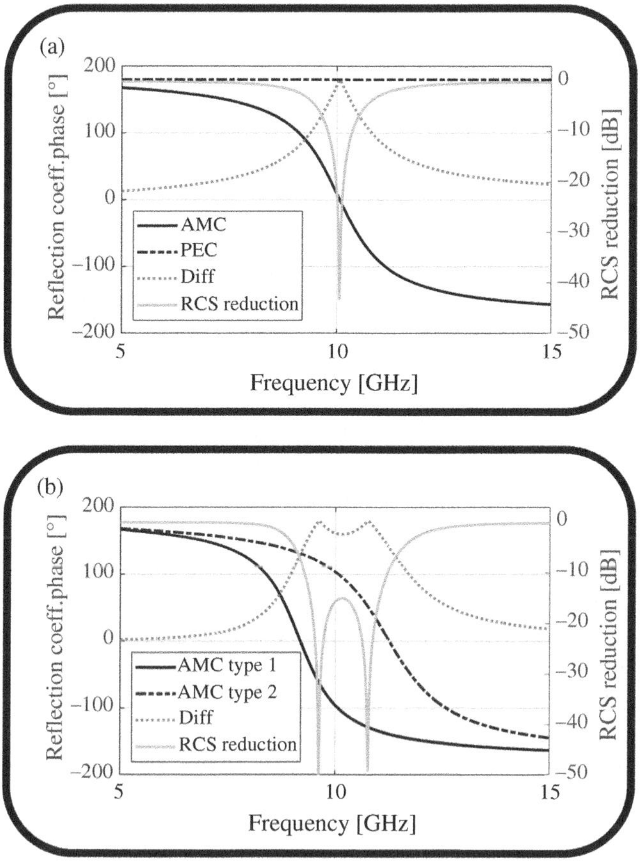

Fig. 4.4 Reflection coefficient phases and RCS reduction: (a) checkerboard made of AMC and PEC; (b) checkerboard made of two kinds of AMC

In the real world, PMCs do not exist, and we cannot build a structure that will provide $\Gamma_{TEM} = +1$ unconditionally. We saw in Section 3.2.3 that AMCs can provide such reflection coefficient at specific frequencies. Thus, opposing PEC squares and squares made of AMC unit cells in a checkerboard surface would lead to an optimal RCS reduction at the very same specific frequencies. As an example, we depict in Fig. 4.4a the reflection coefficient phases of an AMC and a PEC surface, as well as the difference between both. We calculate the AMC reflection coefficient with (2.50) considering an impedance mismatch between a surface impedance Z_s calculated with (3.15) and the free-space intrinsic impedance $\eta_0 \cong 120\pi\ \Omega$. We use $C_{FSS} = 0.25$ pF and $L_{slab} = 1$ nH to obtain a resonant frequency $f_0 = 10$ GHz.

As expected, a phase difference of 180° occurs exactly at the resonant frequency of the AMC. Considering that each pattern occupies half the area of the checkerboard, we can calculate the monostatic RCS reduction by superimposing the reflected fields as [26]:

$$RCS\ reduction = 10\log\left|\frac{e^{j\varphi_{\Gamma 1}} + e^{j\varphi_{\Gamma 2}}}{2}\right|^2 \tag{4.10}$$

where $\varphi_{\Gamma 1}$ and $\varphi_{\Gamma 2}$ are the reflection coefficient phases of the two materials. Accordingly, Fig. 4.4a also shows the RCS reduction for the checkerboard metasurface previously stated using (4.10). Indeed, the reduction is optimal at $f_0 = 10$ GHz and keeps < -10 dB in a narrowband around this frequency, which is not that useful for operational purposes.

Instead of interchanging AMC and PEC patterns, we can exploit two AMC structures with shifted resonant frequencies to enlarge the BW of the checkerboard metasurface. The task is as easy as designing both AMCs so that their phases present the same decay slopes but properly shifted in the spectrum to provide a difference close to 180° in the broadest possible frequency range. Figure 4.4b depicts the reflection coefficient phases of two AMCs whose unit cell capacitances are different. Specifically, the AMC type 1 has $C_{FSS} = 0.30$ pF and type 2 has $C_{FSS} = 0.20$ pF. Both present $L_{slab} = 1$ nH. The range where the monostatic RCS reduction is better than −10 dB significantly outperforms the one in Fig. 4.4a.

Following this idea, W. Chen et al. [26] experimentally demonstrate a checkerboard metasurface made of squared- and circular-patch AMCs. The structure is 6.35 mm thick. They achieve an RCS reduction of up to −25 dB around 4.4 and 7.5 GHz. The structure provide a reduction better than −10 dB in a wide band of 60% for normal incidence (from 4.10 to 7.59 GHz). For an oblique incidence of 30°, this BW degrades to 45%, and for 40° to 34% due to the metasurface angle-dependent behavior we saw in (2.73) and (2.76).

D. Sang et al. [27] propose a metasurface made of two 0.5-mm thick AMCs whose unit cells are a meander cross and a cross contour. They achieve an RCS reduction again up to −25 dB, now at 6.2 and 7.1 GHz, keeping < -10 dB from 3.77 to 10.14 GHz, which represents a BW of 91.5%. Increasing the angle of incidence degrades performance for both TE and TM mainly from 8.0 GHz, which still represents a wide BW.

M.-J. Haji-Ahmadi et al. [28] propose a checkerboard composed of two patterns of pixelated AMC unit cells. Its thickness is 6 mm. The fabricated prototype is shown in Fig. 4.5.

Such metasurface presents an even wider 10-dB BW, 95%, ranging from 3.8 to 10.7 GHz. The RCS reduction is optimal at 7.2 GHz (−35 dB). An oblique incidence degrades BW to 82% at 30°, and 70% at 40° for TM polarization. For TE polarization, $BW = 52\%$ at 30°, and 30% at 40°.

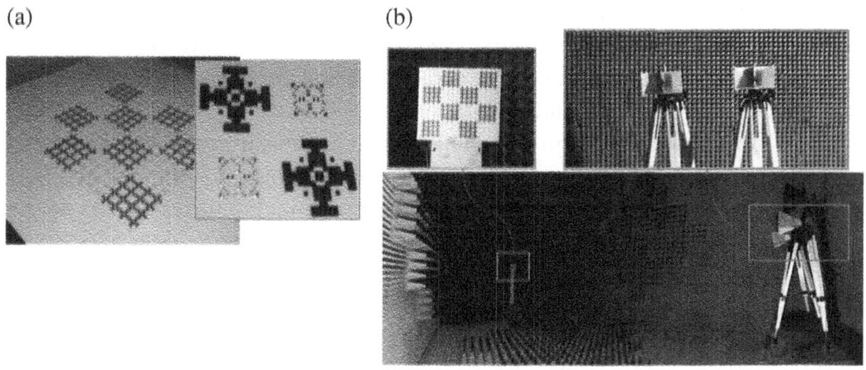

Fig. 4.5 Metasurface of [28]: (a) fabricated prototype; (b) measurement setup. "Reproduced from Haji-Ahmadi et al. [28]/Springer Nature/CC BY 4.0."

We can also use the checkerboard approach in the antenna design. A ground plane disposing two types of squared-patch AMCs in a checkerboard configuration reduces the RCS of the dipole when backing it [29]. The metasurface thickness is 3 mm. The dipole operates at 4.1 GHz, and the metasurface provides an RCS reduction better than < -6 dB out of band, namely from 6.5 to 18.0 GHz, for both polarizations. It also provides optimal reductions of -20 and -30 dB close to 14 GHz in each polarization.

H. Jiang et al. [30] cover a 2×2-slot antenna array with a 4-mm thick checkerboard metasurface made of a metallic grounded slab and a fractal square shape. The metasurface provides an in-band RCS reduction (around 5.9 GHz) better than -11 dB in the co-polarized and -6 dB in the cross-polarized component. Besides, it also provides an out-of-band RCS reduction better than -10 dB for both polarizations in the X-band. An optimal out-of-band result of -35 dB is seen around 9.7 GHz.

S. Varault et al. [31] propose patches and empty space forming a checkerboard over a ground plane to provide a stable gain in the broadside direction with simultaneous RCS reduction. Optimal results occur at 4.5 and 13.5 GHz, where the reduction gets around -25 dB. They achieve a 10-dB BW of $5:1$ for oblique incidence of up to $60°$ and both polarizations, from 3.8 to 19.0 GHz. The overall thickness is 2.4 mm.

4.4.2 Phase-Gradient Metasurfaces

We studied in Section 3.2.7 that the generalized law of reflection [32] implies that a linear phase gradient imposed by a metasurface on reflected waves deflects them according to (3.35). Phase-gradient metasurfaces can exploit this law for RCS reduction purposes. We can divide a metasurface into multiple supercells, each one presenting a phase gradient in a specific direction, as depicted in Fig. 4.3b. In this way, the metasurface spreads reflected waves in multiple directions, directing the skin returns away from the source of radiation and also avoiding their focusing on any particular point around the platform.

This is exactly what K. Singh et al. [33] do with a metasurface divided into multiple squared-patch AMC supercells presenting phase gradients oriented in different directions. Eight types of supercells ensure the spread of the reflected energy. The metasurface achieves a monostatic RCS reduction better than −45 dB at 20.0 GHz. The 10-dB BW is approximately 32% (from 16 to 22 GHz). The thickness of the metasurface is 1.6 mm.

S. Choudhary et al. [34] exploite a similar idea but with unit cells made of patches and loops with circular shapes. Again, the metasurface presents eight different supercells. For both polarizations, it provides monostatic RCS reductions better than −30 dB at 19.5 GHz. Further, the 10-dB BW is 76% (from 9.2 to 20.5 GHz). The thickness is 3 mm.

Y. Azizi et al. [35] propose a 1.6-mm thick phase-gradient metasurface with supercells in a coaxial disposition providing thus a radial diffusion of energy. Figure 4.6 shows the fabricated prototype in an anechoic chamber. The metasurface provides RCS reduction values around −15 dB from 18 to 25 GHz and better than −10 dB from 11.3 to 32.3 GHz ($BW = 96.3\%$).

There is another interpretation of how a phase-gradient metasurface reduces the RCS. We saw in Section 4.4.1 that the principle behind RCS reduction with checkerboard metasurfaces is the destructive interference of waves reflected on two

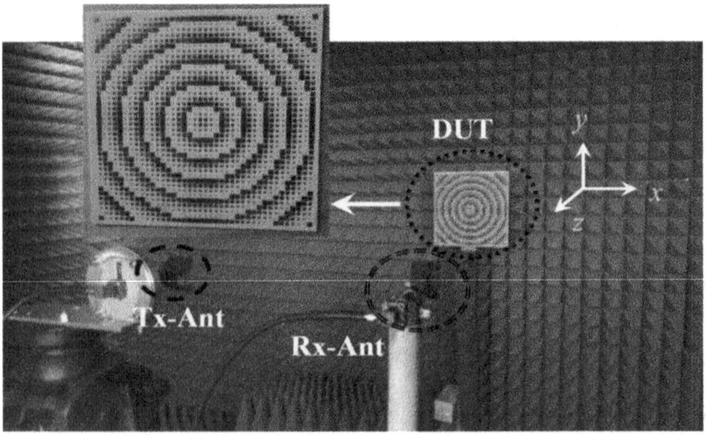

Fig. 4.6 Metasurface of [35]. "Reproduced from Azizi et al. [35]/MDPI/CC BY 4.0."

types of materials, which leads to (4.10). Extending this concept, we can evaluate the monostatic RCS of a phase-gradient metasurface with n unit cells as [35]:

$$RCS\ reduction = 10 \log \left| \frac{\sum_{i=1}^{n} |\Gamma_i| e^{j\varphi_{\Gamma i}}}{\sum_{k=1}^{n} |\Gamma_k|} \right|^2 \tag{4.11}$$

where $|\Gamma_{i,k}|$ is the reflection coefficient magnitude and $\varphi_{\Gamma i}$ the reflection coefficient phase of the ith or kth unit cell in the metasurface.

This interpretation is quite reasonable if we consider the definition of a plane wave. As we saw in Section 2.1, a possible solution to the wave equations (2.29) and (2.31) is a plane wave represented by (2.40). In the specific case of (2.40), a constant magnitude and phase are seen at any moment in the xy plane, whose normal direction $+\hat{\mathbf{a}}_z$ matches the propagation direction. In general, a constant magnitude and phase through any plane that is perpendicular to a fixed direction in space, at any moment, is what characterizes a plane wave.

In a plane wave, all the energy propagates in a single direction. When a plane wave impinges over a flat, conventional surface, Snell's law (2.61) orients all the reflected energy to another plane wave at the specular direction. However, if the surface counts with numerous unit cells, different reflection phases can break the equiphasic wavefronts in the reflected wave. In other words, the reflected energy will be a mixture of weaker plane waves spread in multiple directions instead of a single direction.

This interpretation opens the possibility for the conceiving of metasurfaces whose phase distribution is actually random. J. Chen et al. [36] achieve a monostatic RCS reduction of up to −30 dB at 12.0 GHz by exploiting this idea. Also, they achieve a 10-dB BW of 98% (from 5.5 to 16 GHz) for one polarization and 63% (from 7.5 to 14.5 GHz) for the other. The thickness of the metasurface is 4.0 mm.

M. Feng et al. [37] propose N-shaped unit cells randomly rotated to provide random reflection phases to both polarizations. They obtain measured RCS reductions of up to −15 dB around 15 GHz. The reduction keeps better than −10 dB from 12.0 to 18.4 GHz ($BW = 42\%$), and the metasurface thickness is 3 mm.

Phase-gradient metasurfaces also find applicability in the antenna RCS reduction. W. Zhang et al. [38] obtain an out-of-band RCS reduction of up to −25 dB in a 2 × 2-slot array antenna by adding a phase-gradient metasurface thereover. The 10-dB BW is 82.4%, ranging from 7 to 16.8 GHz. At the antenna operating frequency, 4.2 GHz, the metasurface is transmissive, and the radiation features remain practically unchanged.

B. Li et al. [39] propose a 2.5-mm thick phase-gradient metasurface with H-shaped unit cells for an antenna operating in the X-band, around 10 GHz. They achieve a 6-dB BW of 23.3% from 8.7 to 11 GHz. At 10 GHz, the RCS reduction is −12 dB. Out of the band, the reduction reaches −22 dB for one polarization and −30 dB for the other.

M.I. Mirzapour et al. [40] use a three-layer metasurface with Jerusalem unit cells that not only provides an in-band RCS reduction of −19 dB but also enhances the gain of a patch antenna operating at 9.8 GHz by 12 dB. The phase gradient focuses on the patch antenna the waves transmitted through the metasurface. On the other hand, the distance between the metasurface and the ground plane is such that energy reflected on both interferes destructively. The 10-dB BW is narrow (8.2%, from 9.3 to 10.1 GHz). The metasurface thickness is 3 mm.

H. Umair et al. [41] associate both the phase-gradient and the checkerboard mechanisms (see Section 4.4.1) in the design of a Fabry-Pérot antenna. Two different AMCs surround the antenna aperture, where the phase-gradient metasurface is. As a result, they obtain an RCS reduction better than −10 dB from 4 to 13 GHz (BW = 105.8%), with optimal values around −40 dB for both polarizations (out of band) and −6 dB in the antenna operating frequency (7 GHz). The radiation pattern is squinted in relation to the broadside direction, the antenna gain is 12 dBi and the metasurface thickness is 1.52 mm.

4.4.3 Polarization-Conversion Metasurfaces

Another way of reducing the probability of radar detection is by converting the polarization of waves, as Fig. 4.7a illustrates. The goal is to weight most of the skin returns in (4.6) and (4.7) by the cross-polarized gain $G_{r,n}$ of the Rx antenna. The

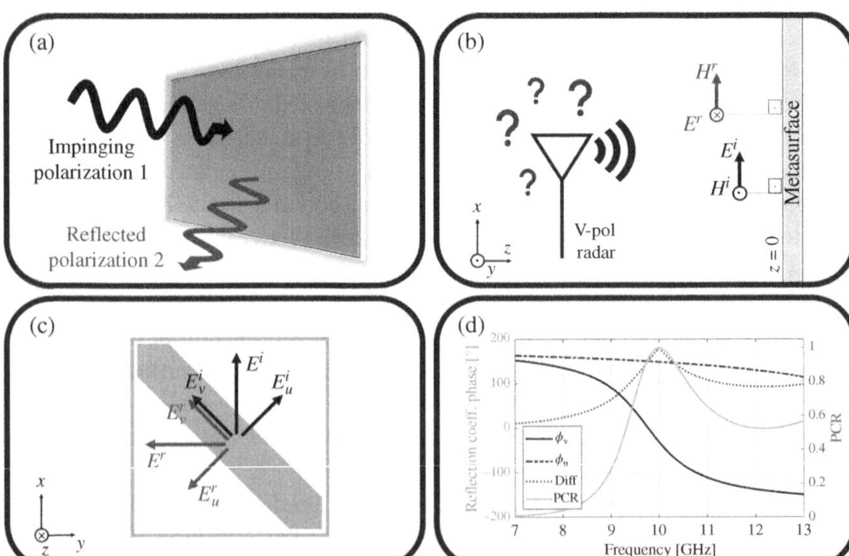

Fig. 4.7 Polarization-conversion: (a) a metasurface converts the polarization of a wave; (b) a linearly polarized radar cannot detect the echoes of its own signals; (c) hypothetical polarization-conversion unit cell; (d) phase shift for each component, phase difference, and PCR

cross-polarized gain of well-designed radar antennas can be 25 or 30 dB below its co-polarized gain [42]. As such, a radar will have significant disadvantages to detect the echoes of its own signals, as Fig. 4.7b represents with a vertically polarized wave converted to the horizontal polarization when reflected on a metasurface.

To highlight the associated mechanism, we will focus here on the conversion of an incident linearly polarized wave to its orthogonal counterpart. Consider in Fig. 4.7c a hypothetical metasurface unit cell composed of a metal inclusion above a ground plane rotated counterclockwise in the z-axis 45° in relation to the x dimension. The metasurface is placed at the plane xy. An x-polarized plane wave with electric field $\boldsymbol{E}^i$ travels at the $+\hat{\mathbf{a}}_z$ direction, being described by (2.40). As such, it impinges upon the unit cell with normal incidence.

For convenience, we can decompose the electric field $\boldsymbol{E}^i$ into two dimensions u and v squinted $\pm 45°$ in relation to the x dimension, leading to two orthogonal vectors $\boldsymbol{E}^i_u$ and $\boldsymbol{E}^i_v$, as Fig. 4.7c also indicates. Considering that the metal inclusion is located at $z = 0$, we can represent the scalar components of the electric field in the u and v dimensions thereon as:

$$E^i_u = \boldsymbol{E}^i \cdot \hat{\mathbf{a}}_{\boldsymbol{u}} = \frac{\sqrt{2}}{2} E_0, \quad E^i_v = \boldsymbol{E}^i \cdot \hat{\mathbf{a}}_{\boldsymbol{v}} = \frac{\sqrt{2}}{2} E_0 \tag{4.12}$$

The asymmetry of the unit cell makes the interactions with the wave differ in the u and v dimensions. If we regard the unit cell in question as an asymmetric AMC and consider the formulation derived in Section 3.2.3, the waves whose electric fields are $\boldsymbol{E}^i_u$ and $\boldsymbol{E}^i_v$ are subjected to different AMC parameters, mainly capacitances C_{FSS}. Therefore, the resonant frequencies $\omega_0 = 1/\sqrt{L_{slab} C_{FSS}}$ in the u and v dimensions will differ and so will the reflection coefficient phases φ_Γ.

Suppose that the unit cell presents $C_{FSS} = 0.1$ pF from the point of view of the wave with $\boldsymbol{E}^i_u$ and $C_{FSS} = 0.265$ pF from that with $\boldsymbol{E}^i_v$. We use $L_{slab} = 1$ nH for both waves considering that the spacing between the metal inclusion and the ground plane is the same for both fields. Now, using these values in (3.15) and then (2.50), we calculate and depict in Fig. 4.7d the reflection coefficient phases φ_u and φ_v related to both dimensions.

As expected, the response of the unit cell differs depending on the dimension u or v. Further, there is a frequency (around 10 GHz) where the coefficient phase in the v dimension is $\varphi_v \cong 0°$ but $\varphi_u \cong 180°$. Putting it differently, the AMC is almost at resonance from the point of view of $\boldsymbol{E}^i_v$ but not from that of $\boldsymbol{E}^i_u$. By using (4.12) and (2.50) in (2.44), the reflected fields over the plane of the metal inclusion ($z = 0$) are:

$$\boldsymbol{E}^r_u = \hat{\mathbf{a}}_{\boldsymbol{u}} \frac{\sqrt{2}}{2} E_0\, e^{j\varphi_u}, \quad \boldsymbol{E}^r_v = \hat{\mathbf{a}}_{\boldsymbol{v}} \frac{\sqrt{2}}{2} E_0\, e^{j\varphi_v} \tag{4.13}$$

At 10 GHz, the difference between φ_v and φ_u is 180° as Fig. 4.7d also shows. Fig. 4.7c also indicate the reflected fields $\boldsymbol{E}_v^r \cong \boldsymbol{E}_v^i$, and $\boldsymbol{E}_u^r \cong -\boldsymbol{E}_u^i$. The scalar components of these fields in the x and y dimensions are:

$$\begin{aligned} E_x^r &= \boldsymbol{E}_u^r \cdot \hat{\mathbf{a}}_x + \boldsymbol{E}_v^r \cdot \hat{\mathbf{a}}_x \\ E_y^r &= \boldsymbol{E}_u^r \cdot \hat{\mathbf{a}}_y + \boldsymbol{E}_v^r \cdot \hat{\mathbf{a}}_y \end{aligned} \tag{4.14}$$

A polarization-conversion ratio (PCR) can represent the effectiveness of the mechanism as [43]:

$$PCR = \frac{\left|E_y^r\right|^2}{\left|E_x^r\right|^2 + \left|E_y^r\right|^2} \tag{4.15}$$

Figure 4.7d also shows the PCR for the illustrated case. The PCR is unitary at 10 GHz, where the difference between φ_v and φ_u is 180°.

We can make a similar analysis concerning the polarization conversion of circularly polarized waves. Most works make similar analyses and, roughly speaking, the task is to develop a unit cell that provides 180° of difference between φ_v and φ_u in the desired frequency range.

M.I. Khan et al. [43] propose a unit cell composed of a rotated-triangle ring loaded with a parasitic metal strip over a ground plane. The metasurface converts linearly polarized waves to their orthogonal counterparts in the X-band, from 8.0 to 11.0 GHz ($BW = 31.6\%$) with a PCR higher than 90% for normal incidence, and higher than 80% for angles of incidence up to 45°. Further, a linear-to-circular polarization conversion occurs in the ranges 7.5–7.7 GHz and 11.5–11.9 GHz. The thickness is 1.6 mm.

Figure 4.8a depicts a unit cell made of a square patch with two corners cut over a ground plane, again with thickness of 1.6 mm [44]. The performance is similar to the previous one [43], but now the linear polarization conversion occurs from 8.0 to 12.0 GHz ($BW = 40.0\%$), and the linear-to-circular polarization conversion takes place in the ranges 7.4–7.6 GHz and 13.0–13.56 GHz. Figure 4.8b shows the fabricated prototype.

M. Karamirad et al. [45] present a 2.0-mm thick metasurface whose unit cell is a 45°-rotated oval pattern split by strip lines over a ground plane to convert linearly or circularly polarized waves into their orthogonal counterparts. They achieve a PCR higher than 90% from 10.2 to 20.5 GHz ($BW = 67.1\%$) for normal incidence. When the angle of incidence is increased to 30°, the upper frequency shifts downward to around 19 GHz.

Some works combine the polarization-conversion technique with the checkerboard approach for RCS reduction [46–48]. J. Chatterjee et al. [46] propose a unit cell composed of two concentric square rings, the outer one split, over a ground plane. The thickness is 3.5 mm. The normal-incidence PCR is better than 90% from

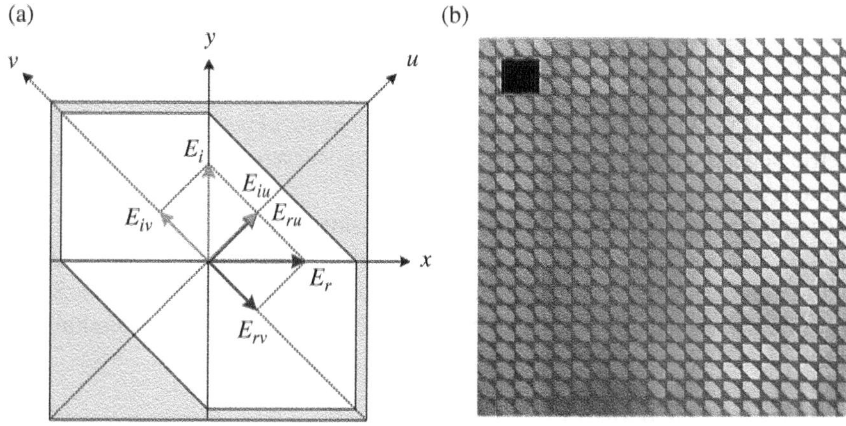

Fig. 4.8 Metasurface of [44]: (a) unit cell; (b) fabricated prototype. "Reproduced from Nguyen et al. [44]/Springer Nature/CC BY 4.0."

6.3 to 20.5 GHz (BW = 106.0%). The PCR degrades to better than 80% for incident angles up to 35°. Four supercells with unit cells rotated toward different directions perform different conversions of polarization and cancel each other at the broadside direction. In this way, they provide an RCS reduction better than 10 dB from 5.5 to 20.5 GHz (BW = 115.4%), with a maximum value around −30 dB at 14.5 GHz for both polarizations. Other works [47, 48] propose unit cells composed of 45°-rotated double-head arrows over a ground plane. Their thicknesses are 3 mm in [47] and 0.25 mm in [48]. Each of the four supercells has arrows pointed to a specific direction. The 10-dB BWs are of the order of 100% for angles of incidence up to 20° for both TE and TM polarizations (from 10 to 30 GHz), with a maximum RCS reduction around −50 dB at 27 GHz in [47] and around −40 dB at 25 GHz in [48].

We can also combine the polarization conversion with the checkerboard approach for the antenna RCS reduction. A. Pesarakloo et al. [49] design firstly a double-head arrow unit cell to provide an RCS reduction better than 10 dB from 5.5 to 20.5 GHz (BW = 115.4%). Then, they introduce an extra resonator to break the polarization conversion in a narrow frequency range around 10 GHz. As such, the metasurface can work as a conventional antenna reflector at this frequency, whereas it reduces the antenna RCS out of band. In the end, this reduction is better than 10 dB in the ranges 4.0–9.0 GHz and 10.8–14.8 GHz. The overall thickness is 7.75 mm.

W. Yao et al. [50] present a 6.5-mm thick dual-layer unit cell composed of a notched-square ring and a notched rectangular patch over a ground plane. They arrange these unit cells in supercells with different orientations in a checkerboard configuration. They achieve an RCS reduction better than 7 dB from 4.3 to 13.5 GHz (BW = 103.4%). Finally, they etch slots in the ground plane, resulting in a 7.30-mm thick antenna operating from 4.8 to 6.0 GHz.

4.4.4 Time-Varying Metasurfaces

Some reconfigurable metasurfaces can dynamically vary their transmission and reflection coefficients with the aid of tunable components, such as PIN and varactor diodes [51]. We can use a time variation in the phase of transmitted or reflected electric fields to perform a signal modulation. Fig. 4.9 represents this mechanism for the reflection of waves.

Using the same notation of Section 2.1.4, consider a metasurface embedded in the free space and located at the plane $z = 0$. An electromagnetic wave with electric field $\boldsymbol{E}^i$ impinges upon it with normal incidence. This scenario presents the same disposition as the one in Fig. 3.8. Suppose that the metasurface transmission and reflection coefficients are of the form:

$$\Gamma_{TEM}(t) = |\Gamma_{TEM}|e^{j\varphi_\Gamma(t)} \tag{4.16}$$

$$\tau_{TEM}(t) = |\tau_{TEM}|e^{j\varphi_\tau(t)} \tag{4.17}$$

By using (2.44) and (2.45), the transmitted and reflected electric fields are:

$$\boldsymbol{E}^r = |\Gamma_{TEM}|e^{j\varphi_\Gamma(t)}\boldsymbol{E}^i e^{+jk_0 z} \tag{4.18}$$

$$\boldsymbol{E}^t = |\tau_{TEM}|e^{j\varphi_\tau(t)}\boldsymbol{E}^i e^{-jk_0 z} \tag{4.19}$$

Now, suppose that $\varphi_\Gamma(t)$ and $\varphi_\tau(t)$ present linear variations with time, that is $\varphi_\Gamma(t) = \omega_{\Gamma m}t$ and $\varphi_\tau(t) = \omega_{\tau m}t$. Here, $\omega_{\Gamma m} = 2\pi f_{\Gamma m}$ and $\omega_{\tau m} = 2\pi f_{\tau m}$ are modulation angular frequencies [rad/s] and $f_{\Gamma m}$ and $f_{\tau m}$ are modulation frequencies [Hz]. By using (4.18) and then (2.16), we can express the reflected electric field in its instantaneous form as:

$$\mathcal{E}^r = \text{Re}\left[|\Gamma_{TEM}|e^{j\omega_{\Gamma m}t}\boldsymbol{E}^i e^{+jk_0 z}e^{j\omega t}\right] \tag{4.20}$$

$$\mathcal{E}^r = \text{Re}\left[|\Gamma_{TEM}|\boldsymbol{E}^i e^{j[+k_0 z + (\omega + \omega_{\Gamma m})t]}\right] \tag{4.21}$$

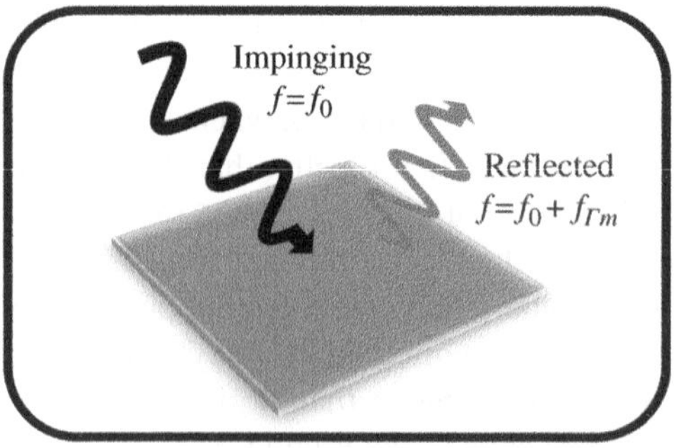

Fig. 4.9 Modulation of reflected waves with a time-varying metasurface

Likewise, we can express the transmitted electric field in the instantaneous form as:

$$\mathcal{E}^t = \mathrm{Re}\left[|\tau_{TEM}|\boldsymbol{E}^i e^{j[-k_0 z + (\omega + \omega_{\tau m})t]}\right] \tag{4.22}$$

We can see in (4.21) and (4.22) that the reflected and transmitted fields are shifted in the spectrum by the angular modulation frequencies $\omega_{\Gamma m}$ and $\omega_{\tau m}$, respectively.

Recently, Wu, Z. et al. [52] exploited this concept in a three-layer metasurface whose transmission coefficient phase could be reconfigured through the biasing of varactors embedded in the unit cells. By modulating the varactor bias voltage with a periodic signal, the metasurface could sweep the transmission phase from 0 to 2π with a linear time variation. Hence, it could shift the frequency of transmitted waves in the spectrum by $f_{\tau m} = 1$ MHz around 10 GHz.

To avoid radar detection, we are mostly interested in shifting the frequency of reflected waves. D. Ramaccia et al. [53] use varactors between the unit cells of an AMC to control its capacitance and thus its reflection phase, rather like we explained in Section 3.2.4. By applying a modulated bias voltage over the varactors, they obtain a linear time variation in the reflection phase. They achieve a frequency shift of $f_{\Gamma m} = 500$ kHz in the reflected waves.

The frequency shift obtained by D. Ramaccia et al. [53] is not sufficient to take reflected signals out of the frequency range of many modern radars, whose BW is typically of the order of 50 or 100 MHz. To this end, a frequency shift $f_{\Gamma m} > 100$ MHz would be desirable. However, modulating varactors in such a rate is complicated because a cutoff frequency exists that depends on the *RC* time constant of the biasing circuit, related to the resistances and capacitances used therein [54]. Nevertheless, many modern radars operate with a moving-target-indication (MTI) mode discriminating targets from clutter through a Doppler frequency shift [55]. In this context, we can use time-varying metasurfaces to cancel the Doppler shift of their aircraft out, making them be considered as clutter and avoiding detection. X.G. Zhang et al. [56] implement this idea with a time-varying metasurface used together with a sensing-feedback time-varying electronic control system to cancel different Doppler shifts in real time, without any human intervention.

Other modern radars operate in a frequency-modulated continuous-wave (FMCW) mode. These radars do not estimate the range through the delay of skin returns. They obtain it rather through an analysis of the spectrum of the received signals [57]. V. Kozlov et al. [58] propose time-varying metasurfaces as a means of deceiving these radars. They experimentally demonstrate a metasurface used against a linear frequency-modulated (LFM) radar.

Another way of exploiting time-varying metasurfaces to avoid radar detection despite the limitation of the *RC* time constant of the biasing circuit is to modulate signals with a random phase reversal. By randomly reversing the phase of reflected signals, the energy of impinging waves is spread over a range of frequencies, thereby providing a spectral camouflage. In fact, the modulating signal does not need to be really random. Repetitive sequences of pseudorandom reversals can already do an excellent job. X. Wang et al. [59] show that a pseudorandom-sequence modulation can spread the energy of a single-tone signal over multiple frequencies with a maximum component of $(N+1)/N$. Here, N is the number of pseudorandom elements in a sequence. Further, X. Wang et al. [60] demonstrate that the frequency range the signal is spread over depends on the switching frequency of the modulating sequence.

M. Liu et al. [61] exploit this concept with a reflective unit cell whose PIN diodes allow the dynamic control of the reflection coefficient over two working states whose phases differ approximately 180°. Vias that lead to a control circuit behind the ground plane bias the diodes. By setting the diodes on or off, the metasurface provides the two reflecting states over a wide frequency range, from around 3.4 to 7.6 GHz. As such, it can perform the pseudorandom-sequence modulation. The metasurface achieves an RCS reduction better than −10 dB in the aforementioned frequency range ($BW = 76.4\%$), with a maximum value around −35 dB at 4.5 GHz for both polarizations with normal incidence. The operation keeps stable for incident angles up to 60° for TM and 40° for TE polarization. The thickness is 6.35 mm.

4.4.5 Space–Time-Modulated Metasurfaces

We can see space–time-modulated metasurfaces as a combination of checkerboard or phase-gradient metasurfaces, seen in Sections 4.4.1 and 4.4.2, and time-varying metasurfaces, seen in Section 4.4.4. They are a hot topic today due to their many applications such as the simultaneous steering and frequency shifting of beams, and nonreciprocal reflection effects, among others [62]. However, since the late 1990s the community have envisioned similar structures for the RCS reduction [63–65].

Figure 4.10a represents a reflective space–time-modulated metasurface. In this example, the unit cells are reconfigurable and can provide different reflection coefficient phases φ_{Γ}. By addressing each unit cell individually, a controller can switch between different phase distributions over the metasurface through time. Figure 4.10b illustrates an electromagnetic wave impinging over this metasurface at the carrier frequency f_0. Ideally, if we repeat a sequence of phase distributions in a loop with period $1/f_{\Gamma m}$, different reflected beams result in the fundamental and harmonic frequencies $f_0 \pm nf_{\Gamma m}$, where $n \in \mathbb{Z}$.

An overly simplified explanation here would assume that each of these beams corresponds to one of the phase distributions within the sequence. However, the response of each unit cell throughout an entire sequence loop should be

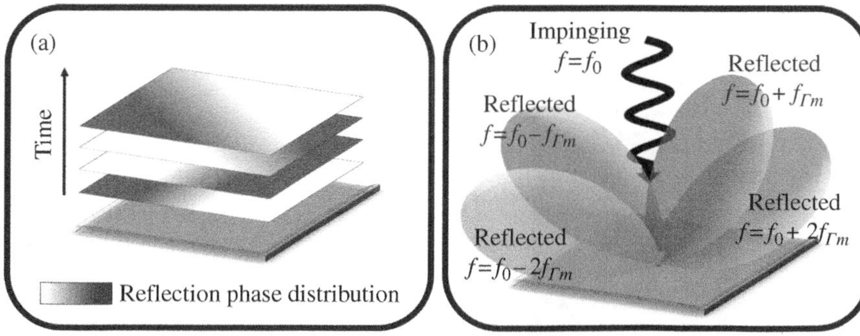

Fig. 4.10 Reflective space–time-modulated metasurfaces: (a) a reconfigurable metasurface periodically switches between many distributions of reflection phase φ_Γ within a period $1/f_{\Gamma m}$; (b) different beams are respectively resulted at each harmonic of $f_{\Gamma m}$

considered in the frequency domain in a precise examination. Then, we can determine the phase distribution for the whole metasurface at each frequency $f_0 \pm nf_{\Gamma m}$.

L. Zhang et al. [66] provide a rigorous derivation in this matter. Further, they propose a reflective unit cell with PIN diodes dynamically providing two working states for the reflection coefficient phase φ_Γ differing 180° at $f_0 = 10$ GHz. Feeding lines controlled by an FPGA bias the diodes. By modulating φ_Γ with the two states in a sequence loop with $f_{\Gamma m}$ = 500 kHz, the metasurface spreads the energy of reflected waves over the frequencies 10.0 GHz $\pm$ n500 kHz. The desired reflected beams are steered versions resulting from the reflection of a normally incident wave. As we saw in Section 3.2.7, we can achieve such steering with an appropriate gradient of φ_Γ according to the generalized law of reflection [32] as expressed by (3.35). L. Zhang et al. [66] obtain the modulations for the unit cells that lead to the desired gradients at each frequency $f_0 \pm nf_{\Gamma m}$ through means of optimization. Figure 4.11a depicts the fabricated metasurface, whose thickness is 2.0 mm. Figure 4.11b illustrates the reflected beams at the harmonic frequencies (the fundamental is not shown due to measurement issues).

L. Zhang et al. [66] also present a modulation scheme where the phase distributions adopt a checkerboard configuration. The metasurface alternate two opposite checkerboard patterns, spreading energy in space (like in Section 4.4.1) and frequency (like in Section 4.4.4) at the same time. Specifically, the metasurface leads to an RCS reduction of $-\infty$ dB for f_0, −9.5 dB for $f_0 \pm f_{\Gamma m}$, −19.1 dB for $f_0 \pm 3f_{\Gamma m}$, and −23.5 dB for $f_0 \pm 5f_{\Gamma m}$. Afterward, an optimization provides the phase distributions that promote the best results, where the RCS reduction keeps better than −21.5 dB for all frequencies $f_0 \pm nf_{\Gamma m}$. Figure 4.11c shows the scattered patterns. These results were only simulated, not measured.

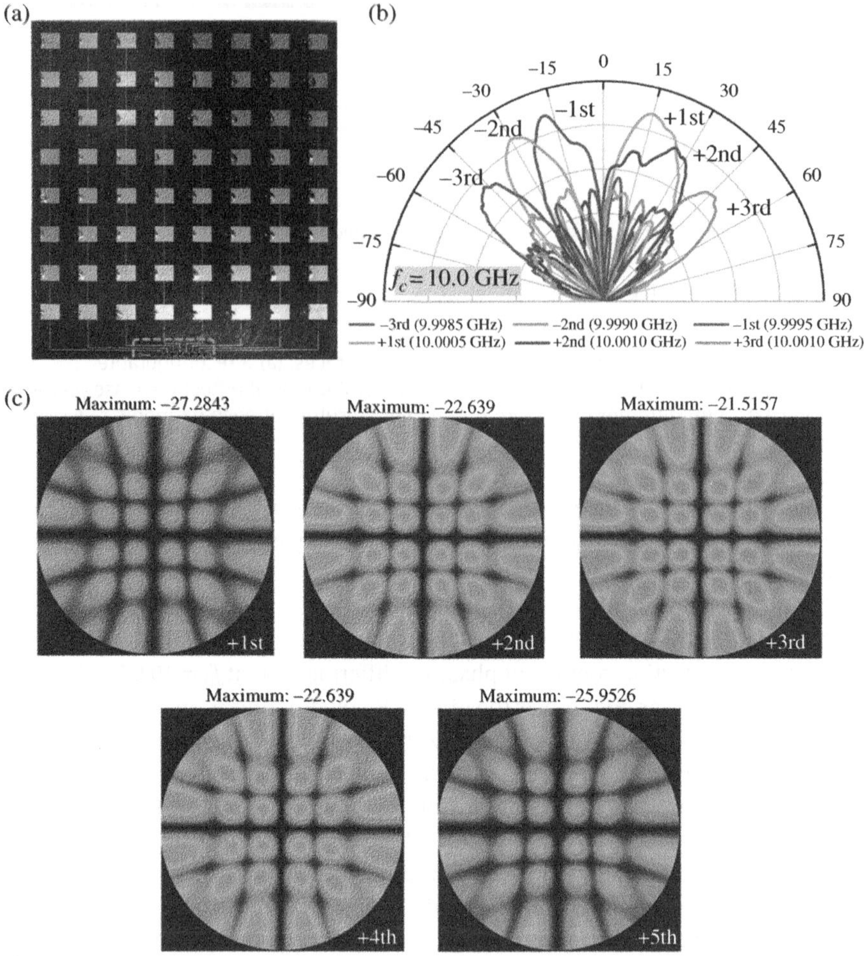

Fig. 4.11 Reflective space–time-modulated metasurface of [66]: (a) fabricated device; (b) reflected beams measured at 10 GHz; (c) scattering patterns for checkerboard modulation. "Reproduced from Zhang et al. [66]/Springer Nature/CC BY 4.0."

4.4.6 Limitation of the Scattering Approach

Each approach we studied throughout this section presents a different mechanism to avoid skin returns from focusing on a radar receiver whose operation we assume to know, whether in space or in the frequency spectrum. These techniques usually consider single-polarized, monostatic radars with matched filters and coherent processing. Rather than for simplicity, we make these assumptions for the sake of generality. However, the technological maturity of radar technology has already reached impressive levels, and complex architectures have already reached the theater of operation.

Spreading energy throughout multiple directions does not ensure that a radar system will not collect this energy and solve for the target. Today, multi-static radar networks are able to coherently process signals arriving at multiple points over the terrain [67–70]. That said, we cannot consider the checkerboard and phase-gradient approaches absolutely effective. Also, we should also call into question the absolute effectiveness of the polarization-conversion approach. Dual-polarization radar antennas can fit the compartment of missile seekers since at least two decades ago [71].

Concerning the time-varying approach, on the one hand, Doppler cloaks can take the target out from the perspective of MTI radar receivers in a given direction. On the other hand, these devices will cause incoherences between velocity and position in other directions, denouncing the use of the technology for multi-static architectures again. Spreading energy throughout the spectrum looks like a fairly robust method to avoid detection for the radars of nowadays. When associated with a spatial dispersion, like in space–time-modulated metasurfaces, this approach becomes even more robust. Nonetheless, the fact that the energy is still in the scenario indicates that some counter technology can appear soon in the cat-and-mouse EW game.

Although all these approaches differ in their related mechanisms, one thing they have in common: they do not destroy the energy, they only redistribute it, being vulnerable to some specific radar architectures. Now, if one intends to avoid detection in a 100% of cases, then they must destroy the energy. Since the energy conservation principle does not allow that, they must at least dissipate it.

4.5 Metasurface Absorbers

Metasurfaces use the same mechanism to absorb energy that conventional absorbers exploit: the dissipation through heat in lossy materials. Since their creation, researchers have improved metasurface absorbers not only in terms of performance but also in functionality, as it will become evident in this section.

4.5.1 Conventional Metasurface Absorbers

In his seminal work, N. Engheta [2] proposed a metasurface absorber composed of an AMC covered by a resistive sheet. The low-profile features of the AMC resulted in a structure with a thickness much smaller than a wavelength λ. Such structure represented a breakthrough in terms of aeronautics and space because the thickness of traditional Salisbury lenses is about a quarter wavelength [72].

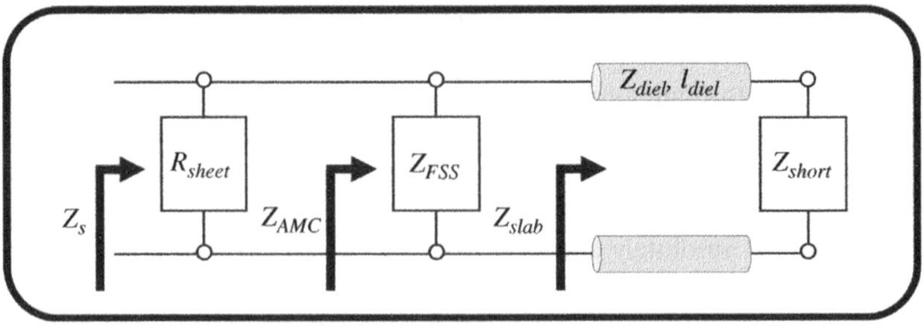

Fig. 4.12 Transmission-line equivalent of the resistive-AMC absorber. "Gonçalves Licursi de Mello [7]/MDPI/CC BY 4.0."

To understand the working principle of that structure [2], let us recall the AMC transmission-line equivalent presented in Fig. 3.5a. We can see the resistive sheet placed over the metal patches as a shunt resistor R_{sheet}. Figure 4.12 represents the new transmission-line schematic [7].

The surface impedance Z_s of the absorber is the parallel between R_{sheet} and Z_{AMC}. Now, recall that an AMC surface impedance $Z_{AMC} = j\infty$ results at the resonant angular frequency $\omega_0 = 1/\sqrt{L_{slab}C_{FSS}}$, as (3.15) indicates. In this condition, $Z_s = R_{sheet}$. If we adjust the resistive layer to match the free-space impedance η_0, the energy of impinging waves is transmitted to the metasurface, and efficient dissipation occurs therein. In this case, we expect a BW of the same order as that of the AMC because the dissipation takes place when the latter resonates.

A whole plethora of metasurface topologies exist for energy dissipation. As a rule of thumb, dissipation occurs only if we employ lossy materials or components. N.I. Landy et al. [73] propose the so-called perfect metamaterial absorber. The structure is designated so because it presents an absorption coefficient of $\alpha = 99\%$ at $f_0 = 11.65$ GHz, where:

$$\alpha^2 = 1 - |\Gamma_{TEM}|^2 - |\tau_{TEM}|^2 \tag{4.23}$$

The dissipation mechanism in this case is a strong resonance that takes place in a lossy dielectric between an electric resonator and a cut wire that compose the unit cell. The metasurface achieves a 10-dB BW of only 4%. A slight increase in the angle of incidence to 16° makes the absorption coefficient drop to $\alpha = 50\%$.

As one may imagine, the above results are not that useful from an operational standpoint. Not coincidently, most works on metasurface absorbers address the frequency- and angle/polarization-dependence problems.

To broaden BW, a common approach is to use two or more layers in the metasurface design [7]. As such, an impedance matching with the free space can be reached in a wider band. The reasoning is similar to traditional multi-section impedance transformers [74]. L. Chen et al. [75] achieve an absorption coefficient $\alpha \geq 90\%$ with a dual-layer metasurface from about 6.0 to 18.0 GHz ($BW = 100.0\%$). X. Begaud et al. [76] achieve a similar performance from 3.7 to 17.5 GHz ($BW = 130.2\%$) with a multilayer metasurface.

As the lossy material, X. Chen et al. [77] use a magnetic absorbing material whose main component is a kind of ferrite. They place this material over a ground plane in a fishnet arrangement. The metasurface provides a monostatic RCS reduction of up to −23 dB at 6.35 GHz. The 10-dB BW is 124.7% (from 4.8 to 20.7 GHz) for normal incidence. For TM polarization, the absorption coefficient keeps satisfactory until 60° of incident angle. However, for TM, it remains reasonable only until 20°. The thickness is 5.0 mm.

X. Begaud et al. [76] address the angle/polarization-dependence limitation with the use of a WAIM. We saw in Section 2.1.6, specifically by comparing (2.73) and (2.76), that the effective impedance changes not only with the angle of the incident wave but also with its polarization. The WAIM is a classical solution that matches the effective metasurface impedance to the free-space impedance η_0 at specific conditions of incident angle and polarization [78]. X. Begaud et al. [76] maximize the number of these conditions while broadening BW by carrying out heavy optimization tasks over a multilayer metasurface with a WAIM. They model structure with the transmission-line equivalence, which allows the optimization to run much faster than it would with full-wave simulations. In the end, the absorber performance remains stable for angles of incidence ranging from 0° to 60° for both TE and TM polarizations. This performance comes at the price of a large thickness of 11.5 mm due to the multilayered design.

P. Tiwari et al. [79] propose an extremely thin metasurface with a 0.2-mm thick unit cell made of four metallic arms. Dissipation happens in eight lumped resistors deployed over the unit cell, depicted in Fig. 4.13a. The metasurface provides a monostatic RCS reduction of about −35 dB at 15.5 GHz. The 10-dB BW is about 78% (from about 11 to 25 GHz) for normal incidence. Measured results for TE polarization show that the performance remains stable until 50° of incident angle. The prototype is shown in Fig. 4.13b,c. They also present a conformal version of the metasurface with a monostatic RCS reduction of about −45 dB at 12.5 GHz and 10-dB BW larger than 106% (larger than 8 to 26 GHz) for normal incidence.

We can also combine the absorption approach with the scattering one in a single metasurface. For instance, S. Leung et al. [80] present a magnetic absorbing material, which is a mixture of barium cobalt ferrite and silicone rubber, displaced over a ground plane. Different shapes for the absorbing material provide a random phase distribution, as previously studied in Section 4.4.2. This combination results in a monostatic RCS reduction of up to −16 dB at about 5 GHz, with a 10-dB BW of 136.5% (from 3.4 to 18 GHz) at normal incidence. The thickness of the metasurface is 1.44 mm.

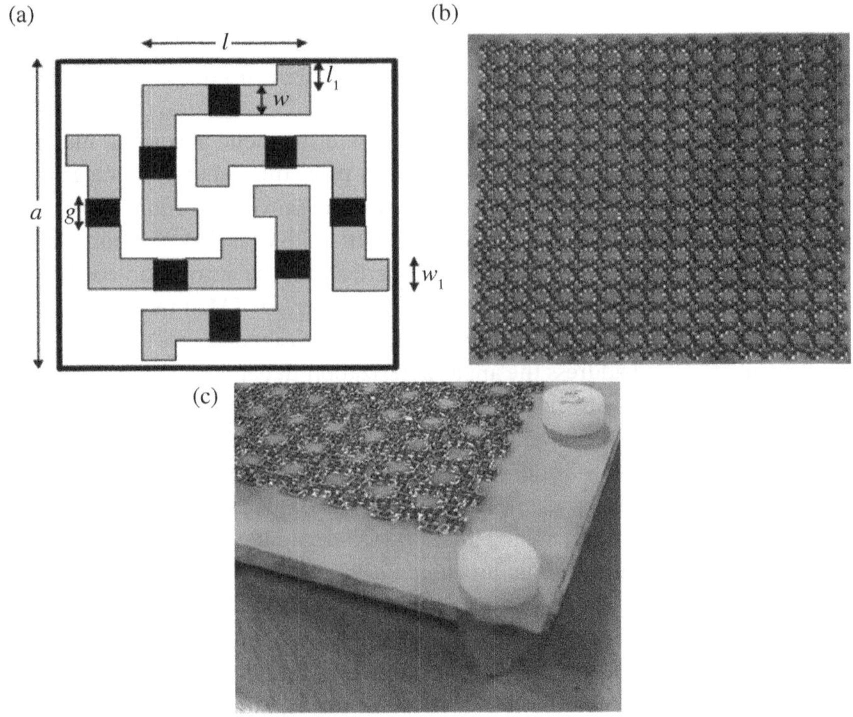

Fig. 4.13 Metasurface of [79]: (a) unit cell; (b and c) fabricated prototype. "Reproduced from Tiwari et al. [79]/Springer Nature/CC BY 4.0."

J. Su et al. [81] also use this combination. First, they model a dual-layer unit cell with resistors with the transmission-line equivalence for the RCS reduction from 1.3 to 13 GHz. Then, they design 3 × 3 supercells presenting unit cells with different spacings between the layers for the phase cancellation from 13 to 40 GHz. The final structure provides a noticeable 10-dB BW of 185.5% (from 1.5 to 40 GHz), which remains stable for angles of up to 50° at both polarizations, except in some narrow frequency ranges. The maximum reduction exceeds −40 dB. Like in a previous case [76], this performance comes at the cost of a large thickness, 20.93 mm here.

We can also exploit the scattering/absorption combination in the antenna design. Y. Jia et al. [82] apply a dual-layer metasurface as the superstrate of a Fabry-Pérot antenna working at 8.4 GHz. Such metasurface presents loops with resistors on one side and a phase-gradient metasurface on the other side. The RCS reduction occurs from 7 to 15 GHz. The in-band reduction is measured as about −18 dB. A reduction better than −10 dB is measured from about 11 to

13.5 GHz ($BW = 20\%$) for one of the polarizations. The radiation pattern is squinted in relation to the broadside direction, the antenna gain is 17.9 dBi and the metasurface thickness is 4.0 mm.

4.5.2 Rasorbers

As previously commented in Section 4.3, we can use bandpass FSSs in the design of antenna radomes to enable the in-band transmission and the out-of-band reflection of waves [20]. The radome shape orients reflected waves away from the source. However, similar to what was discussed in Section 4.4.6 for the metasurface case, when there are multi-static radar networks deployed over the theater [67], this approach does not guarantee nondetection. To increase the furtivity of the platform, an absorption is desired out of band rather than reflections. Metasurfaces that perform this task are called rasorbers [83].

The rasorber idea dates back to times when the community had not defined the term "metasurface" yet [84]. There is an easy way to understand their working principle, and the transmission-line equivalent can help us with that again.

We can design a typical rasorber by stacking a bandpass and a band-stop FSS, the latter loaded with lossy materials/components. The bandpass FSS, which we can model as the parallel of a capacitance C_{FSS_1} and an inductance L_{FSS_1}, resonates at a frequency f_1. The lossy band-stop FSS, which we can model as a capacitance C_{FSS_2}, an inductance L_{FSS_2}, and a resistor R_{FSS_2} in series, resonates at a frequency $f_2 \neq f_1$. A dielectric with thickness l_{diel} and characteristic impedance Z_{diel} separates the FSSs. Figure 4.14 depicts the transmission-line schematic of such rasorber.

When the parallel circuit of the bandpass FSS resonates at frequency f_1, it behaves as an open circuit. On the other hand, the series circuit of the band-stop FSS also behaves as an open at this frequency since it does not resonate ($f_1 \neq f_2$). As such, waves impinging upon the rasorber with frequency f_1 do not see the FSSs and are transmitted.

Conversely, at frequency $f_2 \neq f_1$, the parallel circuit of the bandpass FSS is shorted. In this case, the surface impedance Z_{FSS_1} indicated in Fig. 4.14 appears as a grounded slab similar to Z_{slab} in Fig. 3.5a. Hence, (3.13) expresses Z_{FSS_1}, which is inductive at frequency f_2 if the thickness of the dielectric is $l_{diel} \leq \lambda/4$. At frequency f_2, the band-stop FSS resonates, and its series circuit impedance reduces to R_{FSS_2}. As such, the rasorber surface impedance Z_s is the parallel of R_{FSS_2}, which is purely real, and Z_{FSS_1}, which is purely imaginary. This situation is not ideal for dissipation purposes because it is impossible to perfectly match Z_s with the free-space impedance η_0. Therefore, even though the grounded slab will avoid energy from being transmitted through the metasurface, it will also reflect some energy.

Dissipation is maximized at a frequency where the series impedance of the band-stop FSS is capacitive enough to make its parallel with Z_{FSS_1} resonate. At this

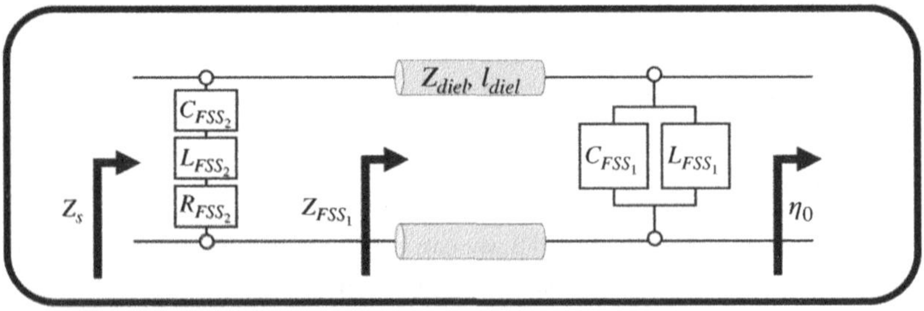

Fig. 4.14 Transmission-line equivalent of a rasorber

frequency, the free-space impedance η_0 can be matched with the proper choice of R_{FSS_2}. In this condition, currents will flow back and forth through the parallel circuit and will be dissipated in R_{FSS_2} in each cycle.

The above example is simplified but not overly. Such an example shows once more the importance of the transmission-line equivalent in the modeling of metasurfaces. F. Costa et al. [85] make a similar analysis in their prominent work. Interdigitated Jerusalem-cross unit cells make up the bandpass FSS. Resistive squared-loop unit cells compose the band-stop FSS. The rasorber provides an insertion loss of only 0.3 dB at 4.6 GHz and an RCS reduction of up to −19 dB at around 19 GHz. The 10-dB BW is around 66.7% (10–20 GHz). For angles of incidence of up to 45°, the absorption band lower frequency shifts to around 11 GHz for TM polarization but keeps stable for TE polarization. The rasorber thickness is 5 mm.

Recently, researchers have taken multiple directions to obtain different filter responses in the design of rasorbers. It is fair to say that, in summary, most of the works in the literature model the rasorber layers as circuits with different levels of complexity we can use in a transmission-line equivalent.

Y. Wu et al. [86] propose a bandpass FSS with a dual-layer metasurface with Jerusalem crosses on one side and square patches on the other. They also propose a band-stop FSS composed of meandered metal squares loaded with resistors. The insertion loss is only 0.4 dB at 3.08 GHz. The rasorber provides a maximum RCS reduction of −19 dB at around 9 and 16 GHz. The 10-dB BW is around 76.9% (8–18 GHz). For angles of incidence of up to 45°, the absorption band lower frequency keeps stable for TE polarization but shifts to 10 GHz for the TM case. The thickness is 3.5 mm.

Some works [87, 88] increase the complexity of the circuits considered in the transmission-line equivalents to obtain different responses. Different from the previously studied works [85, 86], Z. Wang et al. [87] obtain the absorption band at frequencies lower than the transmission band. A slot array plays the role of bandpass FSS, while a dipole array loaded with resistors plays the role of band-stop FSS.

A noticeable insertion loss is simulated as only 0.006 dB and measured as 0.2 dB at 5.6 GHz. The RCS is reduced in a 10-dB BW of 56.4% (2.8–5.0 GHz). The maximum reduction occurs at around 3.8 GHz (−26 dB). When the angle of incidence increases to 45°, the absorption band lower frequency shifts to 3.4 GHz for the TM case. The thickness is 17.0 mm.

R. Shi et al. [88] obtain a transmission band surrounded by two absorption bands with a 6-mm thick rasorber. A square slot unit cell works as the bandpass FSS, and a complex resonator loaded with resistors works as the band-stop FSS. The rasorber provides an insertion loss of 0.25 dB at 10 GHz. It also provides maximum RCS reductions of about −17 and −23 dB around 5 and 13 GHz. The 10-dB BWs are around 82.3% (3.84–9.21 GHz) and 24.1% (10.91–13.90 GHz). For angles of incidence of up to 40°, the absorption BWs keep stable for TE polarization but slightly narrow for TM polarization.

A last type of metasurface worth mentioning for the RCS reduction is the asymmetric transmitting metasurface. These metasurfaces allow waves to enter antenna radomes but not to leak back from them. The idea in this case is to mitigate the scattering from the antenna and keep the total RCS as the one of the metasurface itself. Despite the potential for RCS reduction, prominent works on this kind of metasurface do not study this aspect thereof [89–91]. Table 4.1 summarizes the performance of the metasurfaces studied in this chapter.

4.6 Challenges for Stealth with Passive Metasurfaces

The use of metasurfaces in stealth is promising mainly because, first, they can be more convenient than dissipative paints and RAMs in terms of maintenance and cost, and second, we can design them to work at arbitrary frequencies. However, we should consider some challenges for the use thereof in the modern scenario.

4.6.1 Low Frequencies

In radar technology, a strong trend exists today for operation in the VHF and UHF bands because it is hard to develop stealth technologies that work well in these bands. In particular, traditional RAMs become so bulky when it comes to low-frequency operation that coating platforms with them is unfeasible.

Setting a passive structure to work in a new frequency range is as simple as scaling its size to match the new range of wavelengths. The problem with lowering the frequency using this approach is that the increase in wavelength, of course, requires larger shapes, thicker substrates, and so on, leading to bulkier devices. As an example, if a traditional Salisbury lens absorber [72] needs a thickness of

Table 4.1 Summary of prominent metasurface works in passive RCS reduction

Reference	Mechanism	Max. meas. RCS red.	10-dB bandwidth	Thickness
Chen et al. [26]	Checkerboard	−25 dB	60% (4.10–7.59 GHz)	6.35 mm
Sang et al. [27]	Checkerboard	−25 dB	91.5% (3.8–10.1 GHz)	0.50 mm
Haji-Ahmadi et al. [28]	Checkerboard	−35 dB	95% (3.8–10.7 GHz)	6.00 mm
Zhou et al. [29]	Checkerboard (ant.)	−30 dB, out-of-band	93.9% (6.5–18.0 GHz) (6 dB, not 10 dB)	3.00 mm
Jiang et al. [30]	Checkerboard (ant.)	−11 dB, in-band −35 dB, out-of-band	5.9 GHz 60% (4.10–7.59 GHz)	4.00 mm
Varault et al. [31]	Checkerboard (ant.)	−25 dB, in-band	6 : 1 (3.0–18.0 GHz)	2.40 mm
Singh et al. [33]	Phase gradient	−45 dB	32% (16–22 GHz)	1.60 mm
Choudhary et al. [34]	Phase gradient	−30 dB	76% (9.2–20.5 GHz)	3.00 mm
Azizi et al. [35]	Phase gradient	−15 dB	96.3% (11.3–32.3 GHz)	1.60 mm
Chen et al. [36]	Phase gradient	−30 dB	63% (7.5–14.5 GHz)	4.00 mm
Feng et al. [37]	Phase gradient	−15 dB	42% (12.0–18.4 GHz)	3.00 mm
Zhang et al. [38]	Phase grad. (ant.)	−25 dB, out-of-band	82.4% (7–16.8 GHz)	3.00 mm
Li et al. [39]	Phase grad. (ant.)	−12 dB, in-band −22 dB, out-of-band	10.0 GHz 23.3% (8.7–11 GHz) (6 dB, not 10 dB)	2.50 mm
Mirzapour et al. [40]	Phase grad. (ant.)	−19 dB, in-band	8.2% (9.3–10.1 GHz)	3.00 mm
Umair et al. [41]	Phase grad. + checkerboard (ant.)	−6 dB, in-band −40 dB, out-of-band	7.0 GHz 105.9% (4–13 GHz)	1.52 mm
Chatterjee et al. [46]	Pol. conv. + check.	−30 dB	106% (6.3–20.5 GHz)	3.50 mm
Ameri et al. [47]	Pol. conv. + check.	−50 dB	100% (10.0–30.0 GHz)	3.00 mm
Ameri et al. [48]	Pol. conv. + check.	−40 dB	100% (10.0–30.0 GHz)	0.25 mm
Pesarakloo et al. [49]	Polarization conv. + checkerboard (ant.)	−30 dB, lower band −27 dB, upper band	76.9% (4.0–9.0 GHz) 31.2% (10.8–14.8 GHz)	7.75 mm

Yao et al. [50]	Polarization conv. + checkerboard (ant.)	−12 dB, in-band −30 dB, out-of-band	103.4% (4.3–13.5 GHz) (7 dB, not 10 dB)	7.30 mm
Liu et al. [61]	Time-varying	−35 dB	76.4% (3.4–7.6 GHz)	6.35 mm
Zhang et al. [66]	Space–time mod.	$-\infty$ dB (simulated)	not informed	2.00 mm
Begaud et al. [76]	Absorption	−37 dB	130.2% (3.7–17.5 GHz)	11.50 mm
Chen et al. [77]	Absorption	−23 dB	124.7% (4.8–20.7 GHz)	5.00 mm
Tiwari et al. [79]	Absorption	−35 dB	78% (11–25 GHz)	0.20 mm
Leung et al. [80]	Abs. + phase grad.	−16 dB	136.5% (3.4–18 GHz)	1.44 mm
Jia et al. [82]	Abs. + phase grad. (ant.)	−18 dB, in-band −30 dB, out-of-band	8.4 GHz 20.0% (11–13.5 GHz)	4.00 mm
Costa and Monorchio [85]	Rasorber (ant.)	−19 dB, out-of-band	66.7% (10–20 GHz)	5.00 mm
Wu et al. [86]	Rasorber (ant.)	−19 dB, out-of-band	76.9% (8–18 GHz)	3.50 mm
Wang et al. [87]	Rasorber (ant.)	−26 dB, out-of-band	56.4% (2.8–5.0 GHz)	17.0 mm
Shi et al. [88]	Rasorber (ant.)	−17 dB, lower band −23 dB, upper band	82.3% (3.84–9.21 GHz) 24.1% (10.9–13.9 GHz)	6.00 mm

about 31 mm (about a quarter wavelength) to work properly at 2.4 GHz, scaling the same structure to work at 240 MHz would lead to a thickness of 310 mm.

We can regard metasurfaces as one possible solution in this matter. Building AMC-based absorbers with the mechanism presented by N. Engheta [2] provides great advantages because different approaches exist to reduce the operating frequency of AMCs without directly increasing their thickness. We can also use these low-frequency AMCs to implement checkerboard and phase-gradient scattering metasurfaces such as those of Sections 4.4.1–4.4.2.

However, since the seminal work of Sievenpiper et al. [92], it is well known that, for a given thickness, the lower the operating frequency, the harder it is to obtain reasonable BWs. A panoramic review of the literature corroborates this. For example, resonant frequencies of 15.7 and 8.8 GHz allow BWs of 55.4 and 50.1% [93, 94]. On the other hand, resonant frequencies of 5.7, 6.2, and 6.0 GHz allow BWs of 40.9%, 40.0 %, and 19.4% [95–97]. We consider here the BW of AMCs as the frequency range where the reflection coefficient phase lies in the range $-90° < \varphi_{\Gamma}(f) < +90°$, which is the classical definition [92].

The community acknowledge multilayered AMCs as a means of achieving low-frequency operation [92]. Genetic algorithms are used to generate multilayered unit cells optimized for resonant frequencies down to 1.8 GHz, 337.5 MHz, and even 87.8 MHz [98–100]. In these works, thicknesses of 7.7, 19.4, and 58.5 mm enable BWs of 25.7%, 66.7%, and 9.0%. Thus, we can see that going down in frequency with AMCs is possible but delicate for some applications.

The community has also envisioned the use of ferrites in the AMC substrate as one way of circumventing this problem, with very promising results. Based on this idea, T. Rexhepi et al. [101] propose two AMCs. The resonant frequencies are 300 and 35 MHz. The ferrite-based AMCs provide noticeable BWs of 88% and 154%. The respective thicknesses are quite reasonable, 14 and 35 mm. D.H. Lee et al. [102] present another ferrite-based AMC with a remarkable BW of 25.5 : 1 for operation in the HF, VHF, and UHF bands (from 19 to 485 MHz). The thickness is only 13.0 mm.

B. X. Khuyen et al. present a series of ultrathin metasurface absorbers unrelated to AMCs for operation at low frequencies. In one of them, a 5-mm thick unit cell made of four split-square slots over a ground plane operates at 250 MHz [103]. The absorber provides an absorption coefficient $\alpha \geq 90\%$ for incident angles up to 50° and both polarizations, but the BW is narrow. B. X. Khuyen et al. [104] also propose two 3.6-mm thick absorbers with capacitive loads interconnecting the unit cells. The first presents a single, narrow operation band at 102 MHz with an absorption coefficient $\alpha \geq 90\%$ for TE and TM polarizations and incident angles up to 55°. The second presents stable operation for angles up to 50° and both polarizations at two narrow bands centered around 224.9 and 284.2 MHz. B. X. Khuyen et al. [105] use a similar approach but for dual- and triple-band operation in the

UHF range with a 2.6-mm thick metasurface. When scaled to work at 2.05 and 2.43 GHz, the absorber structure becomes so thin that it provides a flexibility useful for conformal applications. Fig. 1.2 previously depicted such metasurface. B. X. Khuyen et al. [106] present another metasurface absorber but with flexibility even in the VHF range. They achieve wideband absorption better than $\alpha \geq 90\%$ from 200 MHz to 3.6 GHz, which represents a BW of 179.0%. The metasurface covers angles of incidence up to 40° for TE polarization and up to 30° for TM.

These works suggest that, even though the design of low-frequency metasurfaces is not trivial, this technology may be key to tackle VHF/UHF radars in the upcoming years.

4.6.2 Deployment Rate

The metasurfaces we commented on so far in this chapter can lead to stealth platforms, either if we use them in the embedded antennas or if they cover the platform surface. In the real world, we do not expect the RCS reduction of antennas to be so different from the values measured in the works commented in the previous sections. However, when it comes to coating the platform surface, a fundamental challenge arises: the fact that usually we cannot cover platforms in their entireness.

Actually, we do not need to cover the whole platform to make it stealthy. Coating the parts that contribute the most to the RCS is enough. However, this can still be a challenge in some designs and perhaps we will not mitigate every $\widehat{RCS}_i$ term in the summation of (4.8). If the platform is mostly composed of shapes with faceted flat plates due to the reasons mentioned in Section 4.2, the job is easier. Nonetheless, we may require some percentage of the platform surface for other purposes, like the deployment of sensors. Further, complex parts of the platform, such as moving parts, tips, cavities, and corners of the fuselage and embedded weapons, may be difficult to coat.

In this case, it is fair to say that we will not reduce the platform RCS by the exact RCS reduction the metasurface provides. At this point, we assume that each element in (4.2) is made up of three main components which, normalized, are:

$$RCS_{ant} + RCS_{surf} + RCS_{complex} = 1 \tag{4.24}$$

where

- RCS_{ant} represents how much of the RCS comes from the antennas embedded in the platform [unitless];
- RCS_{surf} represents how much comes from the platform regular surfaces [unitless];
- $RCS_{complex}$ represents how much comes from complex parts [unitless].

We will also assume in the simulations of this book that, for some reason, we do not apply the metasurfaces wherever we can on the platform. Costs, weight, or convenience issues are among the possible reasons. Therefore, only a percentage of the metasurface RCS reduction should have effect in the platform RCS. To this end, we define that the reduction of RCS_{ant} and RCS_{surf} should take place with a multiplication with:

$$RCS_{ant}\ reduction = (1 - RCS_{ant\%}) + RCS_{ant\%} RCS_{mts}\ reduction \tag{4.25}$$

$$RCS_{surf}\ reduction = \left(1 - RCS_{surf\%}\right) + RCS_{surf\%} RCS_{mts}\ reduction \tag{4.26}$$

where

- $RCS_{ant/surf}$ *reduction* is the RCS reduction associated with the platform antennas/surfaces [unitless];
- $RCS_{ant\%/surf\%}$ represents the percentage of the platform antennas/surfaces that benefit from the metasurface in question [unitless];
- RCS_{mts} *reduction* is the RCS reduction provided by the metasurface [unitless].

By using (4.24) through (4.26), we can estimate how much of the platform RCS we can reduce with the application of metasurfaces either on the antennas or by coating the surfaces. We consider the complex component $RCS_{complex}$ uncoverable by metasurfaces.

As an example, consider an aircraft that received stealth treatment in the moving parts, tips, cavities, and corners of the fuselage and embedded weapons so as to present an $RCS_{complex} = 5\%$. Moreover, five embedded antennas are responsible for $RCS_{ant} = 50\%$ and the surfaces $RCS_{surf} = 45\%$ of the total aircraft RCS. Now, suppose that we replace four of the embedded antennas, that is, $RCS_{ant\%} = 80\%$, by metasurface-based antennas whose RCS reduction is RCS_{mts} *reduction* $= -15$ dB. Besides, we cover $RCS_{surf\%} = 90\%$ of the aircraft surfaceswith a metasurface that provides RCS_{mts} *reduction* $= -12$ dB. Then, using (4.24)–(4.26), we have:

$$RCS_{ant}\ reduction = 0.2 + 0.8 \times 0.03 = 0.22 \tag{4.27}$$

$$RCS_{surf}\ reduction = 0.1 + 0.9 \times 0.06 = 0.16 \tag{4.28}$$

$$RCS_{total}\ reduction = 0.50 \times 0.22 + 0.45 \times 0.16 + 0.05 = -6.3\ \text{dB} \tag{4.29}$$

As one may notice, the fact that we do not cover the platform entirely with the metasurfaces degrades the RCS reduction from respective values of −15 and −12 dB in the antenna and surfaces to −6.3 dB in the whole platform. This result may look modest at first glance. However, recalling (4.9), an RCS reduction of −6.3 dB provides a reduction in the maximum range r_{max} of radar detection to 0.7 of its original value, which is not neglectable at all.

Now, consider that we cover the whole aircraft with conformal metasurfaces. We can achieve that with off-the-shelf structures [105–108] or adapting other ones with the techniques mentioned in Section 3.4.3. Besides, optically transparent metasurfaces can cover the pilot's canopy as needed [75, 107–110]. Also, consider that the five antennas are metasurface-based. Repeating the procedure performed from (4.27) through (4.29), the new platform RCS reduction is −10.2 dB, which, using (4.9), again leads to a reduction in $r_{\max}$ to 0.55 of its original value.

We will use the RCS reduction values outlined in this section, related to (4.24)–(4.26), in the simulations of Chapter 5.

4.7 Final Remarks of the Chapter

Many people think about stealth as the first application when they listen to the word "metamaterials" or "metasurfaces." In this chapter, we introduced different approaches to reduce the RCS using passive metasurfaces.

We studied scattering metasurfaces, which spread energy either throughout diverse directions or the frequency spectrum. We also saw metasurfaces that reduce the RCS by absorbing energy. We can deploy these structures over the surfaces of the platform. By coating strategic parts of the platform, mainly those that contribute the most to the RCS, we can achieve a reduction in the latter. This coating can be an alternative to dissipative paints or RAMs traditionally used in stealth, allowing for a different maintenance regime and the addressing of specific frequency ranges. Moreover, we can also use metasurfaces in the design of antennas, contributing further to the platform RCS reduction.

The choice of each RCS reduction method will depend on a series of factors such as complexity of implementation, thickness, cost, and durability of the employed solution, and target radars.

In particular, the possibility of choosing the frequency of operation of metasurfaces, until a certain point arbitrarily, makes them an invaluable element for the unfolding cat-and-mouse EW game. The main reason is that radar technology currently has a strong trend to go low in frequency, where traditional dissipative paints and RAMs do not go.

Usually, metasurfaces cannot cover every part of the platforms because the latter can present complex parts, such as corners, tips, cavities, and shaped surfaces. Consequently, it is not fair to directly apply the RCS reduction we see in metasurface works to the platform. Rather, we should consider only a percentage of the platform and its antennas in the total RCS reduction. Nevertheless, according to the derived formulation, reductions of around −6.0 dB are sufficient to reduce the maximum range $r_{\max}$ of radar detection to 0.7 of its original value.

Some applications can demand larger RCS reductions. In this case, we can use other kinds of metasurface, namely power-amplifying ones, as will be seen in the next chapter.

References

1 Schurig, D., Mock, J.J., Justice, B.J. et al. (2006). Metamaterial electromagnetic cloak at microwave frequencies. *Science* 314 (5801): 977–980. https://doi.org/10.1126/science.1133628

2 Engheta, N. (2002). Thin absorbing screens using metamaterial surfaces. *Proceedings of the IEEE Antennas and Propagation Society International Symposium (IEEE Cat. No.02CH37313)*, San Antonio, TX, USA (16–21 June 2002), New York, US: IEEE. https://doi.org/10.1109/APS.2002.1016106

3 Mihajlovic, M., Anicic, D. (2019). *Missilers against the stealth*. Toronto, CA-ON: MSM Publishing.

4 Knott, E.F., Schaeffer, J.F., Tulley, M.T. (2004). *Radar cross section*, 2nd ed. Raleigh, NC: Scitech Publishing.

5 Balanis, C.A. (2012). *Advanced engineering electromagnetics*, 2nd ed. New York, NY: Wiley, Section 13.1, pp. 742.

6 Balanis, C.A. (2012). *Advanced engineering electromagnetics*, 2nd ed. New York, NY: Wiley, Section 2.13, pp. 79.

7 Gonçalves Licursi de Mello, R. (2023). Metasurfaces and blinking jamming: convergent study, comparative analysis, and challenges. *MDPI Micromachines* 14 (7): 1405. https://doi.org/10.3390/mi14071405

8 Vinoy, K.J., Jha, R.M. (1995). Trends in radar absorbing materials technology. *Sadhana* 20: 815–850. https://doi.org/10.1007/BF02744411

9 Savi, P., Giorcelli, M., Quaranta, S. (2019). Multi-walled carbon nanotubes composites for microwave absorbing applications. *MDPI Applied Sciences* 9 (851): 1–10. https://doi.org/10.3390/app9050851

10 Zhang, Z., Cai, Z., Wang, Z. et al. (2021). A review on metal–organic framework-derived porous carbon-based novel microwave absorption materials. *Nano-Micro Letters* 13 (56): 1–29. https://doi.org/10.1007/s40820-020-00582-3

11 Goodall, J.C. (2018). *Lockheed SR-71 Blackbird: the illustrated history of America's legendary Mach 3 spy plane*. Atglen, PA: Schiffer Publishing Ltd.

12 Sweetman, B. (2015). Stealth aircraft-history, technology and outlook. *Proceedings of the Gas Turbine and Aeroengine Congress and Exposition*, Brussels, Belgium (11–14 June 1990), New York, US: American Society of Mechanical Engineers. https://doi.org/10.1115/90-GT-172

13 Cohn, S.B. (1956). Dielectric properties of a lattice of anisotropic particles. *Journal of Applied Physics* 27 (1106): 1106–1107. https://doi.org/10.1063/1.1722190

14 Chen, C.-C. (1970). Transmission through a conducting screen perforated periodically with apertures. *IEEE Transactions on Microwave Theory and Techniques* 18 (9): 627–632. https://doi.org/10.1109/TMTT.1970.1127298

15 DeLyser, R.R., Kuester, E.F. (1991). Homogenization analysis of electromagnetic strip gratings. *Journal of Electromagnetic Waves and Applications* 5 (11): 1217–1236. https://doi.org/10.1163/156939391X00806

16 Lee, S.-W., Zarrillo, G., Law, C.-L. (1991). Simple formulas for transmission through periodic metal grids or plates. *IEEE Transactions on Antennas and Propagation* 30 (5): 904–909. https://doi.org/10.1109/TAP.1982.1142923

17 Whitbourn, L.B., Compton, R.C. (1985). Equivalent-circuit formulas for metal grid reflectors at a dielectric boundary. *Applied Optics* 24 (2): 217–220. https://doi.org/10.1364/AO.24.000217

18 Whites, K.W., Mittra, R. (1996). An equivalent boundary-condition model for lossy planar periodic structures at low frequencies. *IEEE Transactions on Antennas and Propagation* 44 (12): 1617–1629. https://doi.org/10.1109/8.546248

19 Munk, B.A. (2000). *Frequency selective surfaces: theory and design.* Hoboken, NJ: Wiley.

20 Munk, B.A. (2000). *Frequency selective surfaces: theory and design.* Hoboken, NJ: Wiley, Section 1.7.1, pp. 14.

21 Munk, B.A. (2000). *Frequency selective surfaces: theory and design.* Hoboken, NJ: Wiley, Section 1.7.2, pp. 14–16.

22 Munk, B.A. (2000). *Frequency selective surfaces: theory and design.* Hoboken, NJ: Wiley, Chapter 2, pp. 26–62.

23 Munk, B.A. (2003). *Finite antenna arrays and FSS.* Hoboken, NJ: Wiley, Chapter 2, pp. 15–55.

24 Murugesan, A., Selvan, K.T., Iyer, A.K. et al. (2021). A review of metasurface-assisted RCS reduction techniques. *Progress in Electromagnetics Research B* 94: 75–103. http://doi.org/10.2528/PIERB21081401

25 Paquay, M., Iriarte, J.-C., Ederra, I., et al. (2007). Thin AMC structure for radar cross-section reduction. *IEEE Transactions on Antennas and Propagation* 55 (12): 3630–3638. https://doi.org/10.1109/TAP.2007.910306

26 Chen, W., Balanis, C.A., Birtcher, C.R. (2015). Checkerboard EBG surfaces for wideband radar cross section reduction. *IEEE Transactions on Antennas and Propagation* 63 (6): 2636–2645. https://doi.org/10.1109/TAP.2015.2414440

27 Sang, D., Chen, Q., Ding, L. et al. (2019). Design of Checkerboard AMC Structure for Wideband RCS Reduction. *IEEE Transactions on Antennas and Propagation* 67 (4): 2604–2612. https://doi.org/10.1109/TAP.2019.2891657

28 Haji-Ahmadi, M.-J., Nayyeri, V., Soleimani, M. (2017). Pixelated checkerboard metasurface for ultra-wideband radar cross section reduction. *Scientific Reports* 7 (11437): 1–12. https://doi.org/10.1038/s41598-017-11714-y

29 Zhou, L., Gao, M., Cui, X. (2017). Radar cross section reduction of dipole antenna using checkerboard surface. *Proceedings of the IEEE International Symposium on*

Antennas and Propagation & USNC/URSI National Radio Science Meeting, San Diego, CA, USA (09–14 July 2017), New York, US: IEEE. https://doi.org/10.1109/APUSNCURSINRSM.2017.8072085

30 Jiang, H., Xue, Z., Zeng, Q. et al. (2018). High-gain low-RCS slot antenna array based on checkerboard surface. *IET Microwaves, Antennas & Propagation* 12 (2): 145–269. https://doi.org/10.1049/iet-map.2017.0524

31 Varault, S., Soiron, M., Barka, A. et al. (2017). RCS reduction with a dual polarized self-complementary connected array antenna. *IEEE Transactions on Antennas and Propagation* 65 (2): 567–575. https://doi.org/10.1109/TAP.2016.2637860

32 Yu, N., Genevet, P., Kats, M.A. et al. (2011). Light propagation with phase discontinuities: generalized laws of reflection and refraction. *Science* 334 (6054): 333–337. https://doi.org/10.1126/science.1210713

33 Singh, K., Afzal, M.U., Lalbakhsh, A. et al. (2021). Reflecting phase-gradient metasurface for radar cross section reduction. *Proceedings of the IEEE Asia-Pacific Microwave Conference (APMC)*, Brisbane, Australia (28 November–01 Decemeber 2021), New York, US: IEEE. https://doi.org/10.1109/APMC52720.2021.9661928

34 Choudhary, S., Hiremath, K.R. (2019). Random phase gradient metasurface for broadband RCS reduction. *Proceedings of IEEE Region 10 Conference (TENCON)*, Kochi, India (17–20 October 2019), New York, US: IEEE. https://doi.org/10.1109/TENCON.2019.8929656

35 Azizi, Y., Soleimani, M., Sedighy, S.-H. et al. (2022). Wideband RCS reduction by single-layer phase gradient modulated surface. *MDPI Sensors* 22, (7108): 1–11. https://doi.org/10.3390/s22197108

36 Chen, J., Cheng, Q., Zhao, J. et al. (2014). Reduction of radar cross section based on a metasurface. *Progress in Electromagnetics Research* 146: 71–76. http://doi.org/10.2528/PIER14022606

37 Feng, M., Li, Y., Zheng, Q. et al. (2018). Two-dimensional coding phase gradient metasurface for RCS reduction. *Journal of Physics D: Applied Physics* 51 (37): 375103. https://doi.org/10.1088/1361-6463/aad5ad

38 Zhang, W., Liu, Y., Gong, S. et al. (2018). Wideband RCS reduction of a slot array antenna using phase gradient metasurface. *IEEE Antennas and Wireless Propagation Letters* 17 (12): 2193–2197. https://doi.org/10.1109/LAWP.2018.2870863

39 Li, B., Liu, X., Shi, H. (2018). Planar phase gradient metasurface antenna with low RCS. *IEEE Access* 6: 78839–78845. https://doi.org/10.1109/ACCESS.2018.2885356

40 Mirzapour, M.I., Ghorbani, A., Namin, F. A. (2021). A wideband low RCS and high gain phase gradient metalens antenna. *AEU–International Journal of Electronics and Communications* 138 (153887): 1–7. https://doi.org/10.1016/j.aeue.2021.153887

41 Umair, H., Latef, T.B.A., Yamada, Y. et al. (2021). Fabry-Perot antenna employing artificial magnetic conductors and phase gradient metasurface for wideband monostatic RCS reduction and high gain tilted beam radiation. *IEEE Access* 9: 66607–66625. https://doi.org/10.1109/ACCESS.2021.3076913

42 Neri, F. (2018). *Introduction to electronic defense systems*, 3rd ed. Boston, MA: Artech House, Section 5.4.15, pp. 438–441.

43 Khan, M.I., Khalid, Z., Tahir, F.A. (2019). Linear and circular-polarization conversion in X-band using anisotropic metasurface. *Scientific Reports* 9 (4552): 1–11. https://doi.org/10.1038/s41598-019-40793-2

44 Nguyen, T.K.T., Nguyen, T.M., Nguyen, H.Q. et al. (2021). Simple design of efficient broadband multifunctional polarization converter for X-band applications. *Scientific Reports* 11 (2032): 1–12. https://doi.org/10.1038/s41598-021-81586-w

45 Karamirad, M., Ghobadi, C., Nourinia, J. (2021). Metasurfaces for wideband and efficient polarization rotation. *IEEE Transactions on Antennas and Propagation* 69 (3): 1799–1804. https://doi.org/10.1109/TAP.2020.3012828

46 Chatterjee, J., Mohan, A., Dixit, V. (2022). Ultrawideband RCS reduction of planar and conformal surfaces using ultrathin polarization conversion metasurface. *IEEE Access* 10: 36563–36575. https://doi.org/10.1109/ACCESS.2022.3163850

47 Ameri, E., Esmaeli, S.H., Shakouri, A.H. et al. (2020). Ultra-wideband RCS reduction using polarization conversion metasurfaces. *Proceedings of the 28th Iranian Conference on Electrical Engineering (ICEE)*, Tabriz, Iran (04–06 August 2020), New York, US: IEEE. https://doi.org/10.1109/ICEE50131.2020.9261054

48 Ameri, E., Esmaeli, S.H., Sedighy, S.H. (2019). Ultra wideband radar cross section reduction by using polarization conversion metasurfaces. *Scientific Reports* 9 (478): 1–8. https://doi.org/10.1038/s41598-018-36542-6

49 Pesarakloo, A., Oruji, A. (2021). Plasmon resonances management of meta-mirror for combining radar cross section (RCS) reduction and specular reflection with frequency selectivity. *Scientific Reports* 11 (17908): 1–12. https://doi.org/10.1038/s41598-021-97403-3

50 Yao, W., Gao, H., Tian, Y. et al. (2023). Wideband low-RCS linear polarized array based on miniaturized polarization conversion metasurface. *IEEE Transactions on Antennas and Propagation* 71 (7): 5663–5674. https://doi.org/10.1109/TAP.2023.3270383

51 Shao, L., Zhu, W. (2021). Electrically reconfigurable microwave metasurfaces with active lumped elements: a mini review. *Frontiers in Materials* 8 (689665): 1–7. https://doi.org/10.3389/fmats.2021.689665

52 Wu, Z., Grbic, A. (2020). Serrodyne frequency translation using time-modulated metasurfaces. *IEEE Transactions on Antennas and Propagation* 68 (3): 1599–1606. https://doi.org/10.1109/TAP.2019.2943712

53 Ramaccia, D., Sounas, D.L., Alù, A. (2020). Phase-induced frequency conversion and doppler effect with time-modulated metasurfaces. *IEEE Transactions on Antennas and Propagation* 68 (3): 1607–1617. https://doi.org/10.1109/TAP.2019.2952469

54 Malvino, A., Bates, D. (2016). *Electronic principles*, 8th ed. New York, NY: McGraw-Hill Ed., Section 2-14, pp. 47.

55 Schleher, D.C. (2010). *MTI and pulsed doppler radar with MATLAB*. Boston, MA: Artech House.

56 Zhang, X.G., Sun, Y.L., Yu, Q. et al. (2021). Smart doppler cloak operating in broad band and full polarizations. *Advanced Materials* 33, (17): 1–8. https://doi.org/10.1002/adma.202007966

57 Jankiraman, M. (2018). *FMCW radar design*. Boston, MA: Artech House.

58 Kozlov, V., Vovchuk, D., Ginzburg, P. (2023). Radar range deception with time-modulated scatterers. *IEEE Transactions on Antennas and Propagation* 71 (5): 4486–4491. https://doi.org/10.1109/TAP.2023.3255108

59 Wang, X., Yang, G.-M. (2021). Time-coding spread-spectrum reconfigurable intelligent surface for secure wireless communication: theory and experiment. *Optics Express* 29, (20): 32031–32041. https://doi.org/10.1364/OE.437938

60 Wang, X., Caloz, C. (2021). Spread-spectrum selective camouflaging based on time-modulated metasurface. *IEEE Transactions on Antennas and Propagation* 69 (1): 286–295. https://doi.org/10.1109/TAP.2020.3008621

61 Liu, M., Kozyrev, A.B., Shadrivov, I.V. (2019). Time-varying metasurfaces for broadband spectral camouflage. *Physical Review Applied* 12 (054052): 1–8. https://doi.org/10.1103/PhysRevApplied.12.054052

62 Zhang, L., Dai, J.Y., Moccia, M. et al. (2020). Recent advances and perspectives on space-time coding digital metasurfaces. *EPJ Applied Metamaterials* 7 (7): 1–14. https://doi.org/10.1051/epjam/2020007

63 Tennant, A. (1997). Reflection properties of a phase modulating planar screen. *Electronics Letters* 33 (21): 1768–1769. https://doi.org/10.1049/el:19971160

64 Tennant, A., Chambers, B. (1998). Experimental phase modulating planar screen. *Electronics Letters* 34 (11): 1143–1144. https://doi.org/10.1049/el:19980803

65 Tennant, A., Chambers, B. (2009). Time-switched array analysis of phase-switched screens. *IEEE Transactions on Antennas and Propagation* 57 (3): 808–812. https://doi.org/10.1109/TAP.2009.2013448

66 Zhang, L., Chen, X.Q., Liu, S. et al. (2018). Space-time-coding digital metasurfaces. *Nature Communications* 9 (4334): 1–11. https://doi.org/10.1038/s41467-018-06802-0

67 Berry, P.E., Dahal, N., Venkataraman, K. (2022). On the design of an optimal coherent multistatic radar network configuration. *IET Radar, Sonar & Navigation* 16 (5): 869–884. https://doi.org/10.1049/rsn2.12226

68 Li, Z., Xie, J., Liu, W. et al. (2022). Transmit antenna selection and power allocation for joint multi-target localization and discrimination in MIMO radar with distributed antennas under deception jamming. *MDPI Remote Sensing* 14 (3904): 1–17. https://doi.org/10.3390/rs14163904

69 Shi, C., Wang, F., Sellathurai, M. et al. (2020). Low probability of intercept-based optimal power allocation scheme for an integrated multistatic radar and communication system. *IEEE Systems Journal* 14 (1): 983–994. https://doi.org/10.1109/JSYST.2019.2931754

70 Zhang, S., Zhou, Y., Zhang, L. et al. (2021). Target detection for multistatic radar in the presence of deception jamming. *IEEE Sensors Journal* 21 (6): 8130–8141. https://doi.org/10.1109/JSEN.2021.3050008

71 Xiaolin, Q., Ming, J., Lizhong, S. (2003). Design of a new dual-polarization antenna on missile. *Proceedings of the 6th International Symposium on Antennas, Propagation and EM Theory*, Beijing, China (28 October–01 November 2003), New York, US: IEEE. https://doi.org/10.1109/ISAPE.2003.1276653

72 Fante, R.L., McCormack, M.T. (1988). Reflection properties of the Salisbury screen. *IEEE Transactions on Antennas and Propagation* 36 (10): 1443–1454. https://doi.org/10.1109/8.8632

73 Landy, N.I., Sajuyigbe, S., Mock, J.J. et al. (2008). Perfect metamaterial absorber. *Physical Review Letters* 100 (20): 207402–207406. https://doi.org/10.1103/PhysRevLett.100.207402

74 Pozar, D.M. (2012). *Microwave engineering*, 4th ed. Hoboken, NJ: John Wiley & Sons, Inc., Section 5.5, pp. 251–252.

75 Chen, L., Ruan, Y., Luo, S.S. et al. (2020). Optically transparent metasurface absorber based on reconfigurable and flexible indium tin oxide film. *MDPI Micromachines* 11 (1032): 1–11. https://doi.org/10.3390/mi11121032

76 Begaud, X., Lepage, A.C., Varault, S., et al. (2018). Ultra-wideband and wide-angle microwave metamaterial absorber. *MDPI Materials* 11 (10): 1–9. https://doi.org/10.3390/ma11102045

77 Chen, X., Chen, X., Wu, Z. et al. (2018). An ultra-broadband and lightweight fishnet-like absorber in microwave region. *Journal of Physics D: Applied Physics* 51 (28): 1–15. https://doi.org/10.1088/1361-6463/aac907

78 Magill, E., Wheeler, H. (1966). Wide-angle impedance matching of a planar array antenna by a dielectric sheet. *IEEE Transactions on Antennas and Propagation* 14 (1): 49–53. https://doi.org/10.1109/TAP.1966.1138622

79 Tiwari, P., Pathak, S.K., Siju, V. (2022). Design, development and characterization of resistive arm based planar and conformal metasurfaces for RCS reduction. *Scientific Reports* 12 (14992): 1–15. https://doi.org/10.1038/s41598-022-19075-x

80 Leung, S., Liang, C.-P., Tao, X.-F. et al. (2021). Broadband radar cross section reduction by an absorptive metasurface based on a magnetic absorbing material. *Optics Express* 29 (21): 33536–33547. https://doi.org/10.1364/OE.440785

81 Su, J., Li, W., Qu, M. (2022). Ultrawideband RCS reduction metasurface based on hybrid mechanism of absorption and phase cancellation. *IEEE Transactions on Antennas and Propagation* 70 (10): 9415–9424. https://doi.org/10.1109/TAP.2022.3184538

82 Jia, Y., Liu, Y., Zhang, W. et al. (2019). High-gain Fabry-Perot antennas with wideband low monostatic RCS using phase gradient metasurface. *IEEE Access* 7: 4816–4824. https://doi.org/10.1109/ACCESS.2018.2886832

83 Han, Y., Chang, Y., Che, W. (2022). Frequency-selective rasorbers: a view of frequency-selective rasorbers and their application in reducing the radar cross sections of antennas. *IEEE Micrwave Magazine* 23 (2): 86–98. https://doi.org/10.1109/MMM.2021.3125463

84 Arceneaux, W.S., Akins, R.D., May, W.B. (1994). Absorptive/transmissive radome. US Patent 5,400,043, filed 11 May 1994 and issued 21 March 1995.

85 Costa, F., Monorchio, A. (2012). A frequency selective radome with wideband absorbing properties. *IEEE Transactions on Antennas and Propagation* 60 (6): 2740–2747. https://doi.org/10.1109/TAP.2012.2194640

86 Wu, Y., Xiong, J., Deng, F. et al. (2021). A miniaturized frequency selective rasorber with high selectivity passband and wideband absorption properties. *EPJ Applied Metamaterials* 8 (5): 1–7. https://doi.org/10.1051/epjam/2020017

87 Wang, Z., Zeng, Q., Fu, J. et al. (2018). A high-transmittance frequency-selective rasorber based on dipole arrays. *IEEE Access* 6: 31367–31374. https://doi.org/10.1109/ACCESS.2018.2843795

88 Shi, R., Yu, S., Kou, N. (2022). A miniaturized frequency-selective rasorber with absorption bands on two sides of passband for antenna dome. *IEEE Antennas and Wireless Propagation Letters* 21 (11): 2161–2165. https://doi.org/10.1109/LAWP.2022.3187958

89 Chen, K., Feng, Y., Cui, L. et al. (2017). Dynamic control of asymmetric electromagnetic wave transmission by active chiral metamaterial. *Scientific Reports* 7 (42802): 1–10. https://doi.org/10.1038/srep42802

90 Ji, W., Cai, T., Wang, G. et al. (2019). High-efficiency and ultra-broadband asymmetric transmission metasurface based on topologically coding optimization method. *Optics Express* 27 (3): 2844–2854. https://doi.org/10.1364/OE.27.002844

91 Stephen, L., Yogesh, N., Subramanian, V. (2018). Broadband asymmetric transmission of linearly polarized electromagnetic waves based on chiral metamaterial. *Journal of Applied Physics* 123 (3): 1–7. https://doi.org/10.1063/1.5008614

92 Sievenpiper, D.F., Zhang, L., Broas, R.F.J. et al (1999). High-impedance electromagnetic surfaces with a forbidden frequency band. *IEEE Transactions on Microwave Theory and Techniques* 47 (11): 2059–2074. https://doi.org/10.1109/22.798001

93 Yang, W., Hua, G., Hong, W. (2009). Wideband artificial magnetic conductor structure for Ku-band antenna applications. *Proceedings of the 3rd International Symposium on Microwave, Antenna, Propagation and EMC Technologies for Wireless Communications*, Beijing, China (27–29 October 2009), New York, US: IEEE. https://doi.org/10.1109/MAPE.2009.5355800

94 Cheng, Y.-F., Zhao, Y., Peng, F., Liao, C. (2020). Wideband AMC-based phased array with enhanced bandwidth and wide-angle scan. *Proceedings of the International Conference on Microwave and Millimeter Wave Technology (ICMMT),*

Shanghai, China, (20–23 September 2020), New York, US: IEEE. https://doi.org/10.1109/ICMMT49418.2020.9386492

95 Wang, Z., Jiao, Y., Zhang, Y. et al. (2019). Wideband AMC surface and applications to low profile circularly polarized slot antennas. *Proceedings of the International Conference on Microwave and Millimeter Wave Technology (ICMMT)*, Guangzhou, China, (19–22 May 2019), New York, US: IEEE. https://doi.org/10.1109/ICMMT45702.2019.8992239

96 Damaj, L., Lepage, A.C., Begaud, X. (2010). Low profile, directive and very wideband antenna on a high impedance surface. *Proceedings of the 4th European Conference on Antennas and Propagation*, Barcelona, Spain (12–16 April 2010), New York, US: IEEE.

97 Akhoondzadeh-Asl, L., Nourinia, J., Ghobadi, C. et al. (2007). Influence of element shape on the bandwidth of artificial magnetic conductors. *Journal of Electromagnetic Waves and Applications* 21 (7): 929–946. https://doi.org/10.1163/156939307780748995

98 Monorchio, A., Manara, G., Lanuzza, L. (2002). Synthesis of artificial magnetic conductors by using multilayered frequency selective surfaces. *IEEE Antennas and Wireless Propagation Letters* 1: 196–199. https://doi.org/10.1109/LAWP.2002.807956

99 Rayno, J.T., Nakatsu, J.S.K., Huang, G.C. et al. (2013) 3D metamaterial broadband ground plane designed using genetic programming for the long slot array antenna. *Proceedings of the IEEE Antennas and Propagation Society International Symposium (APSURSI)*, Orlando, US (07 July 2013) New York, US: IEEE. https://doi.org/10.1109/APS.2013.6710861

100 Rayno, J., Iskander, M.F., Celik, N. (2013). Synthesis of broadband true-3D metamaterial artificial magnetic conductor ground planes using genetic programming. *IEEE Transactions on Antennas and Propagation* 2014 (62): 5732–5744 https://doi.org/10.1109/TAP.2014.2357416

101 Rexhepi, T., Crouse, D. (2016). A study of composite substrates for VHF and UHF artificial magnetic conductors and their application to a SATCOM antenna. *Progress in Electromagnetics Research C* 64: 1–9. http://doi.org/10.2528/PIERC16030409

102 Lee, D.H., Kwon, O.H., Park, W.B. et al. (2022). A circular sector-shaped dipole antenna with meandered arms and added ferrite-loaded artificial magnetic conductor. *Applied Science* 12 (9149): 1–11. https://doi.org/10.3390/app12189149

103 Khuyen, B.X., Tung, B.S., Yoo, Y.J. et al. (2016). Ultrathin metamaterial-based perfect absorbers for VHF and THz bands. *Current Applied Physics* 16 (9): 1009–1014. https://doi.org/10.1016/j.cap.2016.05.027

104 Khuyen, B., Tung, B., Yoo, Y. et al. (2017). Miniaturization for ultrathin metamaterial perfect absorber in the VHF band. *Scientific Reports* 7 (45151): 1–7. https://doi.org/10.1038/srep45151

105 Khuyen, B.X., Tung, B.S., Kim, Y.J. et al. (2018). Ultra-subwavelength thickness for dual/triple-band metamaterial absorber at very low frequency. *Scientific Reports* 8 (11632): 1–9. https://doi.org/10.1038/s41598-018-29896-4

106 Khuyen, B.X., Tung, B.S., Kim, Y.J. et al. (2019). Broadband and ultrathin metamaterial absorber fabricated on a flexible substrate in the long-term evolution band. *Journal of Electronic Materials* 48: 7937–7943. https://doi.org/10.1007/s11664-019-07637-1

107 Chang, Q., Ji, J., Chen, K. et al. (2024). Transparent and ultra-thin flexible checkerboard metasurface for radar–infrared bi-stealth. *MDPI Sensors* 24 (1531): 1–13. https://doi.org/10.3390/s24051531

108 Dai, H., Li, S., Dong, P. et al. (2023). Design of an ultra-wideband transparent wave absorber. *Materials* 16 (5962): 1–13. https://doi.org/10.3390/ma16175962

109 Gao, Z., Fan, Q., Tian, X. et al. (2021). An optically transparent broadband metamaterial absorber for radar-infrared bi-stealth. *Optical Materials* 112: 1–7. https://doi.org/10.1016/j.optmat.2020.110793

110 Xu, C., Wang, B., Yan, M. et al. (2020). An optically transparent sandwich structure for radar-infrared bi-stealth. *Infrared Physics & Technology* 105: 1–13. https://doi.org/10.1016/j.infrared.2019.103108

5

Active-Cancelation Stealth with Metasurfaces

In the last decades, defense actors have considered stealth a key technology because it enhances the capacity of concealment, increasing the survivability of platforms penetrating within radar zones. We can design stealth platforms by associating a series of techniques that reduce the RCS by the absorption and spread of electromagnetic waves, as we saw in Chapter 4. We can achieve that with the proper shaping and coating of the platform [1].

We can coat platforms with dissipative paints, RAMs, or metasurfaces [2]. By covering the parts that contribute the most to the RCS, we can reduce the probability of radar detection significantly. However, the operating frequency and maintenance regime of dissipative paints and RAMs are not always convenient [3], and passive metasurfaces can be a good alternative thereto. Still, a common limitation to all of these approaches is the fact that usually we cannot cover platforms entirely, which limits the RCS reduction. Complex shapes in the platform also contribute to the RCS and are difficult to coat. Hence, a deployment rate limits the RCS reduction, as we made evident in Section 4.6.2.

One way to reduce further the platform RCS, in this case, consists of transmitting canceling signals that will interfere destructively with the platform skin returns [4–6]. Figure 5.1 illustrates this concept. Some mentions exist about the use of this technique, known as active-cancelation stealth, in the Northrop B-2 Spirit and the Lockheed Martin F-22 Raptor [7]. Nevertheless, despite being envisioned in the 1960s [8], no open report has shown its operational performance to date. Despite this lack of information, we can still see research endeavors all around the world [4–13] because, theoretically speaking, the technique can provide an RCS reduction of $-\infty$ dB. Although this performance is particularly attractive, the implementation challenges must not be ignored.

Ideally, the skin returns and canceling signals must have the same amplitude and be 180° out of phase. This condition can be challenging because radars do not work with single-tone signals. Instead, they transmit pulses, which requires

Metasurface-Driven Electronic Warfare, First Edition. Rafael Gonçalves Licursi de Mello.

Published 2025 by John Wiley & Sons, Inc.
Companion website: www.wiley.com/go/licursidemello/metasurfacedrivenelectronicwarfare1

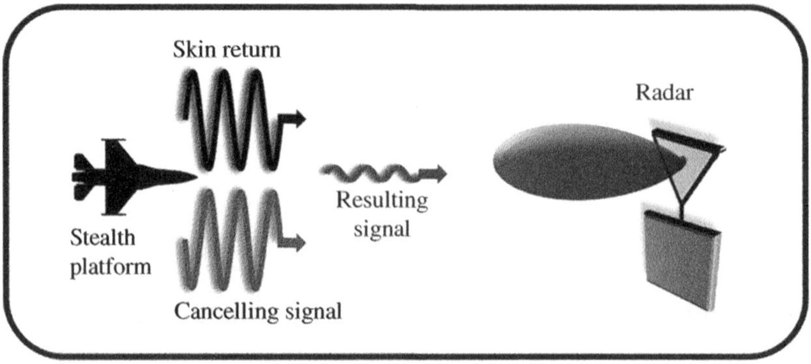

Fig. 5.1 Active-cancelation stealth principle: the stealth platform transmits signals with the same amplitude but out of phase, with its skin returns toward the radar

a superposition of the skin returns and canceling signals in the time domain. This question can be even more critical in the case of intrapulse-modulated signals, such as those of LPI radars [14–19]. To generate canceling signals, the system must have information about the RCS of its own platform. Besides, an ES system should provide the intercepted signals' amplitude and DOA precisely [7]. Last, the time delay between the skin returns and the canceling signals should be neglectable [7, 9–12].

Prominent works suggest the implementation of active-cancelation stealth with DRFMs [7–13]. DRFMs are extensively used to increase survivability, either in active-cancelation stealth or in EA, as we will see in Chapter 7. These devices use antennas and RF front ends to intercept and retransmit radar signals. In between, ADC and DAC circuits allow a processor to modify these signals arbitrarily.

We can envision power-amplifying metasurfaces as a simpler alternative. The low thickness, weight, power consumption, and exemption of an antenna and dedicated compartment make them suitable for active-cancelation stealth in constrained applications. Besides, in the modern EW scenario, it is more reasonable to consider combats between networks of radars and jammers than one-on-one combats. In this context, jamming platforms address multiple radar receivers at once [20–23], and resource allocation is a critical task [24–30]. In these platforms, we can use power-amplifying metasurfaces in parallel to DRFMs to alleviate their tasks.

In this chapter, after a brief introduction to LPI radars and DRFMs, we discuss the theory about active-cancelation stealth and peculiarities related to their implementation with DRFMs. Recently, H. Sing et al. [31] proposed the RCS reduction with active FSSs and metamaterials. However, by "active," that book means reconfigurable circuits. Thus, to the best of the author's knowledge, this is the first study about the application of power-amplifying metasurfaces in active-cancelation stealth. We will introduce the reader to this kind of metasurface, showing how they can mitigate some of the inconveniences of DRFMs. Then, we will also make clear some challenges we should address in a practical application.

This chapter also introduces our EW scenario simulator, which we designed in Matlab [32] and will extensively employ hereafter in this book. Simulations considering the performance of passive metasurfaces, seen in Chapter 4, and the active-cancelation stealth with power-amplifying metasurfaces will be carried out for comparison purposes.

5.1 The LPI Radar

In the cat-and-mouse EW game, LPI techniques were one of the radar responses to traditional ES and EA techniques. Back in the 1980s, advancements in these radars represented a big challenge for EW engineers [14]. Previously, EW systems could simply identify peaks of energy as radar pulses and then transmit noise in a direction that matched the estimated DOA, among other techniques. However, LPI radars avoid detection by transmitting coded, low-amplitude pulses. In this way, they spread energy in time instead of concentrating it in short-period, high-amplitude pulses. In the radar receiver, peaks of energy in the output of a pulse compression filter unveil potential targets when some matching with the waveform of transmitted pulses occurs.

Many types of coded pulses exist such as Frank, Barker, and polyphase [15–18]. Some radars even try to camouflage their signals by adopting a waveform similar to noise [19]. We pick one of these types, the chirp pulse, an LFM pulse, as an example of LPI radar [33]. We will use chirp pulses throughout many simulations in this book. In the complex representation, we can express an ideal pulse of amplitude V_0 by the following formula:

$$V_{pulse} = V_0 \text{rect}\left(\frac{t - TOA}{PW}\right) e^{j2\pi ft} \tag{5.1}$$

where

- rect denotes the rectangular function;
- TOA is the time of arrival of the pulse [s];
- PW is the pulse width [s].

In the absence of intrapulse modulation, we can assume the frequency f to be equal to a carrier frequency f_0. However, chirp pulses present an instantaneous frequency $f(t) = \mu_{chirp}t + f_0$. Here, $\mu_{chirp} = (f_2 - f_1)/PW$ is the chirp rate [Hz/s], f_1 is the starting frequency and f_2 the final frequency. We obtain the instantaneous phase by integrating the frequency f in the time t. Hence, we can express the chirp pulse as:

$$\begin{aligned} V_{pulse} &= V_0 \text{rect}\left(\frac{t - TOA}{PW}\right) e^{j2\pi\left(\frac{\mu_{chirp}}{2}t^2 + f_0 t\right)} \\ &= V_0 \text{rect}\left(\frac{t - TOA}{PW}\right) e^{j\pi\mu_{chirp}t^2} e^{j2\pi f_0 t} \end{aligned} \tag{5.2}$$

Figure 5.2a shows the real and imaginary parts of a chirp pulse in an arbitrary receiver. For the sake of visualization, the pulse is in the baseband, that is, $f_0 = 0$. Further, the instantaneous frequency varies from $f_1 = 0$ to $f_2 = +BW$. Here, $BW =$ 10 MHz, $TOA = 1$ µs, $PW = 5$ µs, $V_0 = 1$ V. We take the receiver thermal noise into account with the addition of a signal with real and imaginary parts randomly and uniformly distributed in the range ±5 mV. This condition characterizes a high signal-to-noise ratio (SNR). Figure 5.2b depicts the spectral power density of this pulse normalized per its highest value. We can see a quasi-rectangular function ranging from 0 to $+BW$.

Now, suppose that the propagation considerably attenuates the chirp pulse before arriving at the receiver. We consider in Fig. 5.2c that $V_0 = 1$ mV. However,

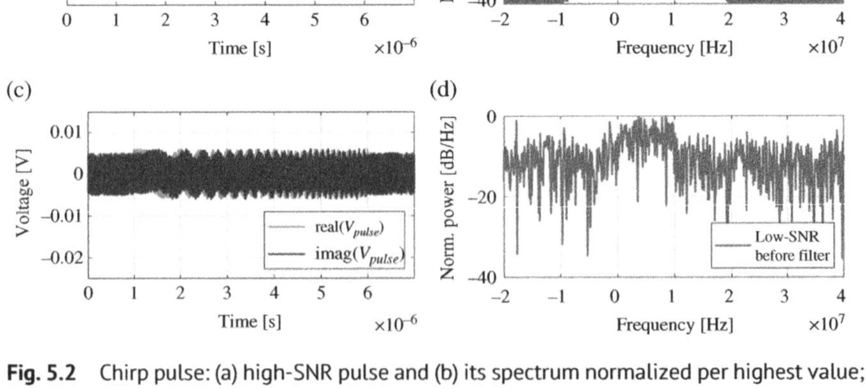

Fig. 5.2 Chirp pulse: (a) high-SNR pulse and (b) its spectrum normalized per highest value; (c) low-SNR pulse and (d) its normalized spectrum

we keep the same noise level as in Fig. 5.2a. Thus, a low SNR occurs. Figure 5.2d shows the new normalized spectral power density. The signal is almost indiscernible from noise in both figures.

If the receiver in question belongs to an ES system that does not have prior information about the pulse, we can consider the latter undetectable (at least by conventional methods). However, the radar has knowledge about the signal and can recover it. A cross-correlation between the received signals and the transmitted waveform will lead to a peak of energy, exposing a potential target when a matching occurs [34]. This operation corresponds to convoluting the received signals $V_{Rx}(t)$ with the impulse response of a pulse compression filter. Such a response should match the complex conjugate, flipped version $V^*_{pulse}(-t)$ of the theoretical waveform of the transmitted pulse:

$$\begin{aligned} V_{match}(t) &= V_{Rx}(t) * V^*_{pulse}(-t) \\ &= \int_{-\infty}^{+\infty} V_{Rx}(t') V^*_{pulse}(t + t') dt' \end{aligned} \tag{5.3}$$

where t' is a dummy variable that we use just to integrate.

Figure 5.3a shows the magnitude of V_{match} considering that V_{Rx} is the high-SNR signal in Fig. 5.2a. A peak of energy happens when the signals are superimposed

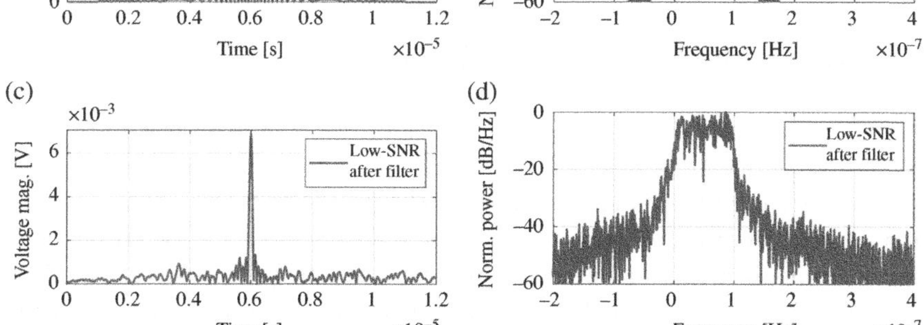

Fig. 5.3 Pulse compression: (a) filter output for a high-SNR pulse and (b) respective spectrum normalized per highest value; (c) output for a low-SNR pulse and (d) respective normalized spectrum

in the time domain. Specifically, $V_{match} = 50$ V at $t = 6$ μs. Note that $V_{match}/V_0 = \sqrt{50} = \sqrt{G_{match}}$, where G_{match} is also known as the signal processing gain associated with the pulse compression. Such gain relates to the time-bandwidth product $PW \times BW$ [33]. Figure 5.3b shows the normalized spectral power density of V_{match} where we see an enhancement in relation to Fig. 5.2b. Now, adjusting (4.9), the maximum detection range $r_{\max}$ of LPI radars is:

$$r_{\max} = \sqrt[4]{\frac{P_t \lambda^2 G_{match}}{(4\pi)^3 SNR_{\min} BW k_b T} G_{t,v}^2(\theta_{plt}, \Phi_{plt}) RCS_{vv}(\theta'_{rdr}, \Phi'_{rdr})} \tag{5.4}$$

The advantage of LPI radars becomes clearer in Fig. 5.3c, which depicts V_{match} again but, now, considering that V_{Rx} is the signal in Fig. 5.2c. The pulse compression leads to a peak of energy despite the low SNR of this signal. The potential target is discernible from noise because vestiges of the former are amplified in the signal, whereas the latter is not. Likewise, the normalized spectral power density in Fig. 5.3d is much better than the one in Fig. 5.2d. The above development shows that:

- LPI radars can recover their signals from a noisy environment due to compression of pulses;
- ES systems without prior information on the waveform cannot do so;
- conventional EA signals arriving at the radar receiver do not benefit from G_{match} in (5.4) because they do not have a similar waveform.

Further, the radar can transmit all the pulses coherently, that is, following a phase reference. This maneuver provides another advantage to it since it allows a coherent pulse integration in the receiving.

5.2 A Brief Introduction to DRFMs

The EA response to LPI radars was the DRFM [35]. The main functionality of DRFMs is the coherent reproduction of intercepted signals with some level of fidelity. As such, copied signals also benefit from G_{match} in (5.4) due to a pulse compression in the radar receiver, as the original ones. DRFMs also allow the introduction of specific modifications to the signals. For instance, a frequency shift can emulate a Doppler deviation associated with the target velocity. Likewise, a time delay can mimic a false target range. Besides, a phase shift can control the interferences between skin returns and the copied signals.

5.2.1 Generic DRFM Architecture

One significant reason why DRFMs have taken so long to enter into service since their first mentions in the literature in the 1970s [36, 37] was the complexity related to implementing one such a system. Fig. 5.4 illustrates the simplified block diagram of a generic DRFM.

In the analog domain, an Rx front end with low-noise amplifiers (LNA) and filters prepares the signals of the addressed frequency band in terms of amplitude and noise for the initial processing. A mixer and a local oscillator (LO) may down-convert signals to the baseband. Then, an ADC circuit digitizes signals, allowing the system to be software defined.

In the digital domain, a memory module handles the intercepted signals for immediate or future operations. A technique generator controls a processor, which may modify signals according to the jamming/stealth operation the DRFM will perform. Next, a DAC circuit converts signals back from the digital to the analog domain, where a Tx front end is used.

In this final stage, a mixer may upconvert signals to their original frequency using the same LO as in the receiving. As such, pulses received at different instants experience the same phase deviation. Putting it differently, coherence is kept. Following, filters attenuate extra components outside the addressed band. Finally, power amplifiers (PA) boost the signals before retransmission.

As one may imagine, each block of Fig. 5.4 offers a series of implementation challenges for a high-fidelity reproduced signal.

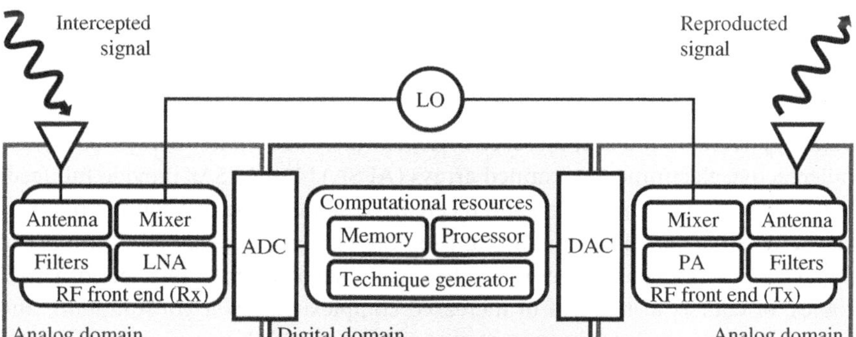

Fig. 5.4 Simplified block diagram of a generic DRFM

5.2.2 Complexity in the Analog Domain

The Rx front end should present a large dynamic range to receive either small or large signals. We accomplish that with LNAs, filters, and mixers that, once in an RF chain, present a low noise figure and good linearity in the addressed band. We often consider the introduction of more complex stages, like an AGC circuit. A compromise between BW and noise exists, opening doors for the use of different architectures. Note that noise is critical here since the signals of LPI radars will be typically low in amplitude when intercepted.

Photonics-based architectures can sweep the whole microwave spectrum with a narrowband filter at an impressive rate. As such, we ensure a high probability of intercept (POI) in frequency with an extremely low noise floor [38–41]. This technology is still complex and expensive today. Conversely, channelized receivers divide the addressed band into smaller ones, instantaneously covering them with 100% of POI in frequency. They present a good compromise between noise level and costs [42–44].

One key element in DRFM-based systems is the antenna because the more energy we focus on the right spots, the more power-efficient and furtive we are. In Fig. 5.4, the antennas that collect and transmit signals can actually be the same if we use a circulator between the RF front ends.

In the modern scenario, addressing multiple distributed radars [20–23] in a kinetically dynamic environment requires electronically steerable antennas. Outside the metasurface world, we can achieve the steering with the dynamic reconfiguration of the amplitudes and phases of the feeding of many antenna elements in an array. As such, we can generate arbitrary field distributions over an aperture, obtaining specific radiation patterns.

Different architectures exist to generate the required amplitudes and phases of currents in these antennas, known as phased arrays [45]. Chapter 8 will explain the working principle behind phased-array antennas. What is important to retain at this point is that active arrays are the intermediate solution in terms of performance and costs in comparison to passive and digital architectures. Consequently, radar applications and DRFM-based systems widely adopt these arrays today, also called active electronically scanned arrays (AESA) [46]. AESAs provide full feeding control in the transmission and receiving of signals. The distribution of transmit/receive modules [47] containing PAs, LNAs, and phase shifters throughout the array at the element level is what provides such control. Here, performance comes, of course, at the cost of increased complexity, power consumption, and costs. J.S. Herd et al. [45] report that the AESA typically represents up to 50% of the costs of a military radar.

Last, phased-array antennas can also generate the multiple independent beams that we need to tackle multiple distributed radars. In the receiving, digital architectures can generate as many equal-gain beams as the available computational capabilities allow [48]. In the transmission, multiple beams at the same frequency result in lower gains to each one of them [49]. That is the reason why antenna resources imposed constraints in DRFM-based systems requiring thus optimized allocation [24–26].

5.2.3 Complexity in the Digital Domain

Taking intercepted signals to the digital domain is an outstanding capability because the system becomes software-defined and reconfigurable in nature there. As such, we can impose arbitrary modifications to the signals like a frequency shift or a time delay [50].

Two intrinsic features of digital systems have a solid impact on the complexity of this domain: the sampling rate f_s, in samples per second [S/s], and the bit resolution n_{bits}, in bits per sample [bit/S]. High sampling rates and bit resolutions require performant ADC/DAC circuits and also powerful processors because the rate of data the system handles is $R_b = n_{bits}f_s$ [bit/s].

The sampling rate f_s relates to the Nyquist frequency f_{Ny}, which in turn is the maximum frequency that a sampler can represent without leading to aliasing distortion. We can calculate the Nyquist frequency as the frequency whose period is two sampling intervals, that is, $f_{Ny} = 0.5f_s$ [51].

In complex sampling, each sample is composed of both a real and an imaginary part, respectively, called the in-phase and quadrature components. This type of representation, also called I/Q sampling, can handle the negative frequencies of the complex representation. As such, it can handle signals whose BW ranges from $-0.5f_s$ to $+0.5f_s$. In this case, the sampling rate matches BW.

Nowadays, modern DRFMs with sampling rates in the order of GS/s are already in operation [52–54]. These rates are more than enough to cover the BW of most of the radars in the theater today, but they are not enough to cover reasonable frequency ranges where opponents can operate.

Representing a signal with a sampling rate much higher than its BW, which is called oversampling, leads to a waste of precious computational power in its processing. To limit this oversampling waste, we can use parallel channels again to cover an addressed band with 100% of POI, but at the cost of increasing size, weight, power consumption, and costs. For example, a DRFM with four channels, each one with ADC and DAC circuits capable of handling a sampling rate of 1 GS/s, can cover the whole X-band (from 8 to 12 GHz). To this end, we should tune the LOs of each channel with a spacing of 1 GHz from one another within this range.

Compressive sensing methods can instantaneously cover wider BWs [55–57]. These techniques can exploit the undersampling of signals that are sparse in the spectrum. They are appropriate for DRFMs because, although they receive signals, for example, either at 2 or 18 GHz, the BW of these signals is far from being that broad. Some of these techniques can make signals from different frequency ranges overlap in the baseband, requiring extra algorithms either to estimate the original frequency [58] or to measure superimposed pulses in the time domain [59].

Now, turning our attention to the bit resolution n_{bits}, this feature relates to the number of amplitude discretization levels representing signals in the digital domain. Let us study how this resolution can impact the spectral content of a signal. To highlight such impact and avoid losing details in the analysis due to the pulse compression, here we are interested in signals before this stage.

Figure 5.5a represents the in-phase component of the pulse of Fig. 5.2a with only $n_{bits} = 2$ bits of discretization levels. We sampled the pulse with a sampling rate $f_s = 1$ GS/s. We can see that the procedure originated a quantization noise outside the pulse interval in the time domain. Figure 5.5b shows the normalized spectral power density. Many spurious components arose around the band of the original signal shown in Fig. 5.2b, with one of them reaching −16.6 dB in relation to the largest value.

In Fig. 5.5c, the in-phase component has a resolution $n_{bits} = 4$ bits. The quantization noise is lower. Fig. 5.5d sketches the respective spectrum, where the

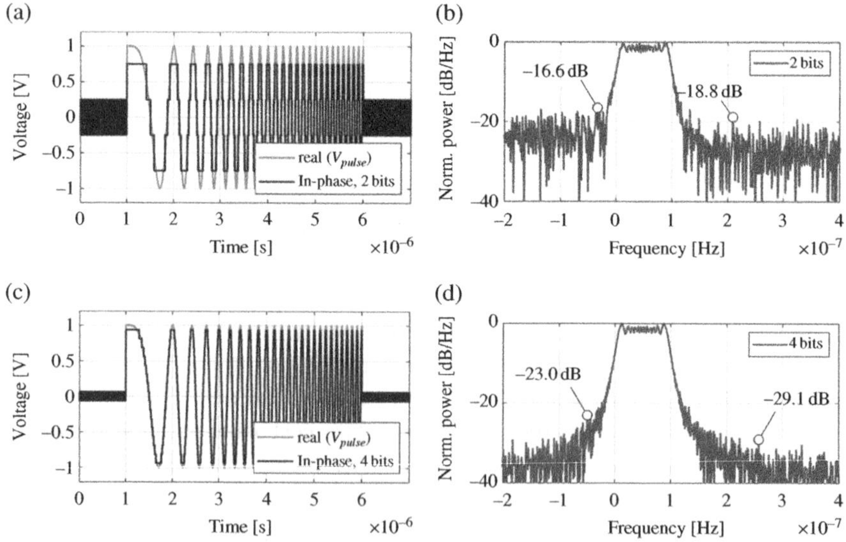

Fig. 5.5 Digitization of a chirp pulse: (a) representation with 2 bits and (b) respective spectrum; (c) 4 bits and (d) its spectrum

spurious components are still present but also at lower levels. Now, the largest one reaches −23.0 dB.

The arise of spurs is intrinsic to the representation of signals in the digital domain. It can trigger EP measures in advanced radar systems receiving the reproduced signals [60]. Besides, the quantization noise contributes to the system dynamic range degradation. It can prevent the detection and processing of LPI radar pulses, whose amplitudes are low. For these reasons, some approaches attempt to replace digital processing by photonic techniques, which presents a high degree of complexity [61, 62].

The above development shows that the better the bit resolution, the lower the level of the spurs and quantization noise. Hence, using high values for n_{bits} can mitigate the problem. Modern DRFMs with $n_{bits} \geq 6$ bits/S are already in operation [52–54]. Unfortunately, the increase of n_{bits} has a direct impact in the data rate R_b that the system handles, exactly like the sampling rate f_s.

High data rates R_b require powerful processors to ensure a small delay in the output of the reproduced signals. If in the software-defined radio early days DSPs could handle most of the tasks, nowadays FPGAs also aid the digital architectures [52, 63–65]. The processing delay is critical to some applications, especially active-cancelation stealth, as it will become more obvious later in this chapter.

From the discussed hitherto in this section, we conclude that wide BWs associated with high-fidelity reproduced signals require large processing capabilities. As a consequence, power consumption, the amount of heat that the system needs to dissipate, bulkiness, complexity, and costs are increased [53]. Allied with the need for an antenna and dedicated compartment, these features prevent the deployment of DRFMs, for example, in small drones [66]. Moreover, they are not aligned with a fundamental military idea, critical in a destituted war effort:

> Everything in war is simple, but the simplest thing is difficult. These difficulties accumulate and produce a friction which no man who has not seen war can imagine exactly.
>
> — Carl P. G. von Clausewitz, military theorist.

5.3 Working Principle of Active-Cancelation Stealth

The DRFM capability of reproducing signals with particular modifications enables a technique to reduce the RCS when other methods are ineffective – or inconvenient. A DRFM can transmit canceling signals that will interfere destructively with skin returns in the radar receiver. As such, we must adjust the complex RCS square root of (4.1) to account for the canceling signal:

$$\begin{aligned}\sqrt{\widehat{RCS}(\theta_s, \Phi_s; \theta_i, \Phi_i)} &= 2\sqrt{\pi} r_{sct} \frac{\boldsymbol{E}^s(\theta_s, \Phi_s) + \boldsymbol{E}^{act}(\theta_s, \Phi_s)}{\boldsymbol{E}^i(\theta_i, \Phi_i)} \\ &= \left|\sqrt{\widehat{RCS}_{plt}}\right| e^{j\varphi_{RCSplt}} + \left|\sqrt{\widehat{RCS}_{act}}\right| e^{j\varphi_{RCSact}}\end{aligned} \tag{5.5}$$

where

- $\boldsymbol{E}^{act}(\theta_s, \Phi_s)$ is the electric field of the active-canceling signal at the observation point [V/m];
- $\left|\sqrt{\widehat{RCS}_{plt}}\right|$ and φ_{RCSplt} are the magnitude and phase of the complex RCS square root of the platform, [m and rad];
- $\left|\sqrt{\widehat{RCS}_{act}}\right|$ and φ_{RCSact} are the magnitude and phase of an equivalent complex RCS square root for the canceling signal, [m and rad].

From the definition of RCS [1] and using (5.5), the RCS a radar receiver sees at the observation point is:

$$\begin{aligned}\widehat{RCS}(\theta_s, \Phi_s; \theta_i, \Phi_i) &= \left|\sqrt{\widehat{RCS}(\theta_s, \Phi_s; \theta_i, \Phi_i)}\right|^2 \\ &= \left|\left|\sqrt{\widehat{RCS}_{plt}}\right| e^{j\varphi_{RCSplt}} + \left|\sqrt{\widehat{RCS}_{act}}\right| e^{j\varphi_{RCSact}}\right|^2\end{aligned} \tag{5.6}$$

which we can express in expanded form as:

$$\begin{aligned}\widehat{RCS} &= \left(\left|\sqrt{\widehat{RCS}_{plt}}\right| \cos\varphi_{RCSplt} + \left|\sqrt{\widehat{RCS}_{act}}\right| \cos\varphi_{RCSact}\right)^2 \\ &\quad + \left(\left|\sqrt{\widehat{RCS}_{plt}}\right| \sin\varphi_{RCSplt} + \left|\sqrt{\widehat{RCS}_{act}}\right| \sin\varphi_{RCSact}\right)^2 \\ \widehat{RCS} &= \left|\sqrt{\widehat{RCS}_{plt}}\right|^2 \cos^2\varphi_{RCSplt} + \left|\sqrt{\widehat{RCS}_{plt}}\right|^2 \sin^2\varphi_{RCSplt} \\ &\quad + \left|\sqrt{\widehat{RCS}_{act}}\right|^2 \cos^2\varphi_{RCSact} + \left|\sqrt{\widehat{RCS}_{act}}\right|^2 \sin^2\varphi_{RCSact} + 2\left|\sqrt{\widehat{RCS}_{plt}}\right| \\ &\quad \left|\sqrt{\widehat{RCS}_{act}}\right| \cos\varphi_{RCSplt} \cos\varphi_{RCSact} + 2\left|\sqrt{\widehat{RCS}_{plt}}\right|\left|\sqrt{\widehat{RCS}_{act}}\right| \sin\varphi_{RCSplt} \sin\varphi_{RCSact} \\ \widehat{RCS} &= \left|\sqrt{\widehat{RCS}_{plt}}\right|^2 + \left|\sqrt{\widehat{RCS}_{act}}\right|^2 + 2\left|\sqrt{\widehat{RCS}_{plt}}\right|\left|\sqrt{\widehat{RCS}_{act}}\right| \cos\left(\varphi_{RCSact} - \varphi_{RCSplt}\right)\end{aligned} \tag{5.7}$$

We can then obtain the RCS reduction by dividing (5.7) by the RCS of the platform in the absence of the canceling signal:

$$\widehat{RCS}\, reduction = 1 + \frac{\left|\sqrt{\widehat{RCS}_{act}}\right|^2}{\left|\sqrt{\widehat{RCS}_{plt}}\right|^2} + 2\frac{\left|\sqrt{\widehat{RCS}_{act}}\right|}{\left|\sqrt{\widehat{RCS}_{plt}}\right|}\cos\left(\varphi_{RCSact} - \varphi_{RCSplt}\right) \tag{5.8}$$

From (5.8), we can see that we obtain an RCS reduction of $-\infty$ dB when the magnitude $\left|\sqrt{\widehat{RCS}_{act}}\right|$ matches $\left|\sqrt{\widehat{RCS}_{plt}}\right|$ and the phase difference $\Delta\varphi_{act/plt} = \varphi_{RCSact} - \varphi_{RCSplt} = (2N+1)\pi$, where $N \in \mathbb{Z}$.

Note that, until now, we have still been adopting the tensorial notation for the RCS because (5.5) through (5.8) consider both vertical and horizontal polarizations. Hereafter, we simplify our analysis for a single, arbitrary polarization.

As one may imagine, achieving the optimal RCS reduction is challenging because it involves transmitting the exact signal that will lead to the aforementioned condition. We can obtain suboptimal reductions if the canceling signal slightly diverges from such a condition. However, large divergences from the ideal signal can even increase the RCS.

Figure 5.6 shows the RCS reduction for diverse values of RCS_{act}/RCS_{plt} and $\Delta\varphi_{act/plt}$. As expected, we see a funnel graph where the RCS reduction indeed tends

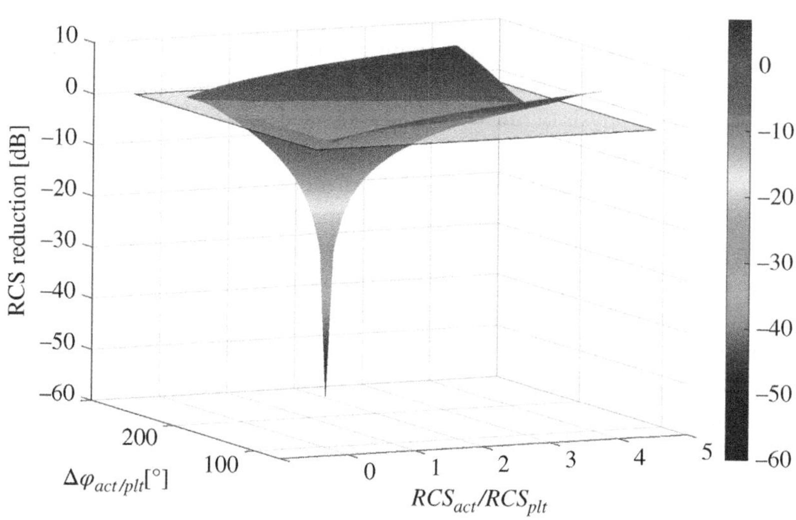

Fig. 5.6 RCS reduction with active-cancelation stealth

to $-\infty$ dB when $RCS_{act}/RCS_{plt} = 1$ and $\Delta\varphi_{act/plt} = 180°$. However, a sharp decrease in the benefits of the technique occurs as soon as these values change. For example, the RCS reduction goes to -10 dB if the phase difference $\Delta\varphi_{act/plt}$ still equals 180° but $RCS_{act}/RCS_{plt} = 0.47$. Likewise, we also obtain -10 dB if $RCS_{act}/RCS_{plt} =$ 1 but $\Delta\varphi_{act/plt}$ diverges 18° from the ideal value.

We can also see in Fig. 5.6 that the RCS actually increases for some combinations of RCS_{act}/RCS_{plt} and $\Delta\varphi_{act/plt}$. For example, for $RCS_{act}/RCS_{plt} = 1$, the RCS reduction is higher than 0 dB when $\Delta\varphi_{act/plt}$ diverges more than 60° from the ideal value. Further, the RCS reduction is also higher than 0 dB whenever $RCS_{act}/RCS_{plt} > 4$, no matter the value of $\Delta\varphi_{act/plt}$.

5.3.1 Impact of the DOA Estimation

The requirements on RCS_{act}/RCS_{plt} and $\Delta\varphi_{act/plt}$ are the reason why an ES system must provide the information about the amplitude and DOA of intercepted signals precisely. In this way, we can predict the electric field $\boldsymbol{E}^s$ of the skin returns since the own-platform RCS is known. As such, we can correctly dimension the electric field $\boldsymbol{E}^{act}$ of the canceling signal. Besides, if we do not have a precise DOA estimation, the active-cancelation system will not point the beam of its Tx antenna precisely toward the radar receiver. Therefore, the electric field $\boldsymbol{E}^{act}$ of the canceling signal over the target will diverge from the desired value.

J. Ding et al. [5] studies the active-cancelation stealth of complex aircraft with full-wave simulations. They put Tx antennas at specific locations over the aircraft aiming at reducing the RCS at the aircraft head-on direction. Indeed, they achieve a reduction of −24.4 dB. However, performance degrades as soon as the direction varies, keeping better than −6 dB in the azimuthal range $\Phi = \pm 10°$. This happens because the platform RCS varies with angle, and so does the radiation pattern of the Tx antenna. The analysis of full-wave simulations of a complex warship allows similar conclusions [9].

H. Zhong et al. [7] investigates this question further for a frequency-modulated signal, namely LFM pulses with a Gaussian envelope. First, they present a stochastic study on the error of TOA measurements as a function of the SNR and PW. Then, they estimate the error deviation in the measurement of DOA with a time-difference-of-arrival (TDOA) method. Such study considers an antenna pattern approximated by a Gaussian function and with 34.5 dBi of gain for the transmission of the canceling signals. They calculated that, under specific conditions, the root-mean-square errors of the DOA estimates cannot exceed values of 0.1°, 0.2°, and 0.5° to achieve 50% of probability of obtaining an RCS reduction of −20, −10, and −6 dB, respectively.

To alleviate this question, we can use less directive antennas in the active-cancelation system because in these antennas the gain variations do not occur so dramatically with angle. On the other hand, these antennas spread more power to other directions, increasing consumption and the probability of detection by third parties.

5.3.2 Impact of Complicated RCS Patterns

When the wavelength is small in comparison to the platform shapes, the RCS may dramatically vary with angle. We can interpret this phenomenon in light of (4.8) because at high frequencies the skin returns of different scattering centers go in and out of phase with small changes in the observation angle, making the RCS pattern complex [67]. As a consequence, it is easier to carry out active-cancelation stealth at low frequencies [8, 67].

Particularly in kinetically dynamic environments, it may be challenging to follow the variations of a complicated RCS pattern in the generation of canceling signals. B. Yang et al. [4] filter the platform RCS with a moving average in the angle domain in order to alleviate this problem. The side effect is that the active cancelation is not optimal. Nonetheless, for an azimuthal range $\Phi = \pm 5°$, average RCS reductions of -14.0 and -16.8 dB are experimentally demonstrated at 8 and 12 GHz, respectively. In the azimuthal range $\Phi = \pm 15°$, the reductions for these frequencies are -7.8 and -10.1 dB.

The good news is that the fact that active-cancelation stealth is difficult to implement at high frequencies makes it complementary to passive stealth, which is hard to perform at low frequencies, as seen in Section 4.6.1.

5.3.3 Impact of the Waveform

In (5.5), we considered the electric fields $\boldsymbol{E}^s$ and $\boldsymbol{E}^{act}$ in the complex representation, per the explanation in (2.16). That means that the above analysis is valid for a sinusoidal signal, at a single frequency. Many components can make up the spectrum of modulated signals. In this case, the cancelation signal should be accurate in terms of waveform in order to cancel each component of the original signal. We saw in Section 5.2.3 that the waveform fidelity of the signals reproduced by a DRFM depends on its bit resolution n_{bits}. Here, we study how the active cancelation of a chirp pulse deviates from the ideal as a function of such feature.

Consider in Fig. 5.7a the in-phase component of signals within a radar receiver related to the instantaneous electric field $\mathcal{E}^s(t)$ of the skin returns of an ideal platform. Here, ideal platform is a platform whose characteristics are non-dispersive within the frequency range of interest. In this way, the skin returns have the same waveform as the impinging waves, whose features in the baseband are the same as described in Fig. 5.5a. An ideal canceling signal arriving at the same receiver has electric field $\mathcal{E}^{act}(t) = -\mathcal{E}^s(t)$. Fig. 5.7a also represents the in-phase component of

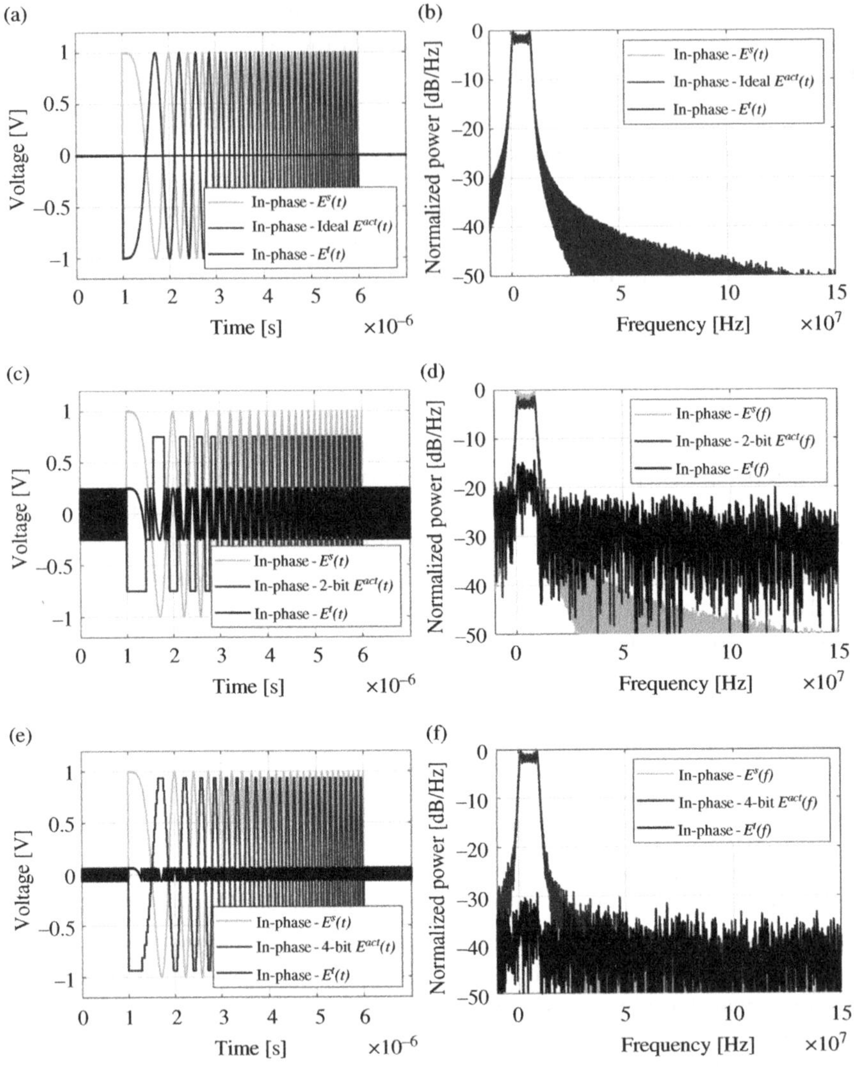

Fig. 5.7 Quantization effects in active-cancelation stealth: time and frequency domain for (a and b) ideal canceling signal; (c and d) 2-bit canceling signal; and (e and f) 4-bit canceling signal

this signal, in baseband. The total field $\mathcal{E}^t(t)$ is this case is zero. All the fields in Fig. 5.7a are normalized per the maximum value of the in-phase component of the skin returns. Figure 5.7b sketches the spectral power density of these signals, normalized per the highest value related to $\mathcal{E}^s(f)$. The skin returns and canceling signals are superimposed. The total field $\mathcal{E}^t(f)$ represents the RCS reduction, which is $-\infty$ dB, as expected.

Now, consider that the canceling signal originated from a DRFM whose bit resolution is $n_{bits} = 2$. Fig. 5.7c,d illustrates the new situation. It is evident that $\mathcal{E}^t(t)$ contains remainders from the DRFM quantization process. The RCS reduction degrades to -14.4 dB in this case. Outside the band of the radar signal, $\mathcal{E}^t(f)$ is much noisier than it would be in the absence of $\mathcal{E}^{act}(f)$, with the existence of spurs that can trigger EP measures in advanced radar systems [60].

Finally, Fig. 5.7e,f illustrates the case where $n_{bits} = 4$. The new RCS reduction is -29.4 dB. Outside the band of the radar signal, $\mathcal{E}^t(f)$ is better than in the case $n_{bits} = 2$, but still with the presence of spurs and a substantial amount of noise at high frequencies.

The above development corroborates with the conclusions taken in Section 5.2.3 concerning waveform fidelity of the signals reproduced by DRFMs, but now for the specific case of active-cancelation stealth. Such analysis indicates that the quality of the technique depends on the bit resolution n_{bits}.

5.3.4 Impact of the Time Delay

Another important question when handling modulated signals is the fact that the cancelation signal should be superimposed to the skin returns in the time domain.

H. Zhong et al. [7] analytically investigate this question for a 3-MHz BW LFM signal at a carrier frequency $f_0 = 1$ GHz. They show that the general performance degrades when the cancelation signal is delayed in relation to the skin return. In order to compensate for that, the performance of other parts of the system should improve. In the studied case, the precision in the DOA estimates should improve (requiring larger SNRs) so that the canceling signals have more precise amplitudes and phases (see Section 5.3.1). Specifically, a 50% probability of obtaining an RCS reduction of -6 dB requires an SNR of 36 dB if the delay was 20 ns. A delay of 50 ns needs 42 dB to achieve the same performance. On the other hand, a delay of 80 ns demands 47 dB, which highlights the criticality of time delay in the technique.

S. Sengupta et al. [68] present many transient analyses related to the active cancelation of radar signals carried out by a sensor and a simple patch antenna. An analog circuit drives the cancelation mechanism, namely an amplifier and a phase shifter. Thus, the delay mainly results from the propagation through the transmission line that connects the sensor, the electronic components, and the patch antenna together.

To obtain further insight, we study again the case of an instantaneous electric field $\mathcal{E}^s(t)$ composed of many spectral components related to the skin returns of an ideal platform. We can use the Fourier transform to represent such field in the frequency domain as:

$$\mathcal{E}^s(f) = \int_{-\infty}^{+\infty} \mathcal{E}^s(t) e^{-j2\pi ft} dt \tag{5.9}$$

To make the total field equal to zero over an Rx antenna, the canceling field $\mathcal{E}^{act}(t)$ should contain the very same spectral components, with the same amplitudes but 180° out of phase. In other words, the ideal canceling field in the frequency domain is $\mathcal{E}^{act}(f) = -\mathcal{E}^s(f)$.

Now, consider that, for some reason, the canceling signal is delayed from the ideal signal by a time t_d. Thus, the actual signal has an electric field $\mathcal{E}'^{act}(t) = \mathcal{E}^{act}(t - t_d)$. Using again the Fourier transform, we can represent such a field in the frequency domain as:

$$\mathcal{E}'^{act}(f) = \int_{-\infty}^{+\infty} \mathcal{E}^{act}(t - t_d) e^{-j2\pi ft} dt \tag{5.10}$$

Making $t' = t - t_d$, noting that the derivative of t' with respect to the time t is $dt'/dt = 1$, and then replacing it in (5.10), we have:

$$\begin{aligned} \mathcal{E}'^{act}(f) &= \int_{-\infty}^{+\infty} \mathcal{E}^{act}(t') e^{-j2\pi f(t' + t_d)} dt' \\ &= e^{-j2\pi f t_d} \int_{-\infty}^{+\infty} \mathcal{E}^{act}(t') e^{-j2\pi f t'} dt' \\ &= e^{-j2\pi f t_d} \mathcal{E}^{act}(f) \end{aligned} \tag{5.11}$$

Examining (5.11) reveals that, if the canceling signal is delayed by a time t_d, each of its spectral components will suffer a phase delay. Moreover, the phase delay increases with increasing time t_d, as well as with increasing frequency f.

Remember that, as we saw in the analysis of Fig. 5.6, phases $\Delta\varphi_{act/plt}$ diverging more than 60° from the ideal value can increase the RCS. Hence, time delays are critical. It is complicated to find a closed formula for the radar performance as a function of the time delay in active-cancelation stealth. Such difficulty is because many waveforms exist. However, (5.11) also reveals that the difference in the phase variation of each component is proportional to the difference in their frequencies. That means that the wider the BW, the larger the difference $\Delta\varphi_{f2/f1}$ between the phase shifts of the extreme components.

As an example, consider a chirp radar operating at $f_0 = 1$ GHz with $BW =$ 40 MHz ($f_1 = 1.00$ GHz to $f_2 = 1.04$ GHz), and $PW = 5$ μs. At $TOA = 1$ μs, a pulse illuminates an ideal platform, which in turn transmits a canceling signal toward the radar receiver. The canceling signal is delayed by $t_d = 12$ ns. For the sake of clarity, we consider that all this delay occurs between the mixers of the Rx and

Tx front ends. In other words, a 1-GHz RF signal is input to the RX mixer, and its copied version is output at 1 GHz but 12 ns later. This allows us to make our analysis only in the range from 1.00 to 1.04 GHz. According to (5.11), the phases of its lowest and highest components will vary by 75.4 rad and 78.4 rad, respectively. As such, the average shift is 76.9 rad, and the phase shift difference is $\Delta\varphi_{f2/f1} = 172.8°$.

Let us see what happens throughout the whole pulse. Figure 5.8 shows the real part of the electric fields $\mathcal{E}^s(t)$ and $\mathcal{E}^{act}(t)$ related to the skin returns and the canceling signal, respectively. For visualization purposes, we use different time ranges in each graph. In this example, we ensured that coherence is kept, that is, a constant phase difference exists between the skin returns and canceling signals when at the carrier frequency $f_0 = 1$ GHz. We can check that in Fig. 5.8a, which depicts the pulse leading edge. The canceling signal succeeds in mitigating the resulted signal from $t = 1.012$ μs because it has approximately the ideal phase. However, the superimposing of the skin return and canceling signal in the trailing edge (Fig. 5.8b) leads to constructive interferences, increasing the amplitude of the resulting signal. This performance degradation happens because a phase mismatch due to the effect of the time delay in (5.11) becomes significant in the highest components even though the coherence related to the carrier frequency is kept. In the chirp case, this effect takes place as the instantaneous frequency changes throughout the pulse, as it is evident in Fig. 5.8c. In Fig. 5.8d, we can see that the canceling signal actually increases the level of some components in the normalized spectral density.

We can try to compensate for the phase deviation expressed in (5.11) by applying a phase shift to the canceling signal. Figure 5.8e,f shows the new scenario for a phase compensation of 86.4°, which is the average shift (76.9 rad) wrapped in the 0–360° range. We can see a perfect cancelation in the middle of the pulse. Still, the final amplitudes are higher than those of the skin returns near the edges. At these regions, the phase of the canceling signal is diverging more than 60° from the ideal value, which leads to increased amplitudes, as discussed in the analysis of Fig. 5.6. This implies that the phase shift difference $\Delta\varphi_{f2/f1}$ cannot exceed 120° if we desire a compensation with a single value of phase. The time t_d plays an important role in this question because it potentializes $\Delta\varphi_{f2/f1}$. On the other hand, this phenomenon does not depend on the carrier frequency. In the end, a constraint related to the product $BW \times t_d$ exists.

Figure 5.8g,h shows the signals for the case when $t_d = 0.55$ ns. The phase shifts of the lowest and highest components are respectively 3.46 and 3.59 rad. The shift difference is only $\Delta\varphi_{f2/f1} = 7.92°$. We carry out a compensation matching the average shift (201.96°) in the canceling signal. In this way, we mitigate the skin return considerably almost in the whole pulse, leading to spectral densities lower than −24 dB/Hz. Thus, we consider −24 dB to be the RCS reduction.

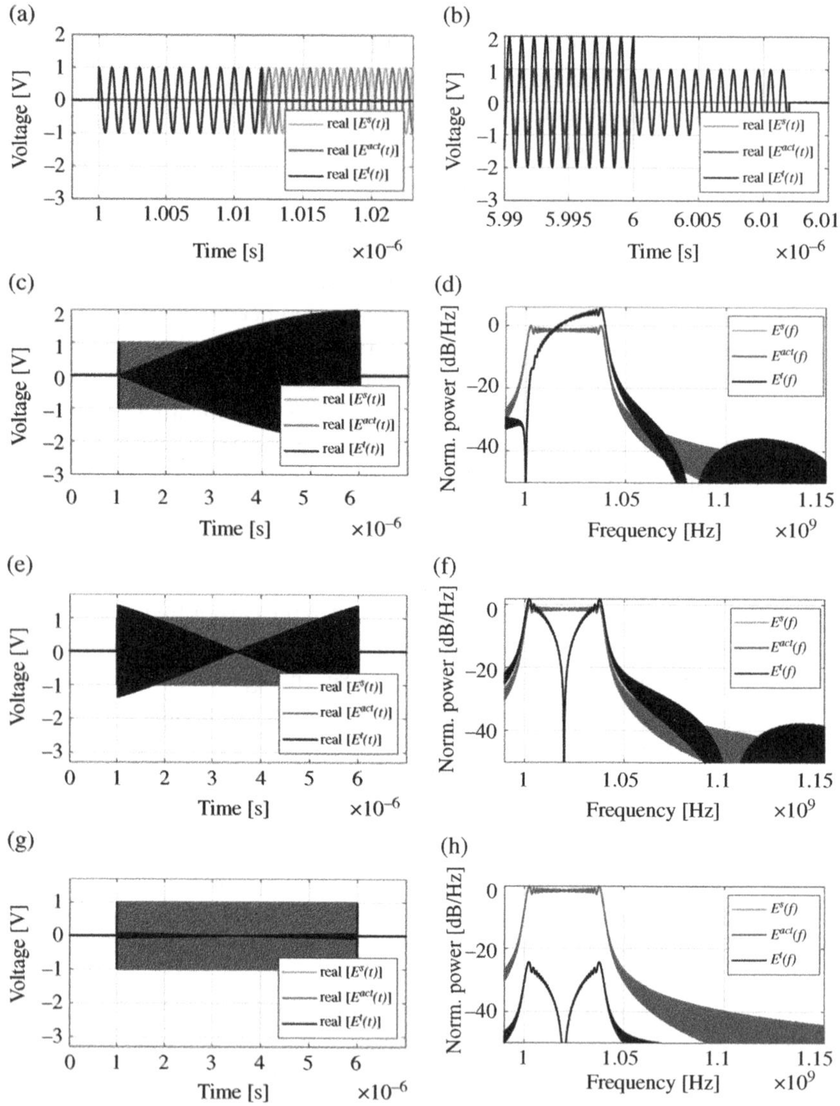

Fig. 5.8 Delay effects in active-cancelation stealth: (a–d) no phase compensation for $t_d = 10$ ns; (e and f) phase compensation for $t_d = 10$ ns; (g and h) phase compensation for $t_d = 0.55$ ns

To date, there is no report about DRFM-based systems with a sub-nano processing delay time. Hence, designers should further implement at least one extra step to use DRFMs in active-cancelation methods. One possibility is to consider the pulse repetition pattern of the target radar, when existent, requiring an ES system to this end. This solution is becoming increasingly unrealistic because, nowadays, highly pulse-dense scenarios may hide exceedingly complicated patterns. Besides, cognitive radars may adjust the PRF according to dynamic environments. For example, G.E. Smith et al. [69] vary the PRF from 2 to 8 kHz in order to maximize the precision in range and velocity estimation of a kinematically dynamic target.

The other option for DRFMs is to handle the delay by applying a dynamic scheme to compensate for the different phases of each signal component. H. Zhong et al. [7] propose a similar solution, by estimating the system delay so as to predict the phase shifts and their respective compensations. However, they indicate that the stability in the system performance (and consequent delay) is a concern in this context. Indeed, in the phase compensation procedure, instabilities in the time delay will make all the scheme also unstable. Ensuring good stability for a system with the complexity of the DRFM, with two RF front ends, cables, and a digital stage, can be challenging. S. Sengupta et al. [68] do not face this difficult because their active-cancelation system is analog. Besides, they make their circuit as short as possible, obtaining a short t_d to simplify this question and use a single-value phase compensation.

5.4 Power-Amplifying Metasurfaces

The waveform and time delay requirements studied in Sections 5.3.3 and 5.3.4 make metasurfaces fit active-cancelation applications like a glove. Instead of using DRFMs to generate signals with all the complexity described in Sections 5.2.2 and 5.2.3, active metasurfaces can simply collect and reradiate impinging waves with a phase shift and power amplification.

The purely analog nature of metasurfaces ensures a high waveform fidelity without quantization noise and spurs, to the limit of the noise figure and linearity of the embedded components. Moreover, the unit cell subwavelength dimensions can provide stable, sub-nano delay times resulting only from the wave propagation through them.

5.4.1 Modeling of Power-Amplifying Metasurfaces

Power-amplifying metasurfaces use active circuits embedded somewhere in the unit cells to provide a gain to impinging waves. As such, it is fair to assume that the power consumption is similar to the PA stage of DRFMs. The consumption related to digital processing is inexistent.

We can illustrate power-amplifying metasurfaces using a structure the reader at this point is already quite familiar with: the AMC without vias, which Section 3.2.3 presented and Chapter 4 extensively exploited. However, consider now that the

AMC has a slotted ground plane behind which circuitry is located, as Fig. 5.9 depicts. In such metasurface, we dimension the slots to couple the energy of impinging waves to the circuits behind the ground plane. There, active components can provide a gain and a phase shift before we couple the energy back to the slots for reradiation. The final effect is equivalent to a power-amplified reflection with a controlled phase [66].

One way of modeling the metasurface of Fig. 5.9 is with the transmission-line schematic approach we depicted earlier in Fig. 5.10. In this model, a quadrupole with input and output impedances Z_{in} and Z_{out} wraps the coupling through the slots, the traveling through the transmission lines of the circuitry, and the amplification and phase shift in the active components. An S_{params} matrix represents the scattering parameters of such quadrupole. Note that the transmission lines surrounding the quadrupole refer to the spacing between the patches and the ground plane, not to the transmission lines of the circuitry. J. Lončar et al. [70] use a

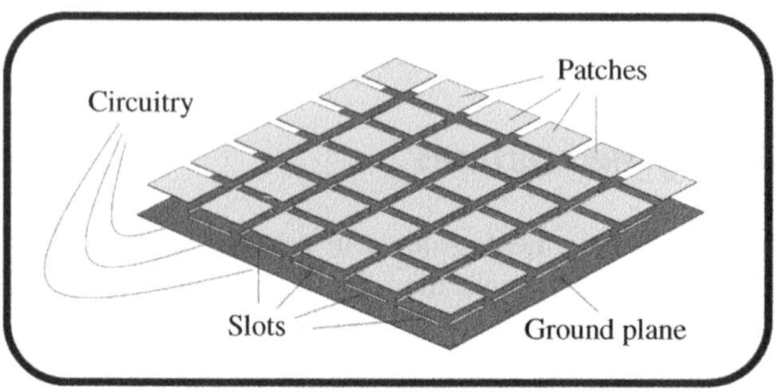

Fig. 5.9 Example of power-amplifying metasurface. "Gonçalves Licursi de Mello [66]/MDPI/CC BY 4.0."

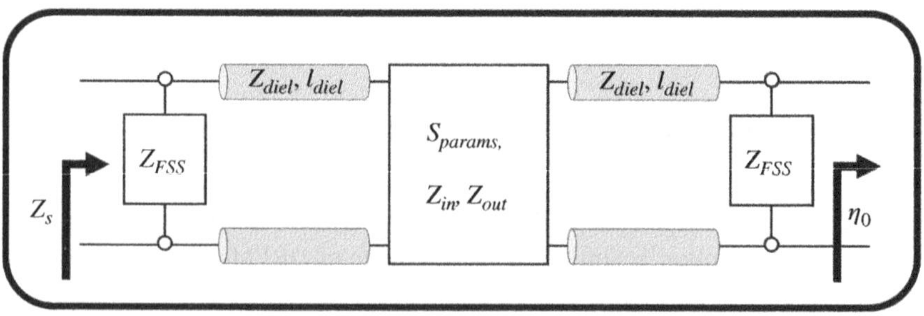

Fig. 5.10 One possibility to model power-amplifying metasurfaces. "Gonçalves Licursi de Mello [66]/MDPI/CC BY 4.0."

similar approach to make an analysis of the challenges related to power-amplifying metasurfaces, including stability issues.

5.4.2 Miniaturization of Active Circuits

The literature reports plenty of power-amplifying metasurfaces. Back in the 1990s, even before the community coined the term "metasurface", an array of amplifiers surrounded by two FSSs provided 11 dB of amplification in the transmission of electromagnetic waves at 3.3 GHz [71]. Since then, many other transmissive power-amplifying metasurfaces have emerged [72–74]. However, in this chapter, we are mostly interested in reflective metasurfaces.

The miniaturization of active circuits has been key for power-amplifying metasurfaces. In the last couple of decades, noticeable advances in this matter allowed the design of high-performance monolithic microwave integrated circuits (MMIC) whose size is of the subwavelength order in the X- and Ka-bands. At first, much of the applications consisted of reflectarrays, a structure with many similarities with metasurfaces. Typically, reflectarrays can be one of the constituent elements of one type of phased-array antenna. In these antennas, we place a well-known radiating element near the reflectarray for the proper illumination thereof. The reflectarray is made of unit cells whose size typically does not exceed half a wavelength and whose reflection coefficient phase we can control. By adjusting the phase of each unit cell, we can translate the illumination from the radiating element into a specific radiation pattern [75].

M.E. Bialkowski et al. [76] and H.J. Song et al. [77] present two power-amplifying reflectarrays for an operation in the X-band: one illuminated by a horn antenna and the other by a four-element antenna array. They use orthogonal slots to couple the energy of both polarizations to MMICs with gallium–arsenide (GaAs) field-effect transistor (FET) amplifiers. The amplifier noise figure is listed as 1.4 dB, the gain as 11 dB, and the 1-dB compression point as approximately 12 dBm. The area occupied by the MMIC is 9.85×9.85 mm^2.

T. Chaloun et al. [78] use an MMIC occupying an area of 4.0×3.85 mm^2 based on silicon–germanium (SiGe) technology in the unit cells of a reflectarray for operation in the Ka-band. They measure the gain as 14 dB, the noise figure as 3.9 dB, and the 1-dB compression point as 7 dBm.

In the late 2010s, N. Landsberg et al. [79, 80] enabled a topology occupying only 90×80 μm^2 by using an advanced 28-nm fully depleted silicon-on-insulator (FD-SOI) complementary metal-oxide–semiconductor (CMOS) process. They embedded this circuit in the unit cells of a 2×2-reflectarray. They obtained a gain of up to 14 dB at millimeter-wave frequencies as high as 125 GHz. They measured a 1-dB compression point of up to −6 dBm along with a noise figure around 11 dB.

5.4.3 Cross-Polarized, Power-Amplified Reflections

For amplification purposes, it does not matter whether the waves are coming from a source nearby, like in reflectarray antennas, or from the free space. At the fundamental level, amplifying reflected waves is a matter of coupling the energy of the waves impinging upon an aperture to a circuitry that can amplify it and couple it back to the aperture for reradiation. We can usually model this procedure with transmission-line equivalents akin to the one in Fig. 5.10.

Some radars can work in a cross-polarized mode, i.e., their Tx and Rx antennas present orthogonal polarizations [81, 82]. In this way, they can detect targets through the RCS_{hv} and RCS_{vh} components in (4.2). Against these radars, we should perform an active-cancelation stealth with the amplification of waves associated to a polarization conversion. This is a particular case. Nonetheless, we pick it to start with because some applicable metasurfaces [83, 84] are easy-understanding examples considering the organization of this book so far.

J. Lončar et al. [83] etch slots in the ground plane of an AMC to couple the energy of impinging waves, in a disposition similar to that in Fig. 5.9. They guide the coupled energy through a circuit in the back of the unit cells, where a right-angle transmission line and an amplifier enable a polarization-conversion with amplification. The unit cell has an area of 7.5×7.5 mm^2. Hence, the delay of a wave traveling through one such circuit is of the order of $t_d = 0.025$ ns. They employ an amplifier with GaAs in enhancement-mode pseudomorph high-electron-mobility transistor (E-pHEMT) technology, occupying an area of 1.0×0.5 mm^2. This amplifier provides a gain of 13.5 dB at 10.0 GHz. Its noise figure is 3.4 dB and the 1-dB compression point is 17 dBm. The thickness of the metasurface is 0.66 mm.

Y. Dong et al. [84] exploit a similar idea with GaAs heterojunction bipolar transistor (HBT) technology for amplified reflection with polarization conversion at 5.44 GHz. The slot-coupling to circuitry behind the ground plane is again present. The unit cell area is 30×30 mm^2 implying a delay time in the circuitry of the order of $t_d = 0.1$ ns. The amplifier provides a gain of 16.4 dB, with noise figure of 4.4 dB and 1-dB compression point of 13.4 dBm. The thickness is 1.07 mm.

J. Yang et al. [85] use a unit cell that couples energy through one metal patch and transmits it to another, both over a ground plane. The thickness is 2.0 mm. The unit cell has an area of 70×70 mm^2 resulting in a delay of the order of $t_d = 0.23$ ns. The patches are orthogonally oriented. They are connected through a right-angle transmission line where an amplifier occupying an area of 2.6×4.6 mm^2 is deployed. This amplifier is based on indium–gallium–phosphorus (InGaP) HBT technology. Its 1-dB compression point is 12.9 dBm and its noise figure is 4.6 dB. At 3.95 GHz, the amplifier gain can reach 12.7 dB, but we can use an external DC-biased voltage for a magnitude-controlled power-amplified reflection at 3.95 GHz.

X. Wang et al. [86] also use two orthogonal patches, but this time in a 3.3-mm thick metasurface for phase-controlled power-amplified reflections. The unit cell area is 45×45 mm^2 leading to a delay of the order of $t_d = 0.15$ ns. It uses an InGaP/GaAs HBT amplifier with an area of 2.3×4.4 mm^2. The 1-dB compression point is 27.5 dBm, the noise figure is 3.9 dB, and the gain is 17.8 dB. In addition, they incorporate a phase shifter with four working states tuned through PIN diodes. In this way, the power-amplified reflections at 4.0 GHz are phase controlled.

If we apply an individual control to the unit cells of X. Wang et al.[86], we can implement a phase gradient such as (3.33) so to perform steering following (3.35) according to the generalized law of reflection [87]. Note that the fact that the unit cell presents only four possibilities for the reflected phase does not avoid the beam steering. It only imposes a certain performance degradation in relation to the case where we have ideal phases [88]. We can dimension one such beam steering to properly provide the desired magnitude in a specific direction. Also, we can control the phase of the reflected signal by choosing the proper reference value for all the reflection phases, that is, the $\varphi_{\Gamma 0}$ term in (3.33). Hence, the unit cell of X. Wang et al. [86] provides simultaneous control of magnitude and phase for reflected waves over a specific direction. This is what the active-cancelation stealth requires.

5.4.4 Co-polarized, Power-Amplified Reflections

Recently, the community have also started to extensively apply amplifiers in metasurfaces used as RISs. If Chapter 9 presents a more detailed study about RISs, we provide next a brief introduction.

In the concept of RIS, we set up the reflection coefficient phase φ_Γ of each unit cell of the metasurface in a specific way, adding multiple, controllable, indirect paths to the wireless propagation channel. From the point of view of the theory of communications, the presence of RISs can tailor the channel and optimize the spectral efficiency [89]. We can use AI algorithms to this end [90].

A series of works have recently proposed using amplifiers in the topology of the unit cells of RISs to compensate for an indirect propagation path larger than the direct one [91–94]. Most of these works express the RIS performance in terms of spectral efficiency, which avoids a direct evaluation of their performance for active-cancelation stealth. One exception reports a remarkable gain of 25 dB at 2.36 GHz associated with a noise figure of 15 dB [94]. Nevertheless, all these works are important for our research because they validate the feasibility of embedding amplifiers in metasurfaces, reinforcing the trends for a higher technological maturity level soon [66].

S. Taravati et al. [95] propose an ultra-thin metasurface with unit cells composed of cascaded rectangular patches, MMIC amplifiers, and tunable phase shifters for a co-polarized power-amplifying reflective beamsteering. The association between amplifiers and phase shifters provides beamforming capabilities. Through the biasing of these components, we can steer the reflected beam to arbitrary directions when the metasurface is illuminated also from arbitrary directions. Figure 5.11a,b illustrates the steering of the reception and reflection beams. The

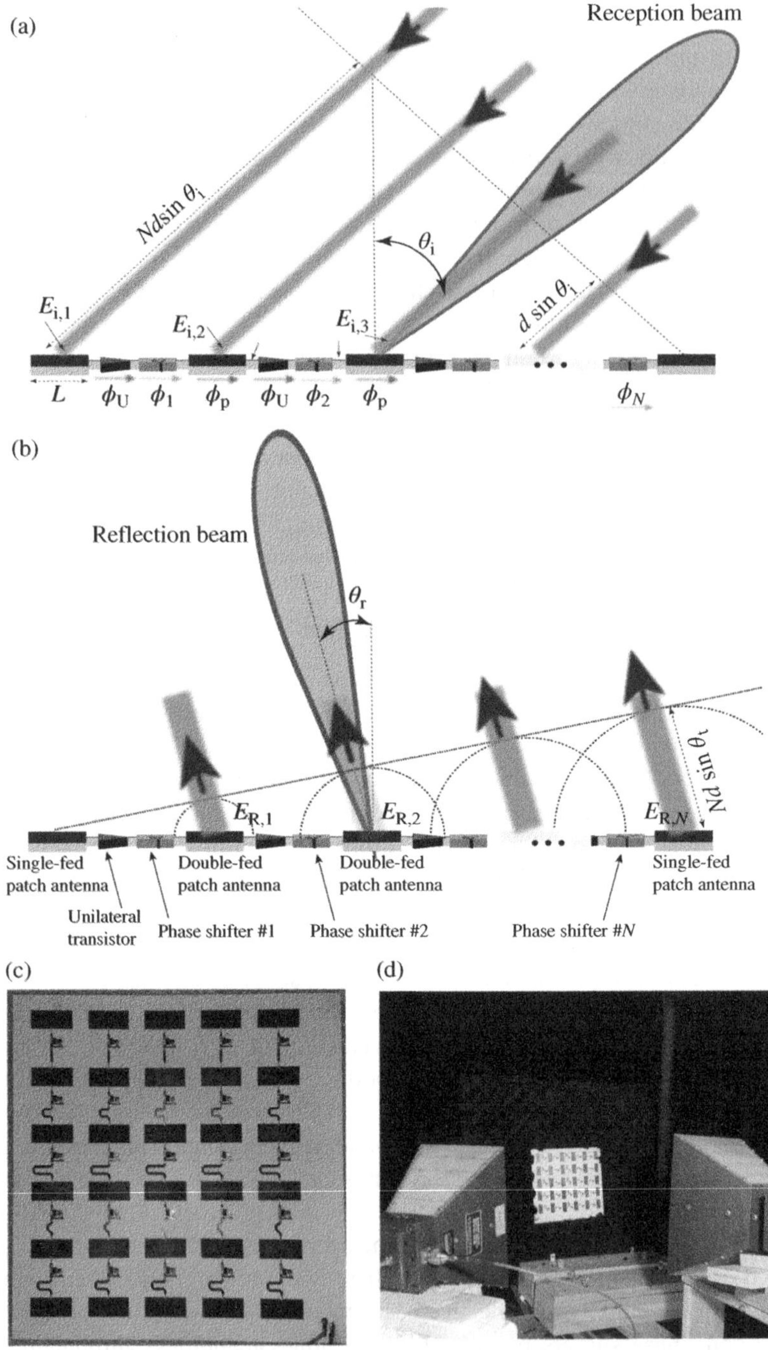

Fig. 5.11 Metasurface of [95]: (a and b) steering of the reception and reflection beams; (c) fabricated metasurface; (d) measurement setup. "Reproduced from Taravati and Eleftheriades [95]/Springer Nature/CC BY 4.0."

thickness is only 0.76 mm. The $30 \times 30\text{-mm}^2$ area unit cell leads to a delay time in the circuitry of the order of $t_d = 0.1$ ns. They employ the same amplifier as J. Yang et al. [85], which allows magnitude-controlled amplification (InGaP HBT technology occupying an area of $2.6 \times 4.6\ \text{mm}^2$, 1-dB compression point of 12.9 dBm, noise figure of 4.6 dB, and gain around 13 dB at 5.81 GHz). Figure 5.11c shows the fabricated prototype. Figure 5.11d depicts the measurement setup, which validated at 5.8 GHz an experimental gain larger than 25 dB even when the angles of incidence and reflection were as high as 45° and 66°, respectively, and larger than 19 dB for angles as high as 70° and 80°.

The metasurface of S. Taravati et al. [95] represents a watershed in co-polarized power-amplified reflections. The fact that it can steer the reflected beam in arbitrary directions allows the right amount of energy to be sent toward a specific spot. Similar to the metasurface of X. Wang et al. [86] for the cross-polarized case, this control is achieved through the implementation of a phase gradient such as (3.33) for steering following (3.35) according to the generalized law of reflection [87]. Besides, same phase shift simultaneously imposed on all the cells enables the controlling of the phase of the signal without disturbing the direction. Therefore, this metasurface accomplishes what the active-cancelation stealth requires for the co-polarized case.

5.4.5 Roadmap to a Metasurface-Driven Active Stealth

In terms of performance, the noise figure values of the active elements of prominent metasurfaces [86, 95] are quite reasonable not to trigger any EP measure in a target radar. One should also note that their reported 1-dB compression points relate to a single amplifier, which ensures a notable dynamic range in terms of waveform fidelity. Their gains allow a good operation even if the deployment rates (see Section 4.6.2) are not large. Last, the delay times for the propagation through the circuit of their unit cells allow single-value phase compensations even for radars whose band is relatively wide (see Section 5.3.4).

In the matter of convenience, we can associate the thicknesses of these metasurfaces [86, 95] with a light weight. Since they amplify reflected waves without the need for an antenna or a dedicated compartment, their deployment over some relatively small surface on an aircraft, vessel, satellite – and even over lightweight drones – is feasible.

A constraint related to these prominent metasurfaces [86, 95] is the fact that they cannot address all the terms of (4.2) at once. Specifically, the metasurface of X. Wang et al. [86] can tackle only RCS_{hv} and RCS_{vh} whereas the one of S. Taravati et al. [95] tackles only RCS_{hh} or RCS_{vv}. However, we should consider that active-cancelation stealth with metasurfaces is an advanced application never studied before this book, at least to the best of the author's knowledge. Hence, to date, no group has put efforts into developing a metasurface dedicated to this task.

The natural way to start such a development would be to decouple the co- and cross-polarization problems. We can use cascaded rectangular patches, MMIC amplifiers, and tunable phase shifters to handle the co-polarized component, as S. Taravati et al. [95] did. Simultaneously, we can implement the orthogonal-slot approach to address the cross-polarized component, as J. Lončar et al. [83] and Y. Dong et al. [84] did. We can also exploit other combinations.

A point to pay attention to in the development of a metasurface dedicated to active-cancelation stealth is the fact that a dispersion behavior in the reflection of waves may degrade the waveform fidelity. One should ensure that the metasurface frequency dependence (see Section 3.4.1) does not distort the reflected waves significantly within the operational band of the target radar.

Another point is that the noise figure values reported in Sections 5.4.3 and 5.4.4 refer to the active components in the metasurface circuitry. According to the Friis equation for the noise figure of cascaded elements [96], an insertion loss before the gain stages can considerably increase the noise figure of the whole circuit. Thus, we should couple the impinging waves properly before they arrive at the circuitry. For the same reason, we should mitigate ohmic losses at points before the gain stages.

As one can see, developing a metasurface to simultaneously cancel all the terms of (4.2) may not be straightforward. However, it is expected that in the upcoming years, as a higher technological maturity level is achieved, the community will feel encouraged to put effort into this task. This will happen mainly when the community realizes that power-amplifying metasurfaces:

a) do not raise spurious components from digitization, proving thus high-fidelity signals, different from DRFMs, as we saw in Section 5.3.3;
b) do not have an intrinsic processing time, providing instead a short, stable delay time due to the signal propagation throughout their tiny unit cells and easing the phase compensation we explained in Section 5.3.4;
c) do not need a dedicated compartment and an antenna to intercept and retransmit signals, different from DRFMs, as we saw in Section 5.2.1;
d) do not present most of the complexity of DRFMs seen in Sections 5.2.2 and 5.2.3, leading to lower weight and costs.

On the other hand, in active-cancelation stealth, power-amplifying metasurfaces:

e) do not exempt the need for information about the DOA of intercepted signals, provided by an ES system, and the RCS of the own platform, as we saw in Section 5.3.1;
f) also need to consume power in the amplification of the reflected signals.

Now, it is time to compare how the techniques presented so far would perform in an EW scenario.

5.5 Introduction to the EW Scenario Simulator in Matlab

We discussed in Section 1.1 how EW has evolved over the decades from a simple clash between static systems to a highly complex combat framework with plenty of reconfigurable systems optimizing their tasks on the fly. This formidable evolution of technologies and techniques was possible due to remarkable advancements not only in electronics, telecommunications, and so forth but also in systems engineering. The latter allowed complex components and phenomena to be packaged together according to their behaviors and relationships. In the end, we can regard advanced systems as a set of small boxes with certain functionalities and interfaces.

One relevant side effect of this evolution is the fact that predicting the performance of such systems in the so-called real world has become increasingly difficult. This is because Electromagnetics is not trivial – and so are not many of its resulting phenomena, as the reader probably recalled/realized in Chapter 2. Hence, whenever we develop a new advanced technique, a question arises about whether it would really work or not in a specific situation. Another reason is that knowledge about the adversary's systems is not always available. Nonetheless, some solutions can circumvent these difficulties.

Simply deploying systems in real combat and collecting as much data as possible is not a possibility because it is not reasonable to take real dangers to risk having unpleasant results in the end. Some governments or big companies could afford air, naval, land, and space measurement campaigns. However, this becomes less interesting when we remember that the rate of innovation of nowadays would constantly require measurement updates. A solution reasonable until recently was the implementation of hardware simulators, known as hardware in the loop (HIL) [52]. Such a solution aims at reproducing realistic operational conditions in controlled environments such as an anechoic chamber. However, the complexity of the modern scenario is not easily reproducible, requiring costly and sophisticated equipment along with a dedicated, specialized team.

Nowadays, a solution the community widely accepts for this problem is the software simulator [97]. This approach allows the quick simulation of a huge diversity of complex scenarios and the obtaining of high amounts of data. Of course, the data quality will be as high as the simulations are realistic. To ease this job, Matlab recently released a series of toolboxes prepared to model realistic phenomena [98]. This book receives support from Mathworks [32] to simulate metasurface-driven EW scenarios at the physics level.

5.5.1 Simulation Modeling

We implement the EW scenario simulator with Matlab and the Radar, Antenna, Phased Array System, and Signal Processing toolboxes. We adopt an I/Q sampling with a rate of $f_s = 100$ MS/s. We can reduce this rate to achieve higher computational speeds. We will define the carrier frequency for each scenario.

We model platforms as *phased.Platform* objects. We model their scattering features as *phased.RadarTarget* objects. They are 2×2 matrices representing the RCS tensor in (4.2). We will always associate systems with platforms, which in turn will present Tx and Rx systems accordingly. For example, we will associate monostatic radars with a Tx and an Rx system in the same platform, whereas we will associate an aircraft equipped only with an ES system with just an Rx system. We model Tx front ends as *phased.Transmitter* objects. Similarly, we model Rx front ends as *phased.Receiver*. In each case, we use specific features taken from the datasheet of components available in the industry. As such, we consider realistic values of transmit power, gain, and noise figure.

For the computation of noise, we associate a base temperature $T = 340$ K in the circuitry (referred to the input) with a specific BW. In general, from 0 to 10,000 ft of altitude, the external temperature lows approximately by 2 K for each 1000 ft. However, we subtract only 1 K for each 1000 ft in the simulations to account for the fact that the circuitry is not in direct contact with the air.

We model realistic antennas with the *phased.CustomAntennaElement*, *phased.Radiator*, and *phased.Collector* objects. We will specify these custom antennas in each case. Phased arrays are a specific case where we model the antenna elements with *patchMicrostrip* objects to be packaged in a *phased.URA* object. Then, we use *phased.Radiator*, *phased.SteeringVector*, and *phased.PhaseShiftBeamformer* objects to radiate and control the beams.

We model LPI radar signals with a *phased.LinearFMWaveform* object with a specific BW, PW, and PRF. For simplicity, we do not consider any sort of agility in this book. In the receiving, we convolute signals with a matched filter we model with a *phased.MatchedFilter* object.

5.5.2 Simulation Algorithm

Figure 5.12 illustrates the flowchart of the simulation algorithm. Simulations start by loading specific parameters as well as all the platforms of the scenario and their respective embedded systems as specified in each case in the book.

In each simulation, we take a Tx system as a reference. We use a for-loop to transmit the signal of the reference system in a specific interval. All the other Tx systems in the scenario also transmit in this loop, but the loop follows the repetition interval of the reference system.

Every signal propagates toward every other platform. There, the signal is collected by the antenna of an eventual Rx system or scattered, propagating toward

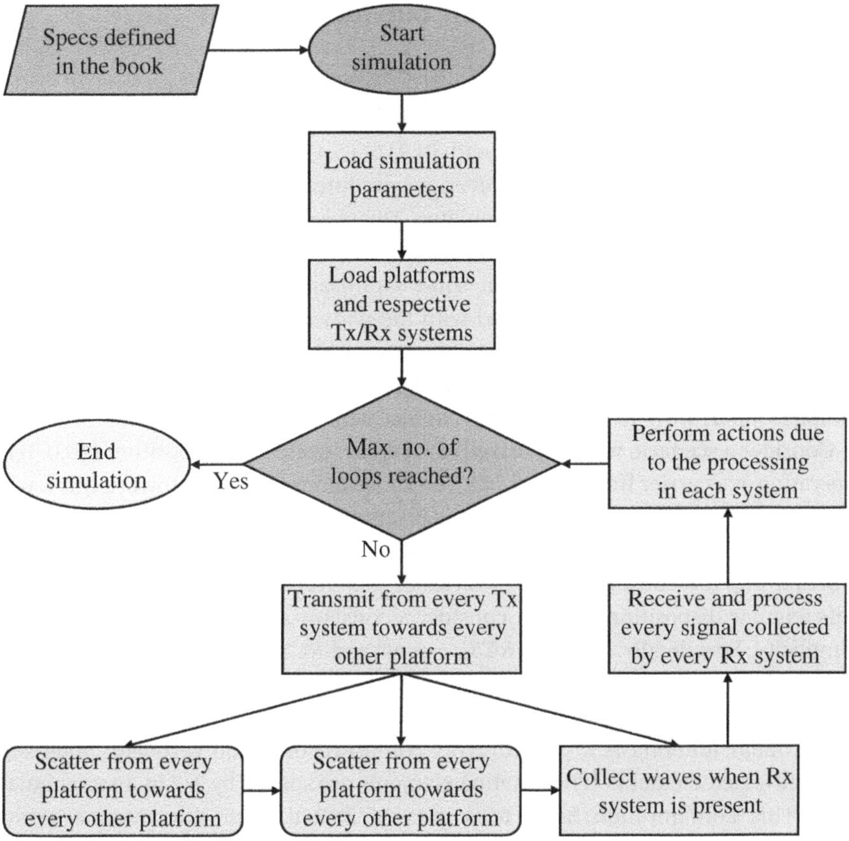

Fig. 5.12 Flowchart representing the EW scenario simulator used in this book

every other platform once again. We implement a multipath propagation of order 2, that is, up to two platforms can scatter signals before we disregard them. In addition, we model the propagation of waves using the *twoRayChannel* object considering reflections over the sea, whose relative permittivity is $\varepsilon_r = 78.4$. Hence, the actual number of paths is doubled.

In the end, we sum up the signals collected by each antenna and process them in each respective Rx system accordingly. At this point, we can take some action, like moving an antenna base system, guiding a missile, performing a jamming, or just showing some information. We use the *phased.ScenarioViewer* object to reveal the geometrical disposition of the scenario.

5.6 Simulation: Comparison Between Stealth Techniques

Predicting the effects of an RCS reduction in the maximum range r_{max} of detection of a target radar is straightforward with (4.9) or (5.4). We can find examples at the end of Sections 4.1 and 4.6.2. However, we are interested at this point in making the first use of the EW scenario simulator. The objective is to show the validity of the simulator rather than the RCS reductions themselves. Besides, it is a good opportunity for the reader to gain some familiarity with this approach. Here, we will only edit the RCSs modeled with the *phased.RadarTarget* objects considering a reduction promoted by some of the metasurfaces studied so far. In the next chapters, more complex scenarios will require the simulator, and there is not a simpler opportunity to having a first contact with it than now.

Consider a scenario where a surveillance radar located at the position $P(0,0,0)$ is operating at a carrier frequency $f_0 = 5.8$ GHz. An aircraft is flying toward this radar at Mach 2 ($v_1 = 686$ m/s) with a dynamic position $P(r_{acf}, \theta_{acf} = 85°, \Phi_{acf} = 7°)$. Note that the elevation of the aircraft is $el_{acf} = 90° - \theta_{acf} = 5°$ from the ground. Further, θ_{acf} and Φ_{acf} are constants but r_{acf} varies as the aircraft approaches the radar. In this angular disposition, we can consider a constant aircraft RCS for the sake of simplicity. Specifically, $RCS_{hh} = RCS_{vv} = 1$ m^2 and $RCS_{hv} = RCS_{vh} = 0$. Figure 5.13 sketches such a scenario together with the rectangular and spherical coordinate systems we use in the simulations.

The radar antenna is a phased array composed of 4×30 vertically polarized squared-patch elements. The antenna elements are spaced by $\lambda/2$ in a rectangular grid. This configuration leads to a gain of 26.3 dBi for the antenna array. Figure 5.14 depicts the radiation patterns in the azimuth and elevation planes. The 3-dB beamwidths in both planes are $\theta_{3dB} = 24.59°$ and $\Phi_{3dB} = 3.38°$, respectively.

The radar works with chirp pulses with $PRF = 2.5$ kHz, $PW = 8$ μs, and $BW =$ 50 MHz. We also consider such BW for noise purposes in the Rx front end, where the listed features of the CA48-2111 LNA from Ciao Wireless, Inc. [99] lead to a gain of 29 dB and a noise figure of 1.3 dB. In the Tx front end, we consider a gain of 22 dB based on the listed gain of the CA26-4114 PA, also from Ciao Wireless, Inc.

If the radar antenna is pointed at $el_{ant} = 90° - \theta_{ant} = 12.3°$ in elevation, a scan pattern is carried out in azimuth from $\Phi_{ant} = +23.66°$ to $\Phi_{ant} = -23.66°$ in 15 steps of 3.38°. The desired probability of detection is $P_d = 0.9$ and of false alarm is $P_{fa} = 10^{-6}$. The number of non-coherently-integrated pulses is 20. To highlight the detection behavior as a function of the RCS, we dimension the peak power to make a target with $RCS_{vv} = 1$ m^2 detectable at a range of around 50 km.

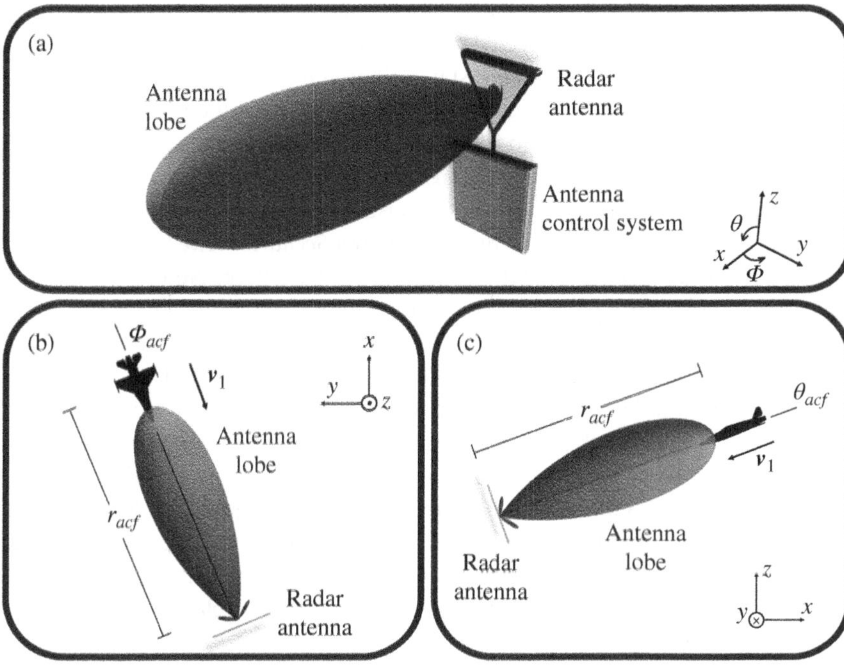

Fig. 5.13 Simulated scenario: (a) surveillance radar and the spherical coordinate system; (b and c) top and side view of the aircraft and the rectangular coordinate system

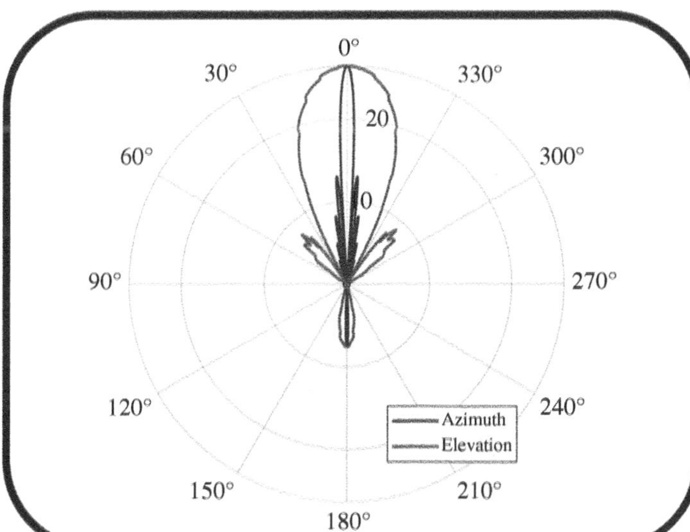

Fig. 5.14 Radiation pattern of the employed antenna array

We model a target tracker using the *radarTracker* object from the Radar Toolbox. We set the tracker with expected velocity Mach 2, with a confirmation threshold of five occurrences out of the six last iterations and a deletion threshold of three occurrences out of the four last iterations. We consider that the radar successfully detects the target when it starts to track it. For all the simulations, the range of the aircraft in relation to the radar at the instant $t = 0$ is $r_{acf} = 50$ km.

The first simulation considers no metasurface stealth technique, that is, *RCS reduction* = 0 dB. Figure 5.15 uses the *phased.ScenarioViewer* object to depict the scenario in a random moment in the first radar sweep. Note how this tool allows a good monitoring of what is happening in the simulation. We observe that the radar tracks the aircraft from $t = 2.16$ s at the range $r_{acf} = 48.6$ km. Figure 5.16 shows the signal intensity associated with the tracking in the radar processing.

In the second simulation, we want to represent one case we saw in Section 4.6.2. Specifically, we consider the case where an aircraft is exploiting metasurface technologies only in some antennas and a percentage of its surface. In this way, the RCS reduction is −6.3 dB as we calculated in (4.27) through (4.29). Running this simulation, we see that the radar tracks the aircraft from $t = 21.5$ s when at the range $r_{acf} = 35.3$ km. These results represent a reduction in the maximum range r_{max} of radar detection to 0.73 of its original value, which is close to what we calculated in Section 4.6.2.

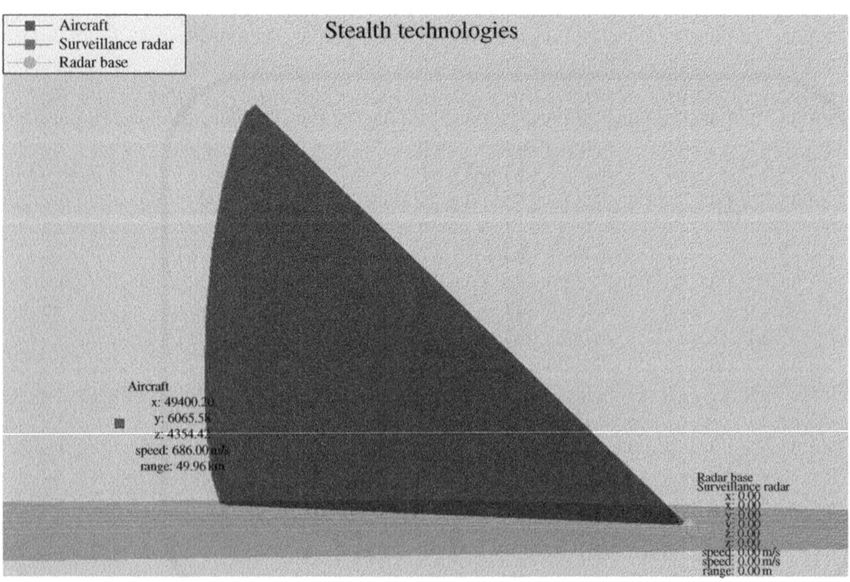

Fig. 5.15 EW scenario represented with the *phased.ScenarioViewer* object

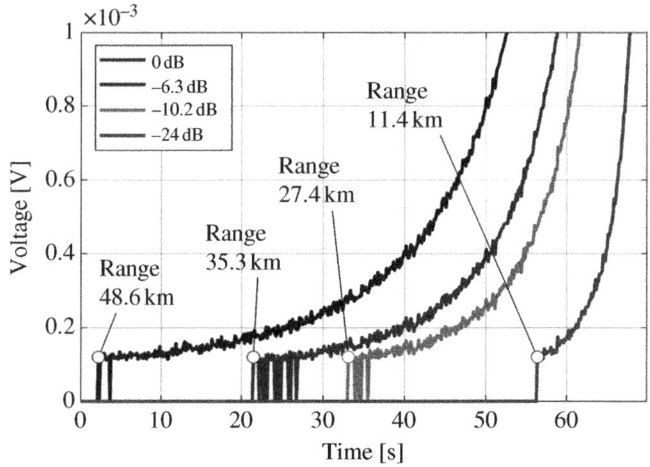

Fig. 5.16 Signal intensity associated with the tracking

In the third simulation, the RCS reduction is −10.2 dB to represent what we obtained at the end of Section 4.6.2. Now, the radar tracks the aircraft from $t =$ 33.1 s when at the range $r_{acf} = 27.4$ km. Therefore, we obtain a reduction in $r_{\max}$ to 0.56 of its original value, which agrees once again with previous calculations in Section 4.6.2.

Finally, we use an RCS reduction −24 dB. We calculated this reduction in the context of Fig. 5.8g,h when we considered a delay time of $t_d = 0.55$ ns in a single-phase compensation active-cancelation stealth technique. Throughout Section 5.4, we saw cross- and co-polarized reflective metasurfaces that can carry out this technique with such a short delay. The radar tracks the aircraft from $t = 56.4$ s with $r_{acf} = 11.4$ km. As expected, the reduction in $r_{\max}$ is the best one so far, specifically to 0.23 of its original value.

We can see in the above simulations that the results are all coherent with what we can calculate using (4.9) or (5.4), which shows the validity of our simulator.

5.7 Exercises

We can use the EW scenario simulator to make simulations much more complex than those of Section 5.6. By referring to the folder entitled *Ch05_stealth* in the companion website, all the reader should do is embed the proper system objects at the right places and run the *main.m* file. The simulator will do its role, which is to propagate the waves from Tx systems until platforms, Rx systems, and so on, followed by the processing of the systems, as Fig. 5.12 illustrated.

In this section, we propose some exercises that provide a good understanding on the working of the simulator. In this manner, the reader can obtain some independence in simulating its own scenarios.

5.7.1 Exploring the EW Scenario Simulator

Difficult level: low.

One invaluable tip in developing either simple or complex electronic systems is to track the signal throughout the different stages of the system. In hardware, we can do this tracking with the aid of an oscilloscope or a spectrum analyzer. In the case of our simulator, a good exercise is to use the Matlab plot function to see how signals look like at different parts of the code.

To start with, place a breakpoint point right after where we call the *wav* vector for the first time in the main code and run the simulation. Then, plot its real and imaginary parts, as well as its magnitude, using *plot(real(wav))*, *plot(imag(wav))*, or *plot(abs(wav))*, respectively. Zoom the signal and observe the chirp waveform and the amplitude of these signals. Next, repeat this procedure after we call *wav* = *pltfms{i}.transmitters{ii}(wav)*. Check if the changes in the magnitude agree with the properties of the Tx front end we previously defined.

Following, you can see how we compute the antenna radiation pattern by plotting *abs(sigtx(1).Z)* and *abs(sigtx(2).Z)* right after *pltfms{i}.radiators{ii}* processes *wav*. The first computes the antenna gain toward the direct path, whereas the second does so toward the indirect path due to reflections on the sea. By changing *Z* for *X* or *Y*, you can check other polarization components. After *prms.txchannel* processes *sigtx*, you can repeat this procedure to check the delay and attenuation due to the propagation as well as the Doppler effect. Continue in this process by plotting *scattered_signal*, *pltfms{iii}.sig_imp*, *sig_imp*, *yc*, and *pltfms{i}.sigrx{ii}* to understand how the simulations consider natural phenomena.

As a last step, put a breakpoint inside the *process_svl_rdr* function and follow the signal within it. We highly recommend getting used to the *V_match* and *int_pulses* vectors here. Such exercise will provide insight into how the compression and integration of pulses work. Last, it will show how we can track the target. Try to change *RCS_rdct* using positive and negative values at the beginning of the main code and see how it affects the signals at this final stage.

5.7.2 Simulating a Realistic DRFM

Difficult level: low.

In Section 5.2.3, we developed a code to digitize the pulse of Fig. 5.2a with n_{bits} quantization levels so as to show in Fig. 5.5 the impact of the bit resolution in the spectral content of a signal. We embedded it in the EW scenario simulator so that the reader can play with a somehow realistic DRFM.

The first step is to look for the *pltfms{2}.pcs{1}.DRFM_state* variable and change it from 0 to 1. We already prepared the *build_acf_DRFM* function to assign proper Tx and Rx systems in the role of DRFM in the aircraft platform. The antenna is a phased array composed of 10×10 vertically polarized squared-patch elements spaced by $\lambda/2$ in a rectangular grid. It presents a gain of 25.3 dBi. In this case, the 3-dB beamwidths are $\Phi_{3dB} = \theta_{3dB} = 10.05°$. To keep the DRFM realistic, we base the features of the Tx and Rx front ends in the same components used in the radar (LNA with 29 dB of gain and 1.3 dB of noise figure and PA with 22 dB). However, we consider a BW of 4 GHz for noise purposes. You can change these parameters as you want. If keeping them in realistic ranges is interesting, exploring exaggerated values can provide extra insight about the system's limitations.

A time, frequency, or phase shift can be chosen for the DRFM by setting *acf_DRFM.pcs{1}.shift* in the end of the *build_acf_DRFM* function. You can also programmatically update these variables in the simulation loop if you want. As such, you can exploit deception capabilities in the DRFM.

In the main code, we set the DRFM to always point its antenna beam over the radar. This comes from the assumption that an ES system provides the estimated DOA of intercepted pulses. More audacious simulations can implement such ES system, e.g., with the technique R.G.L. Mello et al. [59] propose to detect the pulses and with the one R.G.L. Mello et al. [100] propose to estimate their DOAs, as we will see in Chapter 6.

In the *process_DRFM* function, an ideal AGC adjusts the amplitude of intercepted signals to the range ±1 V. Following, signals are digitized with a resolution n_{bits}, defined by the *acf_DRFM.pcs{1}.n_bits* variable.

5.7.3 Simulating a Realistic Active-Cancelation Stealth

Difficult level: medium.

In Section 5.6, we simulated the effects of diverse stealth techniques just by changing the *RCS_rdct* variable at the beginning of the main code. The reader can, of course, try to implement a realistic active-cancelation stealth simulation.

If the reader did the exercise of Section 5.7.2, the first task here starts by setting the DRFM phase shift as π rad while keeping its time delay as 0 in the *build_acf_DRFM* function. You should use the *acf_DRFM.pcs{1}.shift* variable to this end. Then, you must implement a code to translate the amplitudes of the received pulses into the amplitudes to be transmitted by the DRFM. For this, you can use the *V_DRFM* variable. Note that we do not need to care about the aircraft RCS at this point because the aircraft presents a constant angular disposition in relation to the radar and we are using a single value for the quantity.

As a second task, the reader can set a time delay to the DRFM representing its processing time. After that, you can adjust the DRFM phase shift to compensate for this delay, as we explained in Section 5.3.4. This will provide a deep understanding of the active-cancelation stealth operation.

Note that we can emulate the behavior of metasurfaces in active-cancelation stealth if a short delay time is associated with an extremely high bit resolution n_{bits} and the proper values for gain and noise figure in the DRFM front ends.

A more audacious simulation here would introduce an angle-dependent RCS matrix and consider rotations around the pitch, roll, and yaw axes of the aircraft in a curved trajectory. Similar to what we proposed in the previous exercise, you can implement an ES system with the techniques R.G.L. Mello et al. [59, 100] propose, as we will see in Chapter 6. Then, you can associate the estimated DOAs to the RCS matrix to define the amplitude that the DRFM should transmit to achieve maximum RCS reduction.

5.8 Final Remarks of the Chapter

The potential of active-cancelation stealth in reducing the RCS makes it a hot issue in EW today despite its implementation complexity mainly against LPI radars. In this chapter, we proposed a new way of implementing this technique using power-amplifying metasurfaces.

As opposed to DRFMs, metasurfaces do not raise spurious components from digitization and do not have an intrinsic delay time and power consumption related to digital processing. Moreover, the use of metasurfaces exempts the need for a dedicated compartment and an antenna to intercept and retransmit signals. These features make them valuable in the unfolding cat-and-mouse EW game because they enable active-cancelation stealth in applications where DRFMs are not pertinent. Also, power-amplifying metasurfaces can be used in parallel to DRFMs to let them free to carry out other tasks.

On the other hand, the use of metasurfaces in active-cancelation stealth does not alleviate the need for information about the amplitude and DOA of intercepted signals and the RCS of the own platform. Besides, the power consumption related to the amplification of signals still exists.

We identified in the literature candidate metasurfaces to perform the application in question. The reconfigurable nature of their active circuits allows cross- or co-polarized reflections with arbitrary amplitude and phase toward arbitrary directions. The metasurface unit cell subwavelength dimensions, allied to a noise figure of the order of 4 dB, and 1-dB compression point higher than 10 dBm, leads to a low delay and high waveform fidelity. The power-amplified reflections require only a small part of the platform to be coated.

A limitation was also identified in the fact that metasurfaces providing arbitrary, cross- and co-polarized reflections simultaneously cannot be found in the literature yet. This limitation is unlikely to last long since the technological maturity level of metasurfaces is getting higher and higher. This chapter also provided guidelines for the development of one such metasurface.

References

1 Knott, E.F., Schaeffer, J.F., Tulley, M.T. (2004). *Radar cross section*, 2nd ed. Raleigh, NC: Scitech Publishing.

2 Murugesan, A., Selvan, K.T., Iyer, A.K. et al. (2021). A review of metasurface-assisted RCS reduction techniques. *Progress in Electromagnetics Research B* 94: 75–103. http://doi.org/10.2528/PIERB21081401

3 Vinoy, K.J., Jha, R.M. (1995). Trends in radar absorbing materials technology. *Sadhana* 20: 815–850. https://doi.org/10.1007/BF02744411

4 Yang, B., Yang, W., Xiao, W. et al. (2019). Active cancellation stealth technology analysis and verification in experiment. *Proceedings of the International Conference on Identification, Information and Knowledge in the Internet of Things, IIKI 2018*, Beijing, China (19–21 October 2018). Amsterdam, NL: Elsevier. https://doi.org/10.1109/ELMAR49956.2020.9219017

5 Ding, J., Xie, Y., Zhang, J. et al. (2022). Simulation of active echo cancellation effects using the ARCS concept. *MDPI Electronics* 11 (3891): 1–10. https://doi.org/10.3390/electronics11233891

6 Hao, Z., Wang, X., Chen, Y. (2023). Theory and effectiveness analysis of dual-aircraft formation active cancellation slow synchronous blinking jamming. *AIP Advances* 13 (075105): 1–12. https://doi.org/10.1063/5.0153633

7 Zhong, H., Wang, F., Zhou, J. (2019). Performance analysis of active cancellation system based on DRFM. *IEEE 4th International Conference on Computer and Communications (ICCC)*, Chengdu, China (07–10 December 2018). New York, US: IEEE. https://doi.org/10.1109/CompComm.2018.8780693

8 Sheng, X., Yuanming, X. (2014). Analyzing active cancellation stealth. *Microwaves & RF*, August 2014. https://www.mwrf.com/technologies/embedded/systems/article/21845860/analyzing-active-cancellation-stealth (accessed 06 August 2023).

9 Xiang, Y.-C., Qu, C.-W., Su, F. et al. (2010). Active cancellation stealth analysis of warship for LFM radar. *Proceedings of the IEEE 10th International Conference on Signal Processing*, Beijing, China (24–28 October 2010). New York, US: IEEE. https://doi.org/10.1109/ICOSP.2010.5655941

10 Yi, M., Wang, L., Huang, J. (2015). Active cancellation analysis based on the radar detection probability. *Aerospace Science and Technology* 46:273–281. https://doi.org/10.1016/j.ast.2015.07.018

11 Yi, M., Wang, L., Pan, Y. et al. (2015). Active cancellation stealth analysis based on cancellaty. *Tech Science Press CMES* 109 (5): 427–446.

12 Huang, Q., Xu, Y. (2016). Active cancellation stealth analysis based on interrupted-sampling and convolution modulation. *Optik* 127 (7): 3499–3503. https://doi.org/10.1016/j.ijleo.2015.12.129

13 Osman, I.A. (2013). Active cancellation system for radar cross section reduction. *International Journal of Education and Research* 1 (7): 1–6.

14 Schleher, D.C. (1985). Low probability of intercept radar. *Proceedings of the IEEE Radar Conference*, Washington, DC, USA (6–9 May 1985), New York, US: IEEE.

15 Kishore, T.R., Rao, K.D. (2017). Automatic intrapulse modulation classification of advanced LPI radar waveforms. *IEEE Transactions on Aerospace and Electronic Systems* 53 (2): 901–914. https://doi.org/10.1109/TAES.2017.2667142

16 Alrubeaan, T., Albagami, K., Ragheb, A. et al. (2019). An investigation of LPI radar waveforms classification in RoF channels. *IEEE Access* 7: 124844–124853. https://doi.org/10.1109/ACCESS.2019.2938317

17 Qu, Z., Hou, C., Hou, C. et al. (2020). Radar signal intra-pulse modulation recognition based on convolutional neural network and deep Q-learning network. *IEEE Access* 8: 49125–49136. https://doi.org/10.1109/ACCESS.2020.2980363

18 Bhatti, S.G., Bhatti, A.I. (2022). Radar signals intrapulse modulation recognition using phase-based STFT and BiLSTM. *IEEE Access* 10: 80184–80194. https://doi.org/10.1109/ACCESS.2022.3195273

19 Meng, X., Shang, C., Dong, J. et al. (2021). Automatic modulation classification of noise-like radar intrapulse signals using cascade classifier. *ETRI Journal* 43 (6): 991–1003. https://doi.org/10.4218/etrij.2020-0338

20 Berry, P.E., Dahal, N., Venkataraman, K. (2022). On the design of an optimal coherent multistatic radar network configuration. *IET Radar, Sonar & Navigation* 16 (5): 869–884. https://doi.org/10.1049/rsn2.12226

21 Li, Z., Xie, J., Liu, W. et al. (2022). Transmit antenna selection and power allocation for joint multi-target localization and discrimination in MIMO radar with distributed antennas under deception jamming. *MDPI Remote Sensing* 14 (3904): 1–17. https://doi.org/10.3390/rs14163904

22 Shi, C., Wang, F., Sellathurai, M. et al. (2020). Low probability of intercept-based optimal power allocation scheme for an integrated multistatic radar and communication system. *IEEE Systems Journal* 14 (1): 983–994. https://doi.org/10.1109/JSYST.2019.2931754

23 Zhang, S., Zhou, Y., Zhang, L. et al. (2021). Target detection for multistatic radar in the presence of deception jamming. *IEEE Sensors Journal* 21 (6): 8130–8141. https://doi.org/10.1109/JSEN.2021.3050008

24 Xing, H., Xing, Q., Wang, K. (2023). A joint allocation method of multi-jammer cooperative jamming resources based on suppression effectiveness. *MDPI Mathematics* 11 (826): 1–25. https://doi.org/10.3390/math11040826

25 Wang, Q., Yan, X., Liu, F. et al. (2022). An optimal allocation method for cooperative jamming resources based on multi-agent genetic algorithm. *Proceedings of 2nd International Conference on Frontiers of Electronics, Information and Computation Technologies (ICFEICT)*, Wuhan, China (19–21 August 2022), New York, US: IEEE. https://doi.org/10.1109/ICFEICT57213.2022.00056

26 Sun, J., Zhang, D., Yi, W. (2021). Jamming resource allocation for multi-jammer cooperatively suppressing netted radar systems. *Proceedings of the CIE International Conference on Radar (Radar)*, Haikou, Hainan, China (15–19 December 2021), New York, US: IEEE. https://doi.org/10.1109/Radar53847.2021.10028342

27 Zheng, G., Na, S., Huang, T. et al. (2019). A barrage jamming strategy based on CRB maximization against distributed MIMO radar. *MDPI Sensors* 19 (2453): 1–13. https://doi.org/10.3390/s19112453

28 Ke, M., Zhang, H., Wu, H. et al. (2021). Game model in multi-target jamming task allocation. *Proceedings of 2nd International Conference on Electronics, Communications and Information Technology (CECIT)*, Sanya, China (27–29 December 2021), New York, US: IEEE. https://doi.org/10.1109/CECIT53797.2021.00114

29 Shao, S., Zhang, T., Ye, F. (2017). Jamming resource allocation aimed to data link based on simulant annealing algorithm. *Proceedings of the Progress in Electromagnetics Research Symposium - Fall (PIERS-FALL)*, Singapore, Singapore (19–22 November 2017), New York, US: IEEE. https://doi.org/10.1109/PIERS-FALL.2017.8293276

30 Guo, J., Wang, H., Cheng, H. et al. (2022). Multiple methods based resource allocation optimization of jamming system for UAV clusters. *Proceedings of the IEEE International Conference on Unmanned Systems (ICUS)*, Guangzhou, China (28–30 October 2022), New York, US: IEEE. https://doi.org/10.1109/ICUS55513.2022.9986780

31 Sing, H., Jha, R.M. (2015). *Active radar cross section reduction*. Cambridge, UK: Cambridge University Press, Chapter 8, pp. 274–277.

32 The MathWorks Inc.(2023) MathWorks - Maker of MATLAB and Simulink. https://www.mathworks.com/ (accessed 04 June 2023).

33 Skolnik, M.I. (2002). *Introduction to radar systems*, 3rd ed. New York, NY: McGraw-Hill Ed., Section 11.5, pp. 420–434.

34 Skolnik, M.I. (2002). *Introduction to radar systems*, 3rd ed. New York, NY: McGraw-Hill Ed., Section 10.2, pp. 369–375.

35 Roome, S.J. (1990). Digital radio frequency memory. *Electronics & Communication Engineering Journal* 2 (4): 147–153. https://doi.org/10.1049/ecej:19900035

36 Spector, S.C. (1975). A coherent microwave memory using digital storage: the loopless memory loop. *Electronic Warfare* 7 (1): 108.

37 Bittencourt, R.P. (1976). Electronic warfare technology. Master Thesis. Naval Postgraduate School. https://apps.dtic.mil/sti/pdfs/ADA028210.pdf (accessed 12 August 2023).

38 Pace, P.E. (2022). *Developing digital RF memories and transceiver technologies for electromagnetic warfare.* Norwood, MA: Artech House, Section 7.15, pp. 467–485.

39 Gao, G., Liang, Q., Liu, Z. et al. (2021). A photonics-based superheterodyne RF reception approach. *Advances in Astronautics Science and Technology* 4: 121–131. https://doi.org/10.1007/s42423-021-00089-y

40 Ghelfi, P., Scotti, F., Onori, D. et al. (2019). Photonics for ultrawideband RF spectral analysis in Electronic Warfare applications. *IEEE Journal of Selected Topics in Quantum Electronics* 25 (4): 1–9. https://doi.org/10.1109/JSTQE.2019.2902917

41 Ghelfi, P., Onori, D., Laghezza, F. et al. (2015). An RF scanning receiver based on photonics for electronic warfare applications. *European Radar Conference (EuRAD)*, Paris, France (09–11 September 2015). New York, US: IEEE. https://doi.org/10.1109/EuRAD.2015.7346271

42 Mikhailov, Y. (2019). DRFM jammer test solutions. *Microwave Journal.* https://www.microwavejournal.com/articles/33097-drfm-jammer-test-solutions (accessed 12 August 2023).

43 Sun, G., Li, Y., Gao, M. (2009). An improved DRFM system based on digital channelized receiver. *2nd International Congress on Image and Signal Processing*, Tianjin, China (17–19 October 2009). New York, US: IEEE. https://doi.org/10.1109/CISP.2009.5305419

44 Ozdil, O., Ispir, M., Ortatatli, I.E. et al. (2016). Channelized DRFM for wideband signals. *IET International Radar Conference*, Hangzhou, China (14–16 October 2015). New York, US: IEEE. https://doi.org/10.1049/cp.2015.1266

45 Herd, J.S., Conway, M.D. (2016). The evolution to modern phased array architectures. *Proceedings of IEEE* 104 (3): 519–529. https://doi.org/10.1109/JPROC.2015.2494879

46 Pace, P.E. (2022). *Developing digital RF memories and transceiver technologies for electromagnetic warfare.* Norwood, MA: Artech House, Chapter 3, pp. 81–136.

47 Brown, A.D. (2022). *Active electronically scanned arrays: fundamentals and applications.* Norwood, MA: Artech House, Chapter 4, pp. 103–134. https://doi.org/10.1002/9781119749097.ch4

48 Sikri, D., Jayasuriya, R.M. (2019). Multi-Beam phased array with full digital beamforming for SATCOM and 5G. *Microwave Journal.* https://www.microwavejournal.com/articles/32053-multi-beam-phased-array-with-full-digital-beamforming-for-satcom-and-5g (accessed 12 August 2023).

49 Skolnik, M.I. (1990). *Radar handbook*, 2nd ed. New York, NY: McGraw-Hill Ed., Section 7.1, pp. 7.9.

50 Jhansi, V. (2020). Design of delay and doppler module for ECM system based on DRFM technique. *International Journal of Creative Research Thoughts* 8 (2): 1808–1816. http://doi.one/10.1729/Journal.25311

51 Leis, J.W. (2011). *Digital signal processing using MATLAB for students and researchers*. Hoboken, NJ: John Wiley & Sons, Inc., Section 3.8, pp. 82.

52 Davidson, K., Bray, J. (2020). Understanding digital radio frequency memory performance in countermeasure design. *MDPI Applied Sciences* 10 (4123): 1–18. https://doi.org/10.3390/app10124123

53 Neri, F. (2018). *Introduction to electronic defense systems*, 3rd ed. Boston, MA: Artech House, Section 5.2.4, pp. 394–403.

54 De Martino, A. (2018). *Introduction to modern EW systems*, 2nd ed. Norwood, MA: Artech House, Section 5.3, pp. 300.

55 Pace, P.E. (2022). *Developing digital RF memories and transceiver technologies for electromagnetic warfare*. Norwood, MA: Artech House, Section 4.3, pp. 143–153.

56 Vasavada, Y., Prakash, C. (2020). Sub-Nyquist spectrum sensing of sparse wideband signals using low-density measurement matrices. *IEEE Transactions on Signal Processing* 68: 3723–3737. https://doi.org/10.1109/TSP.2020.3000637

57 Almslmany, A., Wang, C., Cao, Q. (2016). Advanced deceptive jamming model based on DRFM sub-Nyquist sampling. *13th International Bhurban Conference on Applied Sciences and Technology (IBCAST)*, Islamabad, Pakistan (12–16 January 2016). New York, US: IEEE. https://doi.org/10.1109/IBCAST.2016.7429963

58 Su, Y. (2020). Digital instantaneous frequency measurement of a real sinusoid based on three sub-Nyquist sampling channels. *Mathematical Problems in Engineering* 2020 (5089761): 1–11. https://doi.org/10.1155/2020/5089761

59 Gonçalves Licursi de Mello, R., De Sousa, F.R. (2018). Precise techniques to detect superimposed radar pulses on ESM systems. *IET Radar, Sonar & Navigation* 12 (7): 735–741. https://doi.org/10.1049/iet-rsn.2017.0563

60 Choi, M. (1996). A comparison of DDS and DRFM techniques in the generation of 'smart noise' jamming waveforms. PhD thesis. Naval Postgraduate School, Section 4.B.1, pp. 75. https://apps.dtic.mil/sti/citations/ADA324218

61 Zhu, D., Chen, W., Liu, S. (2022). Photonics-assisted radio frequency memory. *Journal of Lightwave Technology* 40 (3): 624–631. https://doi.org/10.1109/JLT.2021.3122141

62 Nguyen, T.A., Chan, E.H.W., Minasian, R.A. (2014). Photonic radio frequency memory using frequency shifting recirculating delay line structure. *Journal of Lightwave Technology* 32 (1): 99–106. https://doi.org/10.1109/JLT.2013.2290294

63 Aseeri, M.A.S., Alasows, A.A., Ahmad, M.R. (2017). Design of DRFM system based on FPGA with high resources. *International Design & Test Sympposium (IDT)*,

Hammamet, Tunisia (18–20 December 2016). New York, US: IEEE. https://doi.org/10.1109/IDT.2016.7843036

64 Yin, L., Xie, M., Li, H. et al. (2017). Design and implementation of the digital radio frequency memory system based on advanced extensible interface 4.0. *9th International Congress on Image and Signal Processing, BioMedical Engineering and Informatics (CISP-BMEI)*, Datong, China (15–17 October 2016). New York, US: IEEE. https://doi.org/10.1109/CISP-BMEI.2016.7852918

65 Pace, P.E. (2022). *Developing digital RF memories and transceiver technologies for electromagnetic warfare*. Norwood, MA: Artech House, Section 2.5, pp. 61–63.

66 Gonçalves Licursi de Mello, R. (2023). Metasurfaces and blinking jamming: convergent study, comparative analysis, and challenges. *MDPI Micromachines* 14 (7): 1405. https://doi.org/10.3390/mi14071405

67 Knott, E.F., Schaeffer, J.F., Tulley, M.T. (2004). *Radar cross section*, 2nd ed. Raleigh, NC: Scitech Publishing, Section 7.2.4, pp. 273.

68 Sengupta, S., Council, H., Jackson, D.R. (2020). Active radar cross section reduction of an object using microstrip antennas. *Radio Science* 55 (2): 1–20. https://doi.org/10.1029/2019RS006939

69 Smith, G.E., Cammenga, Z., Mitchell, A. (2016). Experiments with cognitive radar. *IEEE Aerospace and Electronic Systems Magazine* 31 (12): 34–46. https://doi.org/10.1109/MAES.2016.150215

70 Lončar, J., Šipuš, Z. (2020). Challenges in design of power-amplifying active metasurfaces. *Proceedings of the International Symposium ELMAR*, Zadar, Croatia (14–15 September 2020). New York, US: IEEE. https://doi.org/10.1109/ELMAR49956.2020.9219017

71 Kim, M., Rosenberg, J.J., Smith, R.P. et al. (1991). A grid amplifier. *IEEE Microwave and Guided Wave Letters* 1 (11): 322–324. https://doi.org/10.1109/75.93899

72 Taravati, S., Khan, B.A., Gupta, S. et al. (2017). Nonreciprocal nongyrotropic magnetless metasurface. *IEEE Transactions on Antennas and Propagation* 65 (7): 3589–3597. https://doi.org/10.1109/TAP.2017.2702712

73 Chen, L., Ma, Q., Jing, H.B. et al. (2019) Space-energy digital-coding metasurface based on an active amplifier. *Physical Review Applied* 11 (054051): 1–6. https://doi.org/10.1103/PhysRevApplied.11.054051

74 Wang, X., Han, J., Tian, S. et al. (2022). Amplification and manipulation of nonlinear electromagnetic waves and enhanced nonreciprocity using transmissive space-time-coding metasurface. *Advanced Science* 9 (2105960): 1–13. https://doi.org/10.1002/advs.202105960

75 Nayeri, P., Yang, F., Elsherbeni, A.Z. (2018). *Reflectarray antennas: theory, designs, and applications*. New York, NY: Wiley-IEEE Press.

76 Bialkowski, M.E., Robinson, A.W., Song, H.J. (2002). Design, development, and testing of X-band amplifying reflectarrays. *IEEE Transactions on Antennas and Propagation* 50 (8): 1065–1076. https://doi.org/10.1109/TAP.2002.801393

77 Song, H.J., Bialkowski, M.E. (2001). Spatial power combiner using an active reflectarray of dual-feed aperture coupled microstrip patch antennas. *Proceedings of the IEEE Antennas and Propagation Society International Symposium*, Boston, MA, USA (08–13 July 2001), New York, US: IEEE. https://doi.org/10.1109/APS.2001.959578

78 Chaloun, T., Menzel, W., Tabarani, F. et al. (2014). Wide-angle scanning active transmit/receive reflectarray. *IET Microwaves, Antennas & Propagation* 8 (11): 811–818. https://doi.org/10.1049/iet-map.2013.0704

79 Landsberg, N., Socher, E. (2016). An F-band reflection amplifier using 28 nm CMOS FD-SOI technology for active reflectarrays and spatial power combining applications. *Proceedings of the IEEE MTT-S International Microwave Symposium (IMS)*, San Francisco, CA, USA (22–27 May 2016), New York, US: IEEE. https://doi.org/10.1109/MWSYM.2016.7540288

80 Landsberg, N., Socher, E. (2017). A Low-power 28-nm CMOS FD-SOI reflection amplifier for an active F-band reflectarray. *IEEE Transactions on Microwave Theory and Techniques* 65 (10): 3910–3921. https://doi.org/10.1109/TMTT.2017.2695181

81 Caris, M., Brehm, T., Palm, S. et al. (2017). High resolution dual-channel SAR-system for airborne applications. *Proceedings of the 18th International Radar Symposium (IRS)*, Prague, Czech Rep. (28–30 June 2017), New York, US: IEEE. https://doi.org/10.23919/IRS.2017.8008093

82 Reigber, A., Schreiber, E., Trappschuh, K. et al. (2020). The high-resolution digital-beamforming airborne SAR system DBFSAR. *MDPI Remote Sensing* 12 (1710): 1–20. https://doi.org/10.3390/rs12111710

83 Lončar, J., Grbic, A., Hrabar, S. (2019). Ultrathin active polarization-selective metasurface at X-band frequencies. *Physical Review B: Condensed Matter* 100 (075131): 1–7. https://doi.org/10.1103/PhysRevB.100.075131

84 Dong, Y., Kang, W., Sima, B. et al. (2022). Retroreflector with polarization isolation based on nonreciprocal metasurface. *IEEE Antennas and Wireless Propagation Letters* 21 (10): 1940–1944. https://doi.org/10.1109/LAWP.2022.3186334

85 Yang, J., Yang, W., Qu, K. et al. (2023). Active polarization-converting metasurface with electrically controlled magnitude amplification. *Optic Express* 31 (18): 28979–28986. https://doi.org/10.1364/OE.499458

86 Wang, X., Xia, D., Li, G. et al. (2023). Design and waveguide measurement of 2-bit reconfigurable amplification metasurface cell. *IEEE Antennas and Wireless Propagation Letters* 22 (9): 2090–2094. https://doi.org/10.1109/LAWP.2023.3275558

87 Yu, N., Genevet, P., Kats, M.A. et al. (2011). Light propagation with phase discontinuities: generalized laws of reflection and refraction. *Science* 334 (6054): 333–337. https://doi.org/10.1126/science.1210713

88 Wu, B., Sutinjo, A., Potter, M.E. et al. (2008). On the selection of the number of bits to control a dynamic digital MEMS reflectarray. *IEEE Antennas and Wireless Propagation Letters* 7: 183–186. https://doi.org/10.1109/LAWP.2008.920908

89 Okogbaa, F.C., Ahmed, Q.Z., Khan, F.A. et al. (2022). Design and application of intelligent reflecting surface (IRS) for beyond 5G wireless networks: a review. *MDPI Sensors* 22 (2436): 1–24. https://doi.org/10.3390/s22072436

90 Puspitasari, A.A., Lee, B.M. (2023). A survey on reinforcement learning for reconfigurable intelligent surfaces in wireless communications. *MDPI Sensors* 23 (2554): 1–23. https://doi.org/10.3390/s23052554

91 Long, R., Liang, Y.-C., Pei, Y. et al. (2021). Active reconfigurable intelligent surface-aided wireless communications. *IEEE Transactions on Wireless Communications* 20 (8): 4962–4975. https://doi.org/10.1109/TWC.2021.3064024

92 Zhi, K., Pan, C., Ren, H. et al. (2022). Active RIS versus passive RIS: which is superior with the same power budget? *IEEE Communications Letters* 26 (5): 1150–1154. https://doi.org/10.1109/LCOMM.2022.3159525

93 Kang, Z., You, C., Zhang, R. (2023). Active-IRS-aided wireless communication: fundamentals, designs and open issues. *ArXiv (preprint)*. https://doi.org/10.48550/arXiv.2301.04311

94 Zhang, Z., Dai, L., Chen, X. et al. (2023). Active RIS vs. passive RIS: which will prevail in 6G? *IEEE Transactions on Communications* 3 (71): 1707–1725. https://doi.org/10.1109/TCOMM.2022.3231893

95 Taravati, S., Eleftheriades, G.V. (2021). Full-duplex reflective beamsteering metasurface featuring magnetless nonreciprocal amplification. *Nature Communications* 12 (4414): 1–11. https://doi.org/10.1038/s41467-021-24749-7

96 Razavi, B. (2011). *RF microelectronics*, 2nd ed. Upper Saddle River, NJ: Prentice Hall, Section 2.3, Eq. 2.132, pp. 54.

97 Neri, F. (2018). *Introduction to electronic defense systems*, 3rd ed. Boston, MA: Artech House, Section 5.4.20, pp. 449.

98 Fountain, T., Rudolph, M. (2022). Development of cognitive EW datasets. *Association of Old Crows* (16 June 2022). https://www.crows.org/general/custom.asp?page=DevelopmentofCognitiveEW (accessed 08 August 2023).

99 Ciao Wireless, Inc. https://www.ciaowireless.com/ (accessed 08 August 2023).

100 Gonçalves Licursi de Mello, R., De Sousa, F.R. (2019). Non-analytical direction-finding method as a key step in pursuing low size, weight, costs, and computational power in the deinterleaving of radar pulses. *IET Radar, Sonar & Navigation* 13 (11): 1876–1882. https://doi.org/10.1049/iet-rsn.2018.5011

6

Metasurfaces for Electronic Support Applications

The elegance of EW lies in the fact that one should be wise enough to accept the conditions they and their opponents are subjected to. From the moment an emitter starts to operate, its platform is at risk of being detected and shot. Conversely, the designer can exploit in the respective receiver the fact that the adopted waveform is known. This situation allows the use of low-amplitude signals, reducing the probability of interception by adversaries. In LPI radars, for example, we perform this procedure by using coded pulses that can provide a signal processing gain G_{match}, as Section 5.1 explained.

In this scenario, the tactical circumstances under which ES systems find themselves are rather particular. The nature of ES, responsible for monitoring the spectrum to obtain pertinent information for EA and EP purposes [1], allows the use of completely passive, furtive systems. On the other hand, in general, ES systems must handle signals without prior information. These conditions have a series of consequences in the system architecture [2]. An important one is the need for antennas able to cover a wide range of frequencies where opponents will potentially operate [3].

Figure 6.1 illustrates an EW scenario where three radars illuminate a platform with an ES system. In this case, the ES system must cover 1.6 and 10.0 GHz. Further, it receives all the signals concurrently, and must deinterleave them prior to the identification of each radar. Traditional strategies usually start deinterleaving from the DOA of intercepted signals. The main reason is that, even though some radars can vary the parameters of their signals, they cannot vary significantly their angular positions with respect to the ES system in the time frame of the radar pulses [4–8]. In Fig. 6.1, the ES system can clearly discriminate signals of radar 1 from those of radars 2 and 3 using the azimuthal DOA because the former is at $\Phi = 45°$ and the others at $\Phi = 300°$. On the other hand, another method is required to discriminate between radars 2 and 3. The classification of their coded pulses can be one solution since both of them are LPI and operate at the same carrier

Metasurface-Driven Electronic Warfare, First Edition. Rafael Gonçalves Licursi de Mello.

Published 2025 by John Wiley & Sons, Inc.
Companion website: www.wiley.com/go/licursidemello/metasurfacedrivenelectronicwarfare1

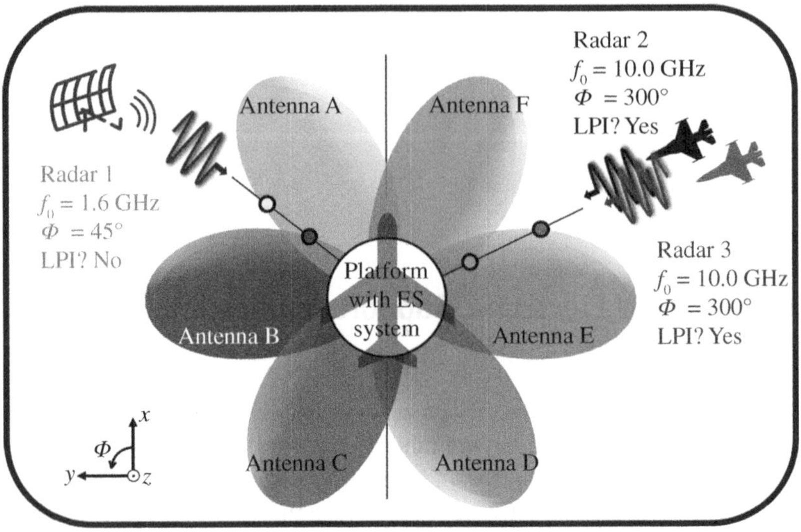

Fig. 6.1 EW scenario with ES system receiving signals from three radars: radar 1 is separated from radars 2 and 3 in both angle and frequency domains

frequency $f_0 = 10.0$ GHz. Prominent AI-based techniques typically demand high SNR levels to this end, as a quick literature review indicates [9–13].

ES systems widely adopt amplitude-based direction-finding methods because they are robust to frequency changes and useful in a wide range. Typically, we require multiple unidirectional antennas with a well-behaved radiation pattern over frequency to achieve a high POI in the angle domain [14–20]. In Fig. 6.1, six antennas receiving signals with different amplitudes, as indicated by the dots, instantaneously cover the whole azimuth, which allows the ES system to resolve the azimuthal DOA. If we desire the elevation DOA, we should deploy antennas covering this dimension.

The theory of frequency-independent antennas organized by V.H. Rumsey [21] can provide devices proper for ES systems. Traditionally, these antennas are backed by a metal cavity filled with absorbers to provide unidirectional radiation patterns in wide frequency ranges. However, the cavity leads to bulky devices that are inappropriate for fitting the platform fuselage in some applications. Besides, this maneuver wastes energy, reducing the radiation efficiency and not contributing to a high antenna gain. This is not aligned with the challenges of detecting LPI radars at all.

This chapter will discuss metasurface-based methods for obtaining low-profile, high-gain antennas appropriate for ES systems. We can obtain unidirectional patterns in multiple, wide frequency ranges by associating wideband radiating elements and multiband reflectors [22–24]. We use multiband AMCs to this end [25–27]. We obtain significant enhancements in the antenna gain, which provides advantages in terms of maximum ES detection range and the accuracy of classification of LPI signals. As a side effect, some blind zones arise between the covered bands. Still, the AMC controls the frequency response of the final device in the proposed method, allowing the designer to avoid leaving radars of interest uncovered. Further, this chapter anticipates the use of reconfigurable AMCs [28–32] to switch between diverse operating states and keep full spectrum coverage (even if not instantaneously).

Here, whether the device is tunable or not, we obtain performance at the cost of POI in frequency–a valuable ES asset. However, from an antenna perspective, there is not much that can be done to face the advancements in LPI radars. Relaxing the full-coverage constraint to specialize in specific missions is in line with current trends in Defense, including the platforms themselves. After the end of the Cold War, Defense funding gradually declined as societies and authorities felt that new large-scale conflicts were unlikely. In this context, multi-role platforms became increasingly common. However, recent conflicts show that the character of war is changing again. Nagorno-Karabakh, Ukraine, and Gaza have employed numerous cheaper, highly specialized unmanned platforms to address specific missions [33–37]. Recently, the dispersal of EW assets rather than their concentration in major platforms has been a strong trend [38]. So interesting this trend may look like, in fact, technology is just meeting a decades-old military principle:

> An army should always be so distributed that its parts can aid each other and combine to produce the maximum possible concentration of force at one place, while the minimum force necessary is used elsewhere to prepare the success of the concentration.
>
> —B. H. Liddell Hart, military theorist

The discussed method is well suited for this scenario. If we cannot relax the frequency POI constraint at all, we can still use metasurfaces. This chapter will also present inhomogeneous reflectors providing moderate gain enhancements without spectral blind zones. First, we will see the role of ES antennas in the radar/ES combat framework. Then, we will review frequency-independent antennas in their traditional form. Next, we will explain traditional and innovative reflection schemes with metasurface reflectors. Our EW scenario simulator will introduce the impact of the antenna performance in the ES problem and then we will propose exercises.

6.1 The One-Way Equation and ES Detection

The one-way propagation from the radar Tx antenna to the Rx antennas of the ES system provides an advantage to the latter. Consider an ES system embedded in an arbitrary platform. The effective area of an ES system Rx antenna can collect the power density (4.3) of a radar signal, which leads to a total power of:

$$S = S_v + S_h \tag{6.1}$$

where

$$S_m = \frac{P_t \lambda^2}{\left(4\pi r_{rdrTx}\right)^2} G_{t,m}\left(\theta_{plt}, \Phi_{plt}\right) G_{r,m}\left(\theta'_{rdr}, \Phi'_{rdr}\right) \tag{6.2}$$

and

- $G_{r,m}$ is the realized gain of the ES system Rx antenna in the horizontal or vertical polarization ($m = h$ or v) [dBi];
- θ'_{rdr}, Φ'_{rdr} is the angular position of the radar Tx antenna referred to the orientation axes of the ES antenna [rad].

Equation (6.2) is also known as the Friis equation for the transmission of signals through the free space [39].

For proper operation, a minimum signal-to-noise ratio $SNR_{\min} = S_{\min}/N_0$ should exist at the heart of the ES processing, where $N_0 = BWk_bT$ in an impedance-matched receiver. For illustration purposes, consider a vertically polarized ES system Rx antenna with perfect polarization purity. From (6.2), we obtain the maximum range $r_{\max}$ of ES detection as:

$$r_{\max} = \sqrt{\frac{P_t \lambda^2}{(4\pi)^2 SNR_{\min} BWk_bT} G_{t,v}\left(\theta_{plt}, \Phi_{plt}\right) G_{r,v}\left(\theta'_{rdr}, \Phi'_{rdr}\right)} \tag{6.3}$$

A first comparison between (6.3) and (5.4) makes the ES advantage evident because a square root is present in the former, whereas a fourth root is in the latter. However, that does not mean that ES systems always detect radars first. The reason is that the other terms in the equations can make all the difference.

First, the lack of prior information about intercepted signals avoids in (6.3) the benefit of a signal processing gain like G_{match} in (5.4). Also, the radar receiver can operate with a BW just wide enough for its own signals. On the other hand, ES systems should cover a range where opponent radars may operate, which may require a wide BW, depending on the chosen ES architecture.

Legacy architectures include crystal video receivers, which use detectors to amplitude-demodulate signals in the whole frequency range of interest instantaneously [40]. The wide BW provides a POI in frequency of 100% but leads in this case to a noise level N_0 unacceptable to tackle LPI radars. An alternative is the

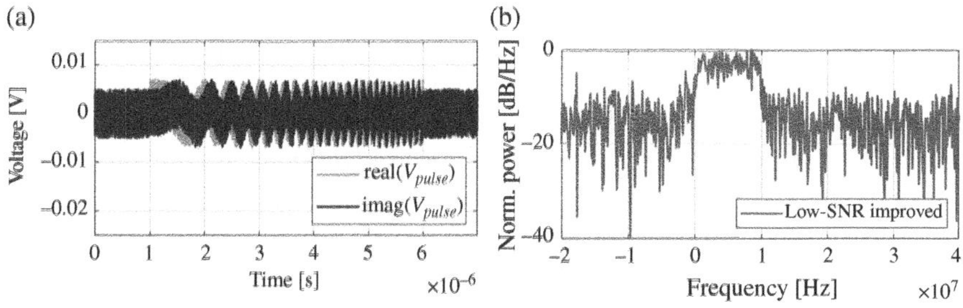

Fig. 6.2 Low-SNR chirp pulse improved by 6 dB: (a) time domain; (b) spectral power density normalized per maximum value

super-heterodyne receiver, which sweeps the frequency range of interest with a narrowband filter. In this way, N_0 is low, but we sacrifice the POI in frequency [41]. Modern trends in this matter are the adoption of receivers based on photonics, which can sweep the spectrum at an impressive rate [42–44]. Fully digital channelized architectures [45] also allow a good noise level and POI in frequency.

Another critical difference between (5.4) and (6.3) is the realized gain of the Rx antennas, which we will exploit. Radars can use extremely directive antennas, allowing for gain values of the order of 30 dBi or better. In the ES case, however, a certain number of antennas must cover a specific angular sector instantaneously, as Fig. 6.1 made evident. Thus, besides being unidirectional, the radiation patterns must have an appropriate directivity so that the system provides a desired angular POI [14–20].

Although these circumstances constrain the directivity of the ES antenna, the antenna gain is still a variable that the designer can work on, especially by avoiding losses from absorbers. We recall Fig. 5.2c,d, where we depicted a chirp pulse considerably attenuated before arriving at an arbitrary receiver. The signal was almost indiscernible from noise. We illustrate in Fig. 6.2a,b the very same signal but considering an improvement of 6 dB on its intensity due to a better gain in the Rx antenna. The noise level here is the same as in Fig. 5.2. Nevertheless, discriminating signal and noise is much easier in Fig. 6.2a,b than in Fig. 5.2c,d, which has a strong impact on the accuracy of AI classification methods [9–13]. Furthermore, increasing the ES antenna realized gain $G_{r,m}$ by 6 dB in (6.3) doubles the maximum ES detection range $r_{\max}$. This can be decisive when it comes to detecting radars before being detected.

6.2 Introduction to Frequency-Independent Antennas

The challenge of designing wideband antennas goes back to the 1940s when typical BWs of 2 or 3 : 1 were not enough for some applications anymore – like ES. At that time, researchers envisioned new antenna geometries, such as the discone

[46], the conical helix [47], the Archimedean spiral [48], and the log-periodic antenna [49, 50]. Since then, the scientific community has created much mystery around these antennas and their extraordinary capabilities. A significant reason is that, in traditional antennas, the relationship between antenna dimensions and wavelength is critical. Hence, at first glance, devices working independently of frequency do not seem reasonable. In this context, the derivations of V.H. Rumsey [21] are enticing in that they give the impression that mathematics is about to outwit physics. Next, we will review two major principles ruling the operation of frequency-independent antennas. Far from a comprehensive review, our goal is rather to provide a basic understanding of some specific antennas that will serve to illustrate concepts later in this chapter.

6.2.1 Scaling Principle

We saw in Section 3.2.5 that we can interchange a set of electric and magnetic radiation sources $\boldsymbol{J}_i$ and $\boldsymbol{M}_i$ and equivalent current densities $\boldsymbol{J}_s$ and $\boldsymbol{M}_s$ over the surface of an enclosing volume v. Then, if we force the fields inside volume v to be $\boldsymbol{E}_v = 0$ and $\boldsymbol{H}_v = 0$, we can calculate the fields $\boldsymbol{E}_0$ and $\boldsymbol{H}_0$ outside the volume from (3.16) and (3.17) using Love's principle [51].

Now, consider a dipole antenna made of two PEC parts spaced by an infinitesimal gap between the terminals. We can enclose the dipole with the aforementioned volume v with surface following the PEC surfaces and running across the gap between the terminals, as Fig. 6.3a illustrates. The configuration of PEC material and infinitesimal gap leads to $\boldsymbol{E}_v = 0$ and $\boldsymbol{H}_v = 0$ inside volume v. Outside it, no sources exist ($\boldsymbol{J}_i = q_{ev} = 0$, and $\boldsymbol{M}_i = q_{mv} = 0$). Moreover, the medium is lossless ($\boldsymbol{J}_c = 0$) and non-dispersive with intrinsic impedance $\eta = \sqrt{\mu/\varepsilon}$.

We are interested in calculating the relationship between $\boldsymbol{E}_0$ and $\boldsymbol{H}_0$ but in a space whose unit of length is the wavelength λ. The nabla operator of this space relates to the nabla of the conventional cartesian coordinate system as:

$$
\begin{aligned}
&\frac{\partial}{\partial x} \xrightarrow{x=\lambda x'} \frac{1}{\lambda}\frac{\partial}{\partial x'}\\
&\frac{\partial}{\partial y} \xrightarrow{y=\lambda y'} \frac{1}{\lambda}\frac{\partial}{\partial y'}\\
&\frac{\partial}{\partial z} \xrightarrow{z=\lambda z'} \frac{1}{\lambda}\frac{\partial}{\partial z'}\\
&\nabla \xrightarrow[\substack{y=\lambda y'\\ z=\lambda z'}]{x=\lambda x'} \frac{1}{\lambda}\nabla'
\end{aligned}
\tag{6.4}
$$

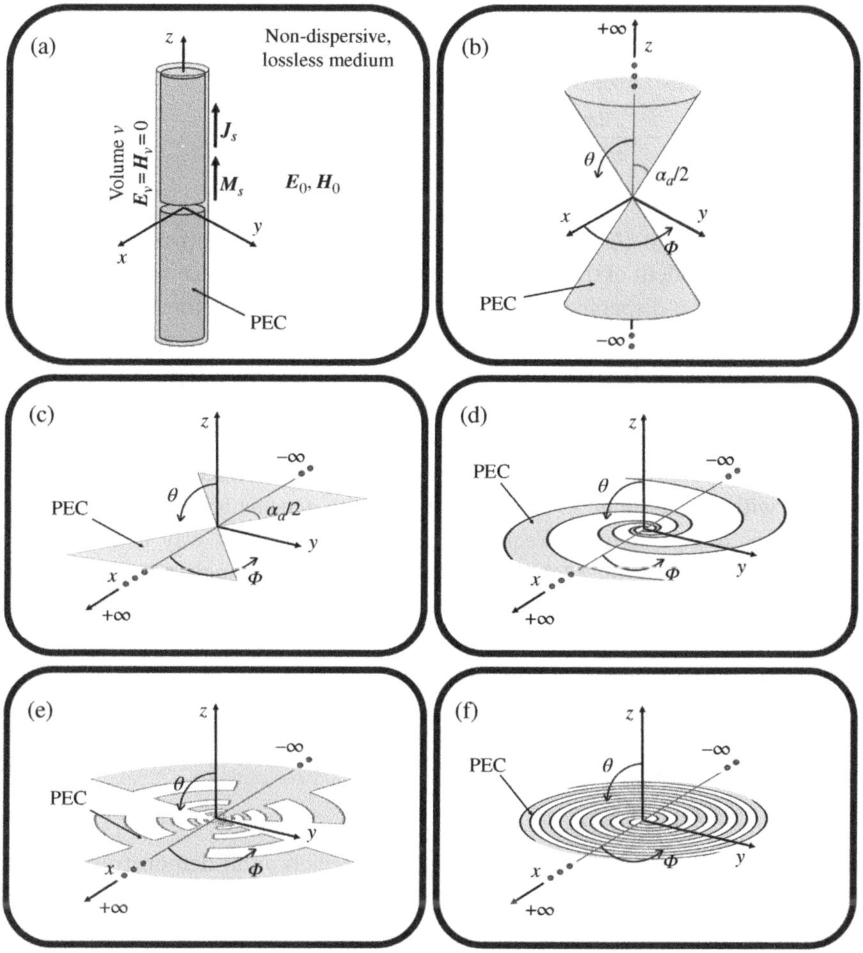

Fig. 6.3 Frequency-independent antennas: (a) dipole surrounded by equivalent currents; (b) biconical antenna; (c) bow-tie; (d) equiangular spiral; (e) log-periodic; (f) Archimedean spiral

Replacing (6.4) together with (2.35) and (2.30) in Ampère's law (2.17), we have:

$$\begin{aligned}
\frac{1}{\lambda}\nabla' \times \boldsymbol{H}_0 &= j\omega\varepsilon \boldsymbol{E}_0 \\
\nabla' \times \boldsymbol{H}_0 &= j\frac{2\pi}{\omega\sqrt{\mu\varepsilon}}\omega\varepsilon \boldsymbol{E}_0 \\
&= \eta^{-1}(j2\pi \boldsymbol{E}_0)
\end{aligned} \tag{6.5}$$

Likewise, Faraday–Lenz's law (2.18) becomes:

$$\begin{aligned} \frac{1}{\lambda}\nabla' \times \boldsymbol{E}_0 &= -j\omega\mu\boldsymbol{H}_0 \\ \nabla' \times \boldsymbol{E}_0 &= -j\frac{2\pi}{\omega\sqrt{\mu\varepsilon}}\omega\mu\boldsymbol{H}_0 \\ &= \eta(-j2\pi\boldsymbol{H}_0) \end{aligned} \tag{6.6}$$

It is evident from (6.5) and (6.6) that the relationship between $\boldsymbol{E}_0$ and $\boldsymbol{H}_0$ becomes independent of the angular frequency ω when the space unit is λ. Hence, in a lossless, non-dispersive medium, if we scale all dimensions of the spatial domain with λ, the relationship between $\boldsymbol{E}_0$ and $\boldsymbol{H}_0$ does not change. Putting more specifically, if we double or triple all dimensions of an antenna together with λ, the performance of the antenna (radiation pattern, polarization, input impedance, etc.) remains the same. This principle is known as the scaling principle [52].

Now, the problem is to find antenna geometries whose scaled versions perfectly overlap with one another so that we can scale the wavelength freely; that is, the behavior is independent of frequency. The straightforward way is to base the antenna development on angled shapes which we call the angle principle. Examples include geometries originating from a common point, like $P(0, 0, 0)$, and described by functions of θ and/or Φ with r as the independent variable. One of them is the infinite biconical antenna, which we define as:

$$\begin{aligned} &r \in [0, +\infty[\\ &\theta = \left\{+\frac{\alpha_a}{2}, \pi - \frac{\alpha_a}{2}\right\} \\ &\Phi \in [0, 2\pi] \end{aligned} \tag{6.7}$$

where $\alpha_a/2$ is the cone flare angle in Fig. 6.3b. Schelkunoff demonstrated much before the work of Rumsey that the input impedance of one such antenna would be independent of frequency [53]. His derivation, however, neglected the current reflections that would occur if the antenna dimensions were not infinite.

The bulkiness of the biconical antenna makes it inconvenient to fit the fuselage of air/spacecraft or embed drones and missile seekers. To this end, we can project its shape over a plane. This maneuver leads to the bow-tie antenna, which may also have a frequency-independent behavior. This is an important antenna for the reader because we will use it to illustrate some concepts later in this chapter. We can define this planar geometry as:

$$\begin{aligned} &r \in [0, +\infty[\\ &\theta = \frac{\pi}{2} \\ &\Phi \in \left\{\left[-\frac{\alpha_a}{2}, +\frac{\alpha_a}{2}\right], \left[\pi - \frac{\alpha_a}{2}, \pi + \frac{\alpha_a}{2}\right]\right\} \end{aligned} \tag{6.8}$$

D. Rutledge et al. [54] show that the input impedance of an infinite bow-tie antenna depends only on the flare angle α_a, indicated in Fig. 6.3c.

Unfortunately, we must truncate practical antennas somewhere before $r = +\infty$, of course. This does not necessarily pose a problem if currents vanish to zero before arriving at the structure's edges because the truncated version will give the same result as the infinite one in this case. This last statement is known as the truncation principle [55].

Typically, practical bow-tie antennas do not meet the truncation principle, which means that, in the size they are used in practice, they do not work as the infinite version. It was not until the mid-1980s that a rigorous derivation showed that the current distribution over an arbitrary-length bow-tie antenna is a superposition of many modes [56]. This derivation is mainly useful for computational approaches and provides little physical insight. To fill this gap, R.G.L. Mello et al. [57] state and test some hypotheses allowing the grouping of all the modes in two major current models. The results show that the shape of the edge of practical bow-tie antennas (triangular, rounded, or grooved) has an impact on performance. Further, these antennas present a good purity of polarization (which in turn is linear) with a gain of around 3 dBi. The radiation pattern is quasi-omnidirectional. These features hold in a BW of the order of 100%.

Figure 6.4a shows a rounded-edge bow tie with a balun and SMA connector for feeding purposes. Fig. 6.4b depicts the normalized radiation patterns in both main planes as simulated and measured at 2.40 GHz [57]. Indeed, we observe a quasi-omnidirectional pattern with low cross-polarization levels. We conclude therefore that we cannot use this kind of antenna, as it is, to estimate the DOA of intercepted signals with methods that require unidirectional patterns [14–20].

One way of forcing currents to vanish so as to meet the truncation principle is to add another feature to the antenna development: curved shapes. We can exploit this feature in applications that allow a rotation in the radiation pattern and polarization with frequency (that is, the antenna will not be totally independent of frequency). V.H. Rumsey [58] explains the general effects of curvatures in the current decay. Scaling an angled curve means that the new curve should coincide with the old one with expansions and rotations, which we can formally write in the plane $\theta = \pi/2$ as:

$$K_1 r(\Phi) = r(\Phi + K_2) \tag{6.9}$$

where K_1 and K_2 are the expansion and rotation factors, respectively. Taking the derivative of (6.9) with respect to Φ and also to K_2 leads to:

$$K_1 \frac{d[r(\Phi)]}{d\Phi} = \frac{d[r(\Phi + K_2)]}{d\Phi} \tag{6.10}$$

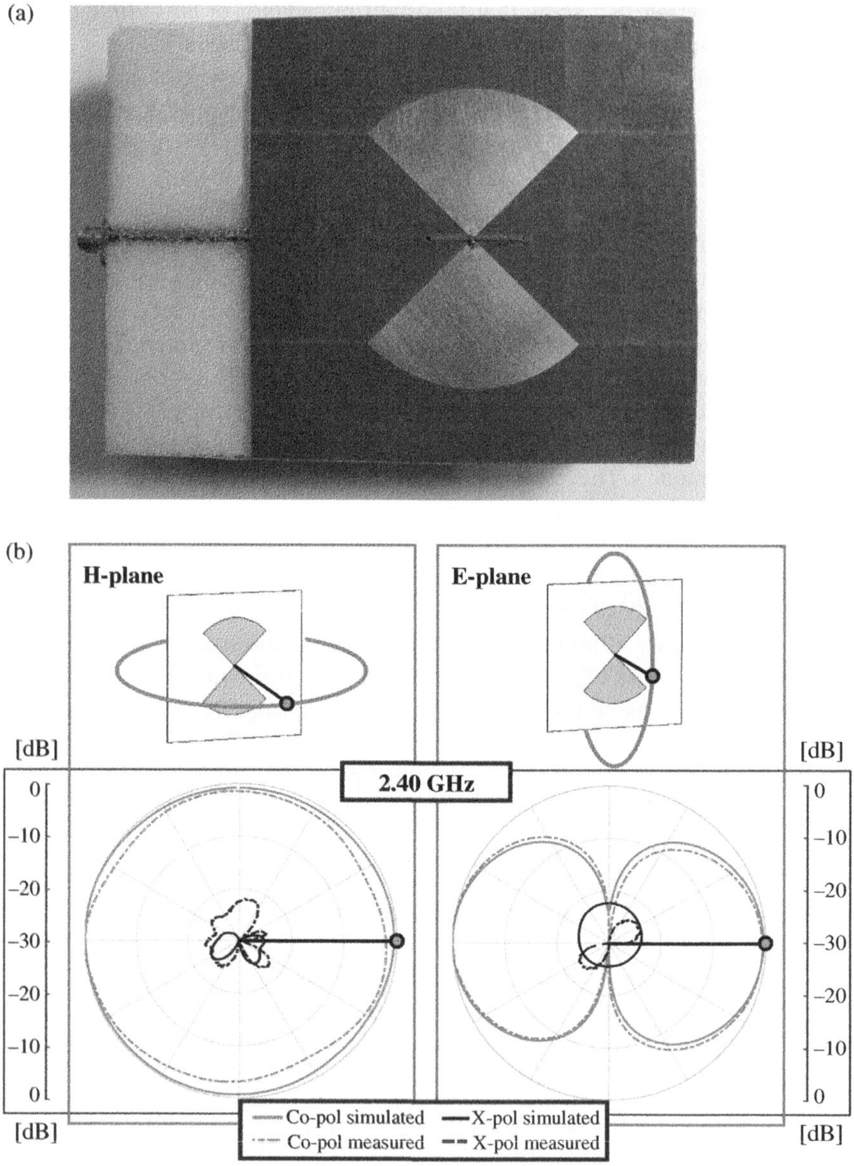

Fig. 6.4 Rounded-edge bow-tie antenna: (a) device fabricated in the framework of [57]; (b) simulated and measured radiation patterns; dots indicate the broadside direction in all the graphs. "Reproduced from Gonçalves Licursi de Mello et al. [57]/John Wiley & Sons/CC BY 4.0."

$$\frac{dK_1}{dK_2} r(\Phi) = \frac{d[r(\Phi + K_2)]}{dK_2} \tag{6.11}$$

However, fundamentally:

$$\frac{d[r(\Phi + K_2)]}{d\Phi} = \frac{d[r(\Phi + K_2)]}{dK_2} \tag{6.12}$$

which allows the following derivation:

$$\begin{aligned} K_1 \frac{d[r(\Phi)]}{d\Phi} &= \frac{dK_1}{dK_2} r(\Phi) \\ \int \frac{d[r(\Phi)]}{r(\Phi)} &= \frac{1}{K_1} \frac{dK_1}{dK_2} \int d\Phi \\ \Phi &= \frac{1}{a} \ln r - \Phi_0 \\ r &= r_0 e^{a\Phi} \end{aligned} \tag{6.13}$$

where

- Φ_0 and r_0 are, respectively, the initial angle and radius, [rad and m];
- $a = dK_1/(K_1 dK_2)$ is the spiral periodicity constant [unitless].

Interestingly, (6.13) represents an equiangular spiral curve. Figure 6.3d illustrates a planar antenna based on this geometry. The antenna is made with four curves with periodicity constant $a = 0.35$, $r_0 = 1.7$ mm, and respective Φ_0 values of $\pi/2 \pm \pi/3$, and $3\pi/2 \pm \pi/3$ over the plane $\theta = \pi/2$. The above development shows why this antenna shape has been a popular choice for wideband applications since at least the late 1950s [59].

In the equiangular spiral antenna, currents follow the curved shapes of the arms, propagating until they reach an "active region" defined by a circle whose circumference is about one wavelength. In this region, the coupling between the energy of adjacent arms is strong, which maximizes radiation [60]. The polarization is circular. The antenna provides typical gain values around 5 dBiC. Its maximum dimensions (which state the lowest frequency where currents vanish before reaching the edges) and the precision of construction of the center terminals (which sets the highest frequency where the currents flow therein according to the intended shape) mainly define its BW. Typically, we can achieve BWs of the order of 20 : 1 with a good axial ratio. Yet, the radiation patterns are bidirectional, and, like the bow-tie antenna, we cannot use the standalone spiral for DOA estimation using methods that require unidirectional patterns [14–20].

Rather than continuous frequency independence across the spectrum, many applications require only a well-behaved performance repeated throughout many smaller frequency bands. If this continuity constraint is relaxed, we can periodically expand an original geometry. Then, each section becomes the "active region" when the operation occurs in a specific band. If the expansion periodicity follows equations similar to (6.13), the geometry is usually called a log-periodic antenna. R. DuHamel et al. [49, 50] study a particular class of these structures, obtaining linearly polarized bidirectional radiation patterns with gain values around 5 dBi with a BW of up to 10 : 1. Figure 6.3e represents the planar version of such a shape.

The antennas seen hitherto are composed of two radiating elements. With four radiating elements in an orthogonal disposition, different signal-combining schemes are possible. In this manner, we can emulate all polarizations [61]. Figure 6.5 shows a dual-polarized log-periodic antenna we developed in the framework of a previous work [8]. R. Sammeta et al. [62] propose a similar antenna, emulating a left-handed circular polarization (LHCP). They measure a bidirectional radiation pattern with a gain of around 5 dBiC with a BW of 6 : 1. Not only log-periodic antennas can adopt this disposition. The literature reports many crossed bow-tie antennas [63, 64].

Among other planar antennas based on the scaling principle, the sinuous antenna represents a linearly polarized version of the spiral antenna [65], which is also suitable for polarization diversity in ES systems [66]. Its radiation pattern is also bidirectional. Moreover, fractal antennas have some similarities with log-periodic antennas. However, they comprise nested repetitive geometries rather than logarithmically expanded shapes. We can design fractal antennas to be wideband and obtain quasi-omnidirectional patterns [67, 68].

Fig. 6.5 Log-periodic antenna developed in the framework of [8]

6.2.2 Self-Complementariness Principle

Another mechanism that can enable a frequency-independent antenna behavior is the use of self-complementary structures. In Fig. 4.3a, we had the first contact with a theoretical checkerboard surface that disposed PEC and PMC squares interchangeably. Consider now that one such a structure has infinite dimensions and is embedded in a source-free ($\boldsymbol{J}_i = q_{ev} = 0$, and $\boldsymbol{M}_i = q_{mv} = 0$), lossless ($\boldsymbol{J}_c = 0$), non-dispersive medium with intrinsic impedance $\eta = \sqrt{\mu/\varepsilon}$.

Figure 6.6 illustrates a connection between adjacent PMC squares with width $w \ll \lambda$ and length $l \ll \lambda$. We impress somehow a surface magnetic current density $\boldsymbol{M}_s$ over both the front and back sides of this connection, as the inlet indicates. Consider that the current distribution is uniform along the surface ($\boldsymbol{M}_s$ is constant over the connection) thereby resulting in a total magnetic current $I_m = 2M_s w$ [V] passing through the connection. According to the boundary condition (2.14) and the vector orientation indicated in Fig. 6.6, a tangential electric field represented by $-\hat{\mathbf{a}}_n \times \boldsymbol{E}_T = \boldsymbol{M}_s$ exists due to the presence of the magnetic current.

Now, we define points A and B on one side of the surface and points A' and B' exactly on the opposite side, respectively. A' and B' are not shown in Fig. 6.6. In this way, we form a surface $S_{AA'B'BA}$ with boundaries defined by these points around the PMC connection. Stoke's theorem [69] states that the line integral of $\boldsymbol{E}_T$ along the path $AA'B'BA$ is equal to the surface integral of $\nabla \times \boldsymbol{E}_T$ over $S_{AA'B'BA}$:

$$\oint \boldsymbol{E}_T \, . \, d\boldsymbol{l} = \oiint (\nabla \times \boldsymbol{E}_T) \, . \, d\boldsymbol{s} \tag{6.14}$$

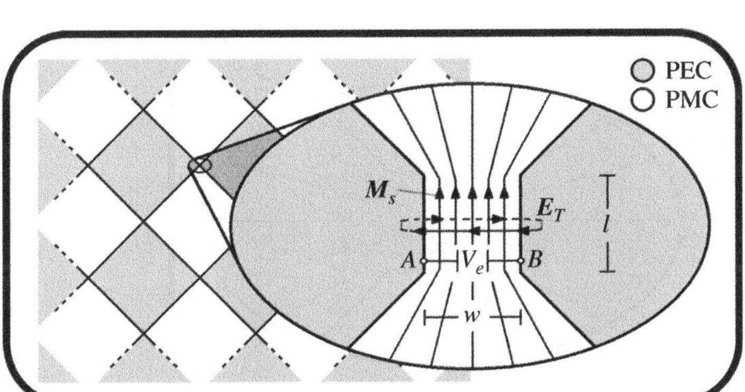

Fig. 6.6 Connection between two PMC squares in a checkerboard surface

From Faraday–Lenz's law (2.18) and considering that the tangential magnetic field is zero over the PMC, we obtain:

$$\oint \boldsymbol{E}_T \cdot d\boldsymbol{l} = -\oiint \boldsymbol{M}_s \,.\, d\boldsymbol{s} \\ = -2M_s w = -I_m \tag{6.15}$$

We are interested now in obtaining the potential difference V_e between A and B which, according to the definition [70], is:

$$V_e = -\int_B^A \boldsymbol{E}_T \cdot d\boldsymbol{l} = -\frac{\oint \boldsymbol{E}_T \cdot d\boldsymbol{l}}{2} \\ = \frac{I_m}{2} = \frac{\eta I_e'}{2} \tag{6.16}$$

where $I_e' = I_m/\eta$ is an electric current equivalent to I_m we define with the duality theorem [71].

Following, Fig. 6.7 shows a connection between two PEC squares with width $w \ll \lambda$ and length $l \ll \lambda$. We impress somehow a surface electric current density $\boldsymbol{J}_s$ over both the front and back sides of the structure. Again, consider a uniform current distribution along the surface. Hence, a total electric current $I_e = 2J_s w$ [A] flows through the connection. Using the boundary condition (2.13), we can represent the tangential magnetic field due to the presence of the electric current as $\widehat{\mathbf{a}}_n \times \boldsymbol{H}_T = \boldsymbol{J}_s$.

We define again points C, D, C', and D' around the connection, which in turn define a surface $S_{CDD'C'C}$. Stoke's theorem leads to:

$$\oint \boldsymbol{H}_T \cdot d\boldsymbol{l} = \oiint (\nabla \times \boldsymbol{H}_T) \cdot d\boldsymbol{s} \tag{6.17}$$

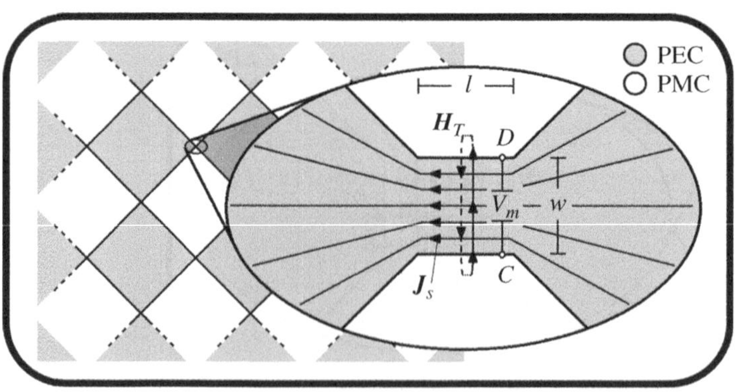

Fig. 6.7 Connection between two PEC squares in a checkerboard surface

From Ampère's law (2.17), and noting that the tangential electric field is zero over the PEC, we have:

$$\begin{aligned} \oint \boldsymbol{H}_T \cdot \boldsymbol{dl} &= \oiint \boldsymbol{J}_s \cdot \boldsymbol{ds} \\ &= J_s w = I_e \end{aligned} \tag{6.18}$$

We can calculate a magnetic scalar potential V_m from C to D as [72]:

$$\begin{aligned} V_m &= -\int_D^C \boldsymbol{H}_T \cdot \boldsymbol{dl} = -\left(-\frac{\oint \boldsymbol{H}_T \cdot \boldsymbol{dl}}{2}\right) \\ &= \frac{I_e}{2} = \frac{V_e'}{\eta} \end{aligned} \tag{6.19}$$

where $V_e' = \eta V_m$ is, in accordance with the duality theorem [71], a magnetic voltage equivalent to V_m. Note that the change of sign in (6.19) is due to the opposite directions of the line integrals.

By multiplying (6.16) and (6.19), we have:

$$\frac{V_e}{I_e'}\frac{V_e'}{I_e} = \frac{\eta^2}{4} = Z_{in}^2 \tag{6.20}$$

If the widths w and lengths l in the connections between both PMC and PEC squares tend to zero, the structure becomes self-complementary. In this case, the equivalent voltages V_e and V_e' converge, as well as the equivalent currents I_e and I_e'. Therefore, we obtain a frequency-independent input impedance $Z_{in} = V_e/I_e' = V_e'/I_e = \eta/2$.

Although the very specific case of the checkerboard was illustrated here, such formulation is valid for every self-complementary antenna. This is because the convergence of the equivalent voltages and currents in the end will lead to $Z_{in} = \eta/2$. H.G. Booker [73] derived this formulation using Babinet's principle from Optics even before V.H. Rumsey [74].

What is fascinating about the above derivation is the fact that an input impedance $Z_{in} = \eta/2$ results if one adjusts the antennas seen in Section 6.2.1 to be self-complementary. For example, we can make the infinite bow-tie antenna of Fig. 6.3c self-complementary by setting its flare angle as $\alpha_a = \pi/2$. Concurrently, we obtain an input impedance $Z_{in} = \eta/2$ by using its dedicated formulation accordingly [54]. Likewise, the equiangular spiral that Fig. 6.3d depicts will be self-complementary if we define its four curves with Φ_0 values of 0, $\pi/2$, π, and $3\pi/2$ which also leads to a $Z_{in} = \eta/2$ [75]. The same is valid for the log-periodic antenna, whose dimensions we can also adjust accordingly [50].

One particular self-complementary antenna that deserves attention here is the Archimedean spiral due to its application in examples later in this chapter. We define this geometry in the plane $\theta = \pi/2$ as:

$$r = a\Phi \tag{6.21}$$

Figure 6.3f sketches an Archimedean spiral antenna with a periodicity constant $a = 0.35$. Discovered in the early EW days as one of the frequency-independent shapes [48], this planar antenna is self-complementary when the width and spacing between the metal arms coincide. Like in the equiangular spiral, currents follow the curved shapes until an "active region". A circle whose circumference is about one wavelength defines such region. There, the strong coupling between adjacent arms maximizes radiation. The polarization is circular. The typical gain is around 5 dBiC with a good axial ratio over BWs of the order of 20 : 1. Again, the radiation patterns are bidirectional.

At this point, the reader may be wondering why the ES designer should choose a specific antenna geometry over the others. Not only the type of radiation pattern, gain, and polarization matter here. Some long, curved shapes (such as the equiangular/Archimedean spirals and the sinuous antenna) may require more computational power in simulations than antennas based on short, straight forms (like the bow-tie and some kinds of log-periodic). Besides, a vast literature refers to some antennas whereas few unclassified works exist about others. These points impact the development time in the context of a cat-and-mouse EW game where the innovation rate is paramount, as we discussed in Section 1.1.1. Nonetheless, one thing is common to all antennas we saw so far in this chapter: none of them presents a unidirectional radiation pattern. Hence, we cannot use them for the DOA estimation with amplitude-based methods requiring these kind of pattern [14–20]. We need to associate an extra structure with these radiating elements to this end.

6.3 Traditional Methods to Obtain Unidirectional Patterns

In this section, we will see two traditional ways of obtaining a unidirectional radiation pattern from planar radiating elements whose standalone versions provide omni- or bidirectional patterns. Unfortunately (for non-metasurface antenna designers), we could also call this section "Traditional methods to obtain bulky, lossy, and/or frequency-dependent antennas from frequency-independent ones" as the reader will realize soon.

6.3.1 Lossy Cavities

Shaping omni- or bidirectional patterns into unidirectional ones is straightforward by backing a planar radiating element with a metal cavity completely filled with absorbers. Consider that the antenna plane divides the whole space into two hemispheres. Then, this maneuver aims at mitigating the electromagnetic energy in one hemisphere while keeping it unaffected in the other. Here, we do not pursue any improvement in the antenna gain.

For example, C.A. Balanis [76] reports a cavity-backed log-periodic antenna with a unidirectional radiation pattern and gain of around 3 dBi from around 2.4–15.0 GHz. In comparison to the standalone version [49, 50], it is fair to say that the cavity suppresses one lobe of the bidirectional pattern and keeps the other roughly unchanged, except for a BW reduction from 10 : 1 to around 6 : 1. The cavity height is 44.4 mm, which represents 0.36 λ at the lowest operating frequency (2.4 GHz in this case). M. Hamza et al. [77] propose a cavity-backed Archimedean spiral antennathat provides a gain of around 3 dBiC from 4.0 to 15.0 GHz. Again, roughly speaking, the lossy-cavity suppresses one lobe of the circularly polarized bidirectional pattern and keeps the other unchanged. Now, however, we observe a reduction in BW from a typical ratio of 20 : 1 to around 4 : 1. The cavity height is 17.5 mm or 0.23 λ at the lowest frequency of 4.0 GHz.

We could list many other antennas backed by lossy cavities here. As a rule of thumb, they provide a unidirectional pattern with approximately the same polarization and gain as the standalone version. However, the lossy cavity degrades BW. This is because a thickness-to-BW ratio exists for physically realizable absorbers [78]. In other words, the larger the desired BW, the thicker the required cavity, which is not convenient for either fitting the fuselage of air/spacecraft or embedding lightweight drones and missile seekers.

Another inconvenience in lossy cavities is that we dissipate energy that we could use somehow. From an antenna engineering perspective, the use of absorbers reduces radiation efficiency. Thinking about the receiving mode, if we have a bidirectional antenna, we do not want to lose one of the lobes and use only the other. In a scenario populated by LPI radars, we want to simultaneously use both of them to detect these challenging adversaries before they detect us.

6.3.2 PEC Antenna Reflectors

As we discussed in Section 3.2.1, the reflection coefficient of a PEC surface is $\Gamma_{TEM} = -1$ for waves normally incident from the free space, that is, the surface reflects the incident electric field with same magnitude but with a phase shift $\varphi_\Gamma = \pm\pi$. We can exploit such property to obtain unidirectional patterns with planar radiating elements without the use of lossy elements.

Consider an infinitely small planar radiating element backed by a reflective surface of infinite extent positioned in parallel to it. According to the image theory [79], the electric field at the broadside direction is the superposition of the field directly coming from the radiating element and the field reflected over the surface. These fields have a phase difference $\varphi_t = \varphi_2 - \varphi_1$ that accounts

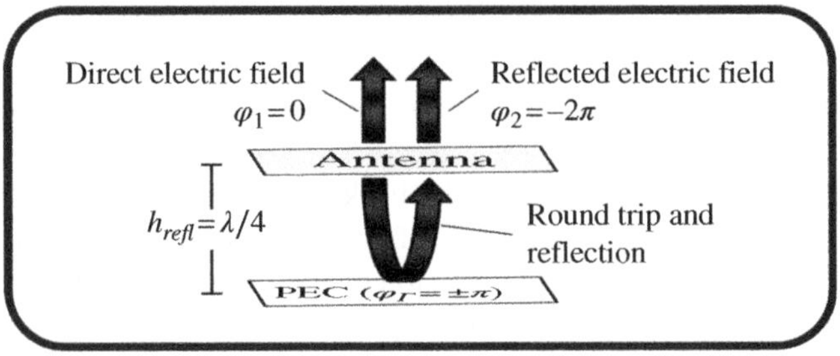

Fig. 6.8 Radiating element distanced by $h_{refl} = \lambda/4$ from a PEC reflector. "Gonçalves Licursi de Mello et al. [22]/MDPI/CC BY 4.0."

for the reflection coefficient phase φ_Γ of the surface, and the path phase delay $\varphi_p(f) = -2kh_{refl}$, related to the round trip between the radiating element and the reflector.

A plate made of a metal whose behavior is close to a PEC maximizes the antenna gain at the broadside direction when spaced by $h_{refl} = \lambda/4$ in this configuration [22]. The reason is that, in this case, $\varphi_t = \varphi_\Gamma + \varphi_p = -\pi - 2kh_{refl} = -2\pi$. Putting it differently, the interferences between direct and reflected electric fields at the broadside direction are perfectly constructive. Figure 6.8 illustrates this case. In fact, we can consider interferences constructive whenever:

$$-120^\circ + 2N\pi < \varphi_t < +120^\circ + 2N\pi \tag{6.22}$$

where $N \in \mathbb{Z}$. Equation (6.22) is the mathematically demonstrated condition for the enhancement of the broadside gain of a reflector-backed antenna [80].

For a PEC reflector and $h_{refl} = \lambda/4$, the path phase delay φ_p depends on the frequency f and imposes limitations on the antenna BW. For example, N.E.A. Rahman et al. [81] use metal cavities with and without an absorber to back an Archimedean spiral. The height of the cavities is 20.4 mm or 0.27 λ at 4.0 GHz. The version with absorber presents a well-behaved gain at the broadside direction around 5 dBiC from 4 to 18 GHz. On the other hand, the pure metal cavity improves the broadside gain to up to 7 dBiC at some frequencies but degrades it to values as low as 1 dBiC at others. What happens there is that the increase of frequency causes fluctuations in $\varphi_p(f)$ and thus in φ_t throughout the whole range 0–2π. Hence, constructive and destructive interferences between direct and reflected electric fields take turns at the broadside direction within this frequency range. This phase variation deforms the radiation patterns with maxima and nulls alternating in this direction as f increases. H. Nakano et al. [82] state that backing an Archimedean spiral antenna with a metal plate dramatically limits its well-behaved performance to a BW of 1.2 : 1.

J.O. Yang et al. [83] demonstrate similar oscillations when they back a log-periodic antenna with a conventional metal cavity, obtaining a gain that varies in the range ±8 dBi. Also, S. Varault et al. [84] show that the broadside gain of a checkerboard antenna backed by a metal plate is not independent of frequency, approaching the theoretical one only in a limited BW.

Improving the gain by 3 dB in (6.3) increases the maximum ES detection range r_{max} by 1.4 times, an advantage not neglectable at all. However, the important point here is that, even if we can consider the radiating element independent of frequency, we cannot consider so the final PEC reflector-backed antenna. The reason is that the path phase delay $\varphi_p(f)$ depends on the wavelength. Therefore, these antennas are not useful in estimating the DOA with amplitude-based methods that require stable, unidirectional patterns in wide frequency ranges [14–20]. Besides, the requirement of $h_{refl} = \lambda/4$ does not contribute to the design of low-profile devices.

6.4 Obtaining Unidirectional Patterns with Metasurfaces

After studying Section 3.4.1, trying to improve the antenna BW by replacing a traditional, frequency-dependent strategy with a metasurface-based one may look at least paradoxical. However, as it will become more obvious in the sequel, the fundamental difference between traditional methods and metasurfaces is that the latter allows the tailoring of the device response at will. This enables a series of innovative reflection strategies that the antenna designer can exploit for ES applications.

6.4.1 Single-Band AMC Reflectors

Section 3.2.1 also made evident that a PMC surface provides a reflection coefficient $\Gamma_{TEM} = +1$, meaning that this theoretical surface reflects all the energy of a normally incident electric field but with a phase shift $\varphi_\Gamma = 0°$. Consequently, such surface provides a frequency-independent phase difference $\varphi_t = \varphi_\Gamma + \varphi_p \cong 0°$ if spaced by $h_{refl} \ll \lambda$ from a parallel radiating element. Figure 6.9 represents such reflection scheme. However, we can obtain the PMC behavior only in special conditions. For example, we can obtain it from AMCs at specific frequencies as thoroughly studied throughout Sections 3.2.1–3.2.3. In this way, (6.22) also limits the improvement on the broadside gain and the antenna BW once again [80]. Some works studied different AMC unit cells to widen BW. Some examples of BWs and their associated resonant frequencies are found in Section 4.6.1.

The reflection scheme in Fig. 6.9 has been exploited since the very first work about AMCs when Sievenpiper et al. [85] put a wire antenna and a mushroom-like

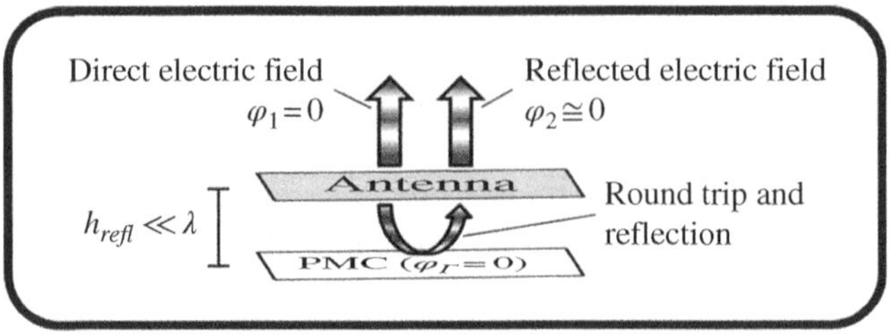

Fig. 6.9 Radiating element distanced by $h_{refl} \ll \lambda$ from a PMC reflector. "Gonçalves Licursi de Mello et al. [22]/MDPI/CC BY 4.0."

metasurface closely spaced in parallel for operation at 13.5 GHz. There, the big advantage is the design of a low-profile, unidirectional antenna with a spacing between these two elements of $h_{refl} = 1$ mm, that is, 0.04 λ instead of $\lambda/4$ as PEC-based schemes require.

The literature about wideband antennas is extensive. A. Thior et al. [86] back a triangular bow-tie antenna with a squared-patch AMC. The total thickness is only 4.8 mm or 0.05 λ at 3.3 GHz. The final device presents a well-behaved, unidirectional radiation pattern with gain of around 7 dBi from this frequency until 5.4 GHz (BW of 48.3%). Thus, in comparison to the standalone version, whose typical gain and BW are around 3 dBi and 100% (see Section 6.2.1), the AMC-backed antenna presents an improvement in the antenna gain but with a BW reduction.

L. Damaj et al. [87] associate another bow-tie antenna with an AMC. This time, a four-arm spiral slot etched within a square patch makes up the AMC unit cell. The total thickness is 8.4 mm or 0.07 λ at 2.5 GHz. Again, well-behaved, the final device provides unidirectional radiation patterns throughout a wide range of frequencies (2.5–5.4 GHz, that is, a BW of 73.5%). They provide a comparison between the gain of the standalone antenna and the AMC-backed structure. The former presents a gain of around 2.5 dBi, whereas the latter 7 dBi in an equivalent BW.

J.M. Bell et al. [88] propose an Archimedean spiral radiating element backed by a mushroom-like metasurface, which leads to a 1.9-mm thick antenna (or 0.06 λ at 10.3 GHz). Results show once more well-behaved, unidirectional radiation patterns in the addressed band (10.3–16 GHz, that is, BW of 43.3%). The gain is 9 dBiC, a significant improvement in relation to the typical one for the standalone Archimedean spiral (5 dBiC as stated in Section 6.2.2).

Again, we cannot neglect the advantages in the maximum ES detection range r_{max} resulted from a gain enhancement in (6.3). However, similar to the PEC case, the AMC-backed antennas reported above depend on frequency despite that their radiating elements follow all the principles stated by Rumsey [21]. Now, the dependence comes from the AMC reflection coefficient phase $\varphi_r(f)$, not from the path phase delay φ_p which the spacing $h_{refl} \ll \lambda$ actually mitigates. Thus, one may conclude that the advantage of a traditional AMC reflector in relation to PEC ones is the low thickness. Nevertheless, neither the former nor the latter can provide stable, high-gain, unidirectional patterns covering reasonable BWs for ES applications. A claim for innovative approaches exists.

6.4.2 Frequency-Reconfigurable AMCs

The reflection scheme of Fig. 6.9 depends mainly on the AMC reflection coefficient phase φ_{Γ} (and not from the path phase delay φ_p). Hence, an interesting functionality is possible: frequency reconfigurability. By controlling the resonant frequency of an AMC, we can shift in the spectrum where the near-PMC behavior occurs and condition (6.22) holds. In this way, we control the operating band of the final device. We can perform such control with varactor diodes. Through the biasing of varactors installed between adjacent unit cells, we can control the equivalent capacitance of the unit cells, as Section 3.2.4 explained. Such maneuver will work as long as the AMC is properly illuminated in the whole frequency range of interest, requiring thus for wideband radiating elements. Figure 6.10 illustrates this concept.

F. Costa et al. [28] place a bow-tie radiating element over an AMC whose patches are connected through varactors. The total thickness is 5.3 mm or 0.04 λ at 2.36 GHz. The device can shift a tuning frequency associated with a well-behaved, unidirectional radiation pattern from 2.36 to 2.82 GHz, which represents a tuning ratio of 1.2 : 1. The broadside gain is around 5 dBi.

S. Yang et al. [29] also use the combination bow-tie and AMC. A five-ring pattern connected to the ground plane through a varactor and a metal via makes up the unit cell. The total thickness is 5.2 mm or 0.03 λ at 1.51 GHz. The tuning ratio is 1.4 : 1 (1.51–2.12 GHz), and the gain is also around 5 dBi.

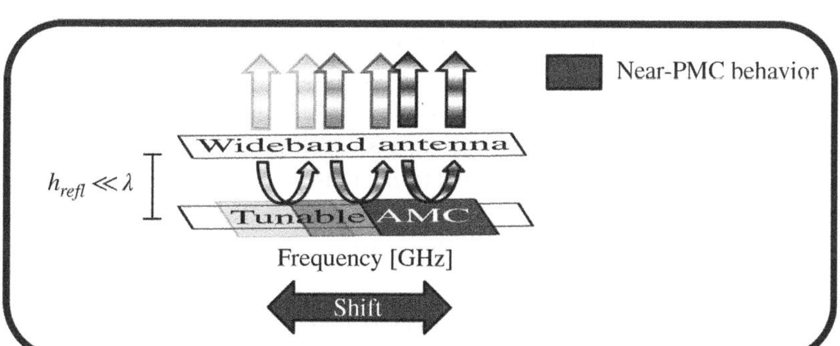

Fig. 6.10 Shifting of the near-PMC behavior throughout the frequency spectrum with a reconfigurable AMC.

B. Sanz-Izquierdo et al. [30] associate a coplanar waveguide-fed radiating element with an AMC whose unit cell is a dual-layer square patch over a ground plane. They set four varactors orthogonally over the square corners, obtaining well-behaved, circularly polarized, unidirectional radiation patterns in a tuning ratio of 1.4 : 1, from 1.15 to 1.60 GHz. By setting different bias voltages to the varactors, the AMC provides different reflection coefficient phases φ_{Γ} for different polarizations. As such, it can perform a polarization conversion, as we previously saw in Section 4.4.3. The gain achieves up to 9 dBiC for both LHCP and right-handed circular polarization (RHCP) at the highest frequencies but presents a dramatic degradation at the lowest ones. This performance drop is due to strong dissipations in the varactors in the proposed AMC topology.

D. Chen et al. [31] associate a U-shaped dipole with another polarization-conversion AMC. Again, four varactors connect neighbor patches over a ground plane but now in a single layer. The final device provides well-behaved, unidirectional radiation patterns for LHCP, RHCP, and linear polarization in a tuning ratio 1.56 : 1, from 1.75 to 2.73 GHz. This time, the gain remains stable at 7.5 dBi/dBiC for all frequencies and polarizations.

The reconfiguration speed for the AMCs mentioned above depends on the *RC* time constant of the biasing circuit, which relates to the resistances and capacitances used therein [89]. We saw in Section 4.4.4 that metasurfaces with a reconfiguration frequency of the order of 1 MHz already exist. Although such tuning speed provides good agility for the reconfigurable antenna, the tuning ratios the above works provide are far from what modern ES systems would require. Nonetheless, the approach we presented here will be useful, together with the one of Section 6.4.3, to understand in Section 6.4.5 how we can conceive powerful metasurface-based ES antennas.

6.4.3 Multiband-AMC Reflection Scheme

One interesting capability of AMCs the community recognized early on is the possibility of a multiband operation [25–27], when we obtain the PMC behavior at multiple frequencies, as we first discussed in Section 3.2.4.

Traditionally, we can obtain unidirectional radiation patterns in multiple bands by exploiting the PMC behavior associated to each resonance of a multiband AMC. Of course, we need a radiating element covering every band of interest. Some works [90–92] exploit this idea with coplanar waveguide-fed radiating elements and dual-band AMCs.

However, we want to exploit the AMC capacity in its entireness to maximize the spectral coverage of the ES antenna. Figure 6.11a depicts a dual-band AMC unit cell as well as its parameters [22]. Two nested metal squares without vias over a ground plane make up the shape. This unit cell is the one that composes the finite-array AMC Fig. 1.11 depicted earlier.

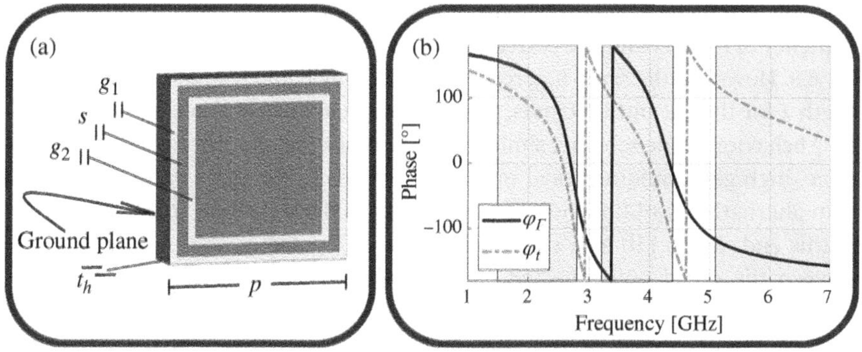

Fig. 6.11 Dual-band AMC unit cell: (a) double square over a ground plane; (b) reflection coefficient phase in the plane of the unit cell (φ_Γ) and in a plane distance h_{refl} = 10 mm from the unit cell (φ_t). "Gonçalves Licursi de Mello et al. [22]/MDPI/CC BY 4.0."

Figure 6.11b shows the reflection coefficient phase φ_Γ obtained with a commercial software (refer to [22] for detailed parameters) that simulates that a normally incident TEM wave illuminates the unit cell. The unit cell emulates a near-PMC behavior around two frequencies where it resonates and $\varphi_\Gamma(f) = 0°$, namely $f_1 = 2.7$ and $f_2 = 4.3$ GHz. Far from resonances, that is, for $f \ll f_1$ or $f \gg f_2$, the reflection coefficient phase $\varphi_\Gamma(f)$ asymptotically goes to $\pm 180°$. Well, this is the reflection coefficient phase of a PEC reflector. Hence, it is reasonable to say that the AMC mimics a near-PEC behavior far from resonances.

The above paragraph enables a glimpse of a reflection scheme exploiting both near-PMC and near-PMC behaviors of an AMC. Fig. 6.12 illustrates such scheme, which is an amalgam of the ones in Figs. 6.8 and 6.9. We can address three bands with a dual-band AMC and a wideband radiating element. We refer to these bands

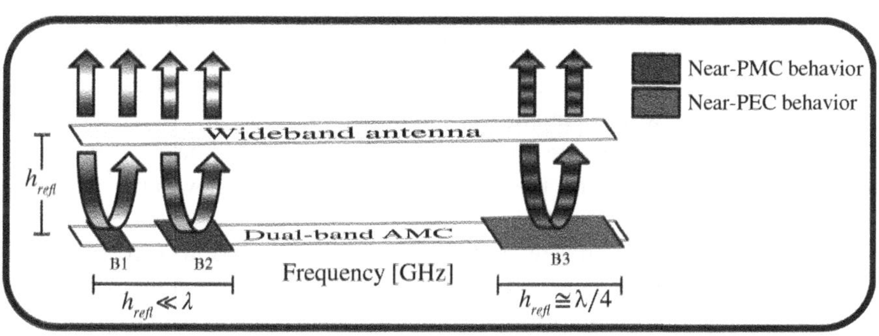

Fig. 6.12 Metasurface-based triple-band reflection scheme: the dual-band AMC works as a near-PMC reflector around two resonances and as a near-PEC reflector far from resonances. "Gonçalves Licursi de Mello et al. [22]/MDPI/CC BY 4.0."

as B1, B2, and B3 bands hereafter. By choosing the spacing h_{refl} as $\lambda/4$ at the central frequency of the B3 band, we correctly exploit the near-PEC behavior, exactly as in Fig. 6.8. However, the same h_{refl} represents a distance much smaller than a wavelength λ for the B1 and B2 bands. That means that we can also exploit the near-PMC behavior in the scheme, similar to the case of Fig. 6.9. Therefore, we obtain stable, high-gain, unidirectional radiation patterns in the three bands.

An alternative workflow here is to simply consider condition (6.22) as a north. To this end, Fig. 6.11b also shows the phase difference $\varphi_t = \varphi_2 - \varphi_1 = \varphi_{\Gamma} + \varphi_p$ between the direct and reflected electric fields in a reference plane distanced $h_{refl} = 10$ mm from the unit cell of Fig. 6.11a. The highlighted areas indicate the ranges where condition (6.22) holds.

R.G.L. Mello et al. [22] exploit this idea backing a grooved bow-tie radiating element with a dual-band AMC based on the unit cell of Fig. 6.11a. The spacing we adopt is $h_{refl} = 10$ mm, which leads to a total thickness of 14.9 mm (or 0.10 λ at 2.0 GHz). Figure 6.13a shows such device. For comparison purposes, we simulate both the AMC-backed bow-tie antenna and the standalone radiating element with commercial software. Fig. 6.13b depicts the broadside realized gain. We define the antenna 3-dB BW as the BW where the measured broadside realized gain is less than 3 dB lower than the maximum value in the band. The AMC-backed bow-tie antenna has three well-defined operating bands (highlighted in Fig. 6.13b). The B1 band has a 3-dB BW of 34.4%, extending from 2.00 to 2.83 GHz. The 3-dB BW of the B2 band is 19.7% (from 3.20 to 3.90 GHz) whereas that of the B3 band is 31.0% (4.90–6.70 GHz). Hence, the antenna besides multiband is also wideband. The ratio between the highest and lowest operating frequencies is $f_h/f_l = 3.4:1$.

For comparison purposes, we take into account the simulated results of both the AMC-backed antenna and the standalone version. The broadside gain of the AMC-backed antenna keeps around 7.5, 8.5, and 10.0 dBi in the B1–B3 bands, which represent enhancements around 6.0, 6.5, and 7.5 dB in comparison to the standalone version. By using (6.3), these values improve the maximum ES detection range $r_{\max}$ in 2.0, 2.1, and 2.4 times, respectively. Fig. 6.13b also shows the broadside realized gain measured in an anechoic chamber. The simulated and measured results agree well. Fig. 6.14 sketches the normalized radiation patterns at some chosen frequencies, where simulations and measurements agree well once again. Throughout all the bands, the antenna indeed presents well-behaved unidirectional patterns. Furthermore, the cross-polarization level is low, and the front-to-back ratio is large.

The presented results indicate that we can use the multiband-AMC reflection strategy in the design of antennas appropriate for the DOA estimation with amplitude-based methods that require stable, unidirectional patterns in wide frequency ranges [14–20]. Doubling the maximum ES detection range $r_{\max}$ is a noticeable improvement if one intends to tackle LPI radars. The ratio $f_h/f_l = 3.4:1$ is much

(a)

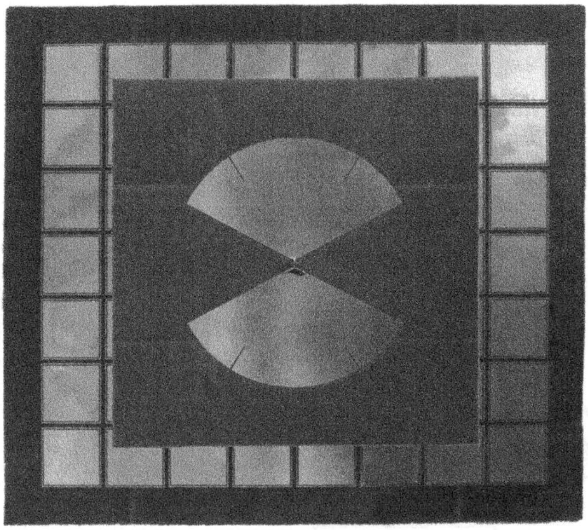

(b)

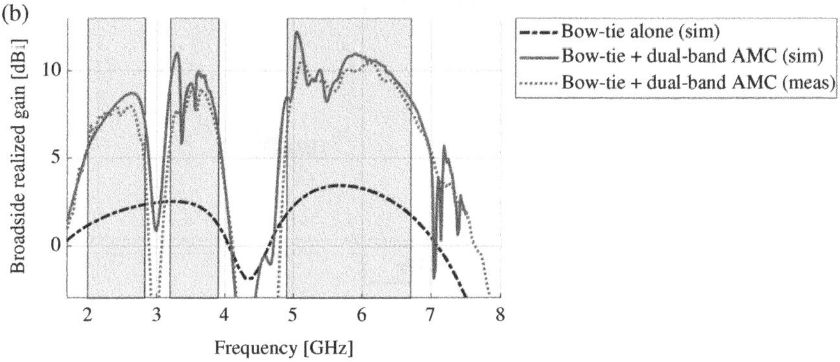

Fig. 6.13 Triple-band antenna using dual-band AMC: (a) fabricated device; (b) broadside realized gain. "Reproduced from Gonçalves Licursi de Mello et al. [22]/MDPI/CC BY 4.0."

wider than the BW of the antennas based either on traditional AMC schemes, seen in Section 6.4.1, or on PEC reflectors, seen in Section 6.3.2. Such a ratio is comparable to the BW of the absorber-backed antennas seen in Section 6.3.1. However, besides the improvement in r_{max}, the proposed strategy allows a much thinner device in the order of 0.10 λ at the lowest operating frequency. On the other hand, some blind zones have emerged in the spectrum that do not exist for standalone/absorber-backed antennas.

To summarize, in the multiband-AMC reflection strategy, a reduction in the POI in frequency is the price to pay for a much thinner antenna and significant improvement in r_{max}. Nevertheless, we can wisely choose the phase difference φ_t per the center frequency of multiple, specific, desired bands, differently from

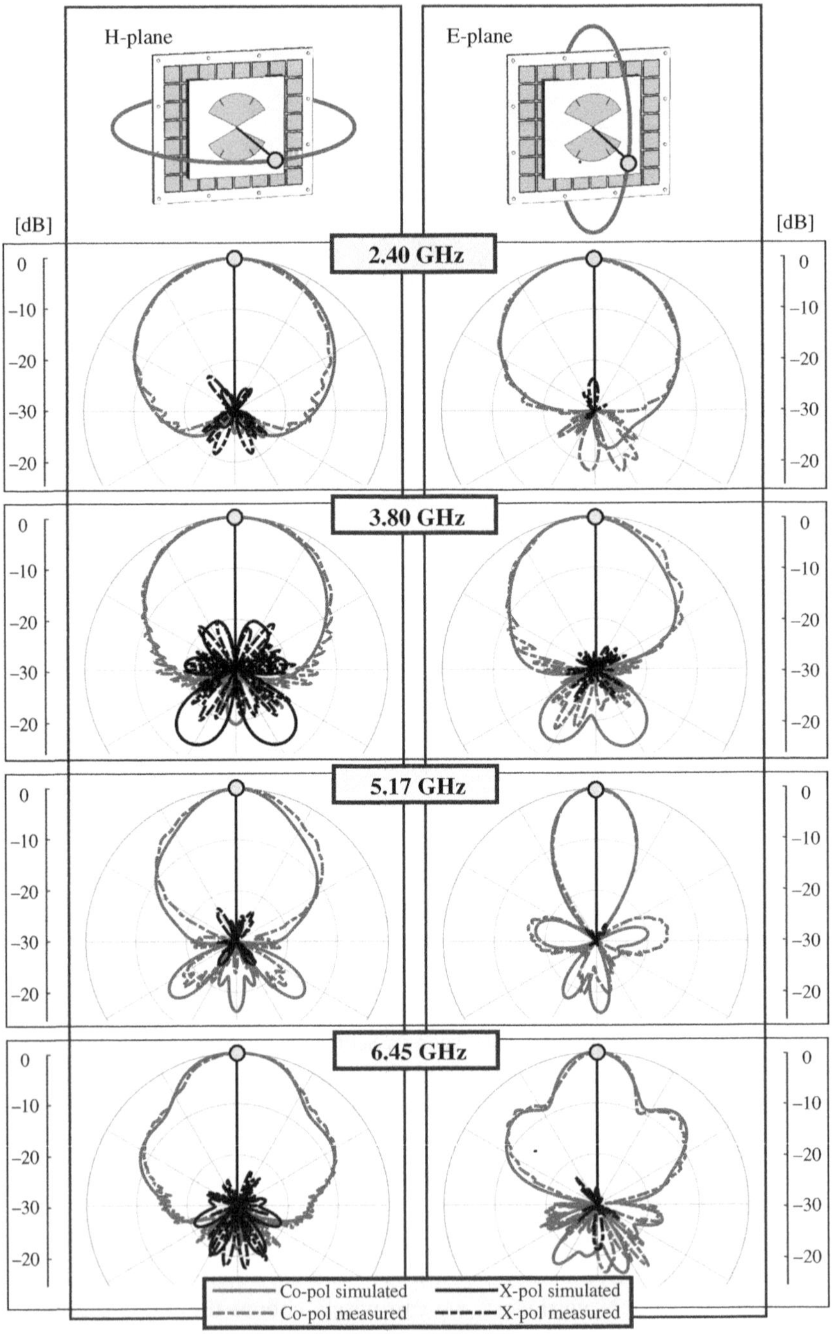

Fig. 6.14 Simulated and measured radiation patterns of the triple-band antenna; dots indicate the broadside direction in all the graphs. "Gonçalves Licursi de Mello et al. [22]/ MDPI/CC BY 4.0."

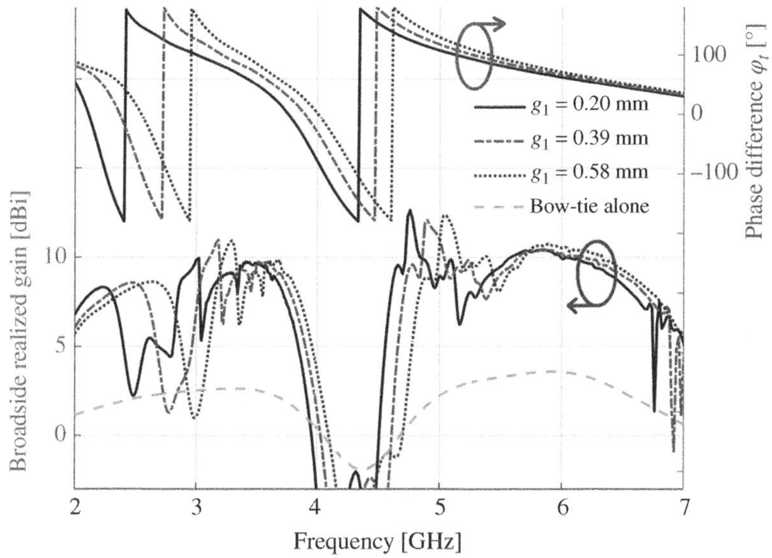

Fig. 6.15 Simulated phase difference φ_t in a plane distance h_{refl} = 10 mm from the unit cell and broadside gain when g_1 is varied. "Gonçalves Licursi de Mello et al. [22]/MDPI/CC BY 4.0."

the PEC case. This choice is possible because both spacing h_{refl} and the reflection coefficient phase φ_Γ of the AMC define such phase, and the engineer can design the latter. To demonstrate this possibility, we depict in Fig. 6.15 the phase difference φ_t and the broadside realized gain when we vary the parameter g_1 of the AMC unit cell. By decreasing g_1, mainly the first resonance of the reflection coefficient phase φ_Γ shifts downward in frequency. Fig. 6.15 does not show the phase φ_Γ. It shows the phase difference φ_t directly impacted in the process. An extra consequence is the shift of the B1 band downward in frequency simply because this process affects the range where (6.22) holds accordingly. The other bands suffer a lower impact which we can compensate with the other AMC parameters indicated in Fig. 6.11a (g_2, s, p, and t_h). R.G.L. Mello et al. [22] also provide a methodology to do so, but we can also use conventional optimizers. In the end, we can correctly set all the bands to avoid target radars remaining uncovered.

The results in Fig. 6.15 are important because they make evident that it is the AMC that controls the frequency response of the final device in the proposed strategy, similar to the works discussed in Section 6.4.2. Likewise, at the fundamental level, there is no difference whether the AMC response is tuned by the proper choice of g_1 in the design or through the real-time biasing of varactors. This insight indicates that we can shift the blind zones throughout the spectrum with an on-the-fly AMC reconfiguration.

6.4.4 Extending the Number of Bands of Antenna Reflectors

We can exploit the multiband-AMC reflection scheme in wider frequency ranges. Triple-band AMCs may work as good, quad-band reflectors as long as (6.22) holds. As an example, Fig. 6.16a shows a triple-band AMC unit cell and its parameters [24]. Now, three nested metal squares without vias over a ground plane make up the unit cell. Figure 6.16b shows the reflection coefficient phase φ_{Γ} (refer to [24] for detailed parameters). The unit cell mimics the near-PMC behavior around three frequencies now, specifically $f_1 = 2.6$, $f_2 = 4.0$, and $f_3 = 6.5$ GHz. Again, far from resonances, the reflection coefficient phase $\varphi_{\Gamma}(f)$ asymptotically goes to $\pm 180°$, emulating a near-PEC behavior. Figure 6.16b also shows the phase difference φ_t of this unit cell but for an $h_{refl} = 5.0$ mm. The highlighted areas indicate the ranges where condition (6.22) holds.

To check this idea, we need a radiating element that can illuminate this AMC throughout the whole range. We use a grooved bow-tie shape surrounded by a metal ring to this end. The spacing between radiating element and AMC is $h_{refl} = 5.0$ mm and the total thickness is 10.8 mm (or 0.08 λ at 2.34 GHz). Fig. 6.17a sketches the complete model including the balun and connector used for feeding. We simulated both such model and the standalone radiating element with commercial software. Fig. 6.17b shows the broadside realized gain And highlights the four operating bands. The BWs are 22.0%, 16.2%, 31.1%, and 21.2% for respective ranges of 2.34–2.92 GHz, 3.28–3.86 GHz, 5.02–6.87 GHz, and 8.28–10.24 GHz. For these BWs, we took into account the 3-dB BW condition and a well-behaved radiation pattern (not shown here). The ratio f_h/f_l is 4.4 : 1.

The broadside realized gain of the AMC-backed antenna keeps around 5.5, 7.0, 9.0, and 10.0 dBi in the four bands. In comparison to the radiating element alone,

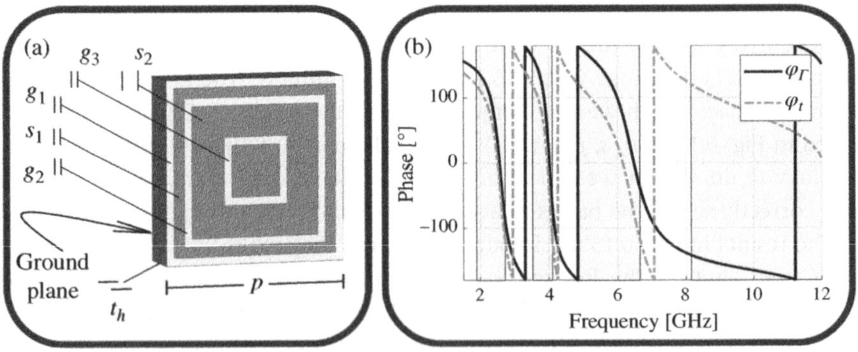

Fig. 6.16 Triple-band AMC unit cell: (a) triple square over a ground plane; (b) reflection coefficient phase in the plane of the unit cell (φ_{Γ}) and in a plane distance h_{refl} = 5.0 mm from the unit cell (φ_t). "Gonçalves Licursi de Mello et al. [22]/MDPI/CC BY 4.0."

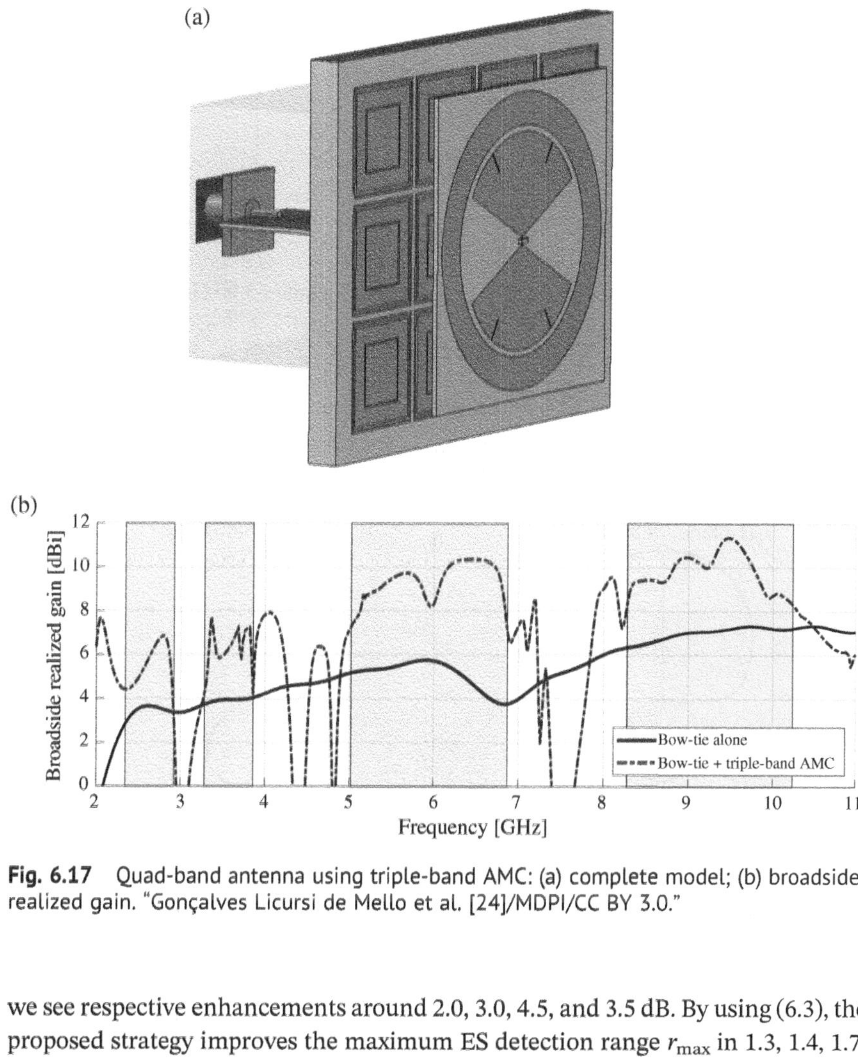

Fig. 6.17 Quad-band antenna using triple-band AMC: (a) complete model; (b) broadside realized gain. "Gonçalves Licursi de Mello et al. [24]/MDPI/CC BY 3.0."

we see respective enhancements around 2.0, 3.0, 4.5, and 3.5 dB. By using (6.3), the proposed strategy improves the maximum ES detection range r_{max} in 1.3, 1.4, 1.7, and 1.5 times, respectively. Fig. 6.18 illustrates the normalized, simulated radiation patterns at some chosen frequencies. Once again, the antenna has well-behaved unidirectional patterns. The cross-polarization levels are low, mainly in the E-plane, where they are out of the scale and do not appear. The front-to-back ratio is large in all patterns.

The proposed antenna shows that extending the technique with new AMC resonances can also provide high-gain, unidirectional, low-profile antennas appropriate for ES systems. The ratio between the maximum and minimum operating frequencies of 4.4 : 1 is even higher than the previous version. The blind zones that

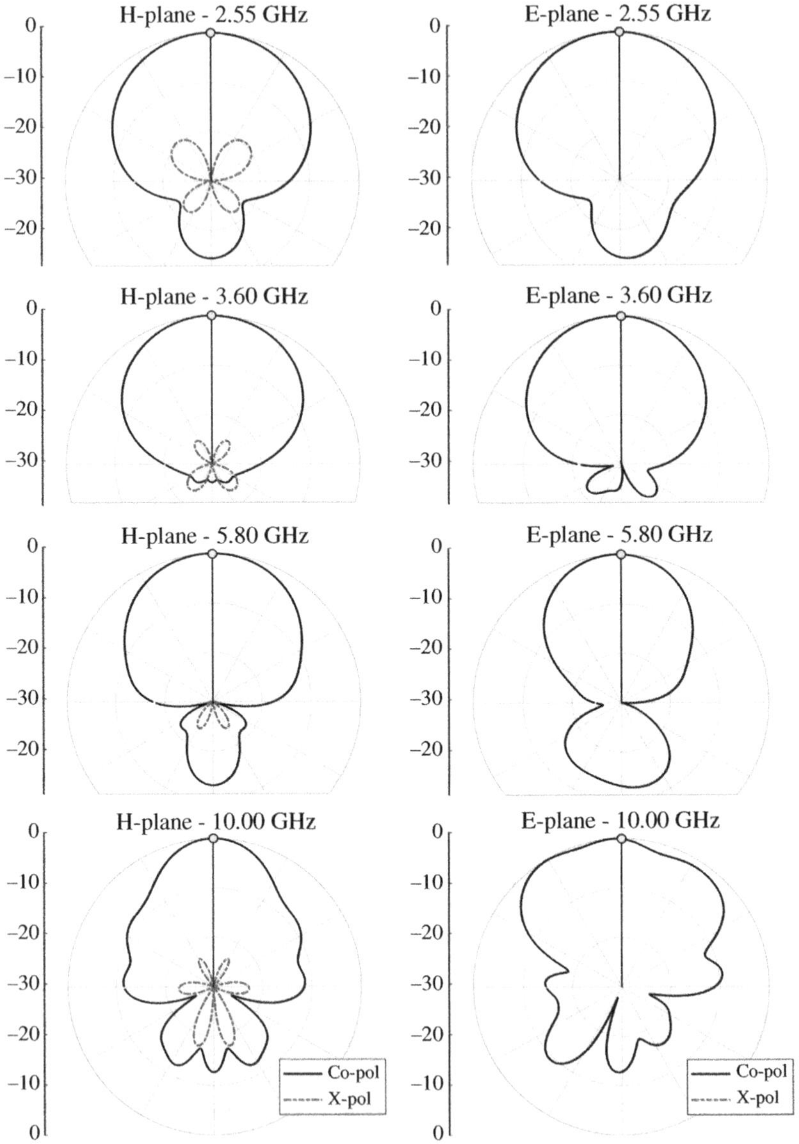

Fig. 6.18 Simulated radiation patterns of the quad-band antenna; dots indicate the broadside direction. "Gonçalves Licursi de Mello et al. [24]/MDPI/CC BY 3.0."

do not exist for standalone/absorber-backed antennas are again present. Last, the thickness is also lower than the previous version, only 0.08 λ at 2.34 GHz (10.8 mm).

6.4.5 Roadmap to Powerful Metasurface-Based ES Antennas

The improvements in the antenna gain that the multiband-AMC reflection scheme provides may be decisive against LPI radars. The accuracy of prominent AI-based approaches [9–13] for the classification of LPI signals in ES systems is really sensitive to the SNR, and every dB counts. Besides, the frequency-stable, unidirectional patterns in wide ranges are appropriate for estimating the DOA with traditional amplitude-based methods [14–20].

In terms of convenience, the low thicknesses of the complete devices, composed of the AMC and radiating element, are appropriate for modern ES systems installed on air/spacecraft, missile seekers, or lightweight drones.

We can manage the constraint related to the spectral blind zones because we can design the AMC to avoid leaving radars of interest uncovered. Nevertheless, we cannot ignore the reduction in the frequency POI. The passive nature of ES systems does not allow adversaries to discover by SIGINT means which frequencies one's forces are monitoring. Nonetheless, we can obtain this information through other intelligence resources. If that is the case, opponents can deploy new radars operating exactly in the spectral blind zones of their adversaries – or quickly reconfigure existing ones accordingly.

The ES response in the cat-and-mouse EW game could be the adoption of new antennas, with different blind zones. However, reconfiguring existing antennas provides a much better innovation rate with the possibility of adapting the ES system within micro or nanoseconds.

Similar to the antennas we saw in Section 6.4.2, reconfiguring the blind zones of AMC-backed antennas such as those of Sections 6.4.3 and 6.4.4 is as easy as reconfiguring the AMC itself. This is due to the fact that, it is the AMC that controls the frequency response of the final device in this kind of antenna, as evident in Fig. 6.15. Of course, the radiating element should properly illuminate the AMC in all of the frequency ranges of interest.

An interesting point is that slight shifts in the AMC resonant frequencies are enough to provide full spectrum coverage. Actually, all one needs is to set two operating states with blind zones in excluding parts of the spectrum. To the best of the author's knowledge, no work dedicated to investigating the details of one such functionality exist to date and we should consider some details thereupon.

As in any antenna project, the first step is to identify the specs in terms of frequency range, polarization, gain, and so forth. Then, we can define an antenna topology, including the proper radiating element, AMC arrangement, and reconfigurable AMC unit cell.

We must wisely choose the radiating element according to a series of factors. For example, R.G.L. Mello et al. [22, 24] choose the bow-tie shape as the radiating element not only due to its wideband behavior but also due to the following reasons:

a) The vast literature on the bow-tie shape allows the understanding of its working principle, behavior, and distribution of currents;
b) Its simple geometry eases the huge number of simulations needed all through the project;
c) Such shape is acknowledged to work well with AMCs in a cartesian disposition [86, 87];
d) As we saw in Section 6.2.1, we can use a couple of bow-tie radiating elements in an orthogonal disposition to emulate different polarizations [63, 64], which may provide ES advantages [61].

Above, AMC in cartesian disposition is an AMC whose periodicity we can defined with two independent variables of the cartesian coordinate system, e.g., squared unit cells disposed side by side. Also, some works [93–95] validate the use of crossed bow-ties with AMCs.

The reader may wonder why we specify that the bow-tie geometry works well with AMCs with a cartesian disposition. The point is that not every radiating element matches every AMC. It turns out that a radiating element may not excite the unit cells of an AMC placed in its near-field region properly. An important factor is due to the current distribution of the radiating element. For instance, in equiangular/Archimedean spirals, currents follow the curved shapes of the arms, establishing an “active region” that maximizes radiation, as stated in Section 6.2. As frequency varies, the active region rotates with the arms, properly illuminating some patches but mismatching others. That is why R.G.L. Mello et al. [23] obtain an unstable broadside gain, albeit the gain improvement resulting from the multiband-AMC reflection scheme.

Some projects will require extremely wideband radiating elements with circular polarization. Despite the comments in the last paragraph, there will be no escape in this case, and the equiangular/Archimedean spirals will probably be the most reasonable radiating element options. Under these circumstances, radial AMCs are more appropriate [96–99]. In this arrangement, we define the AMC periodicity as a function of Φ and r in the plane $\theta = \pi/2$. Such a solution provides a good matching with currents following the spirals' curved shapes at the cost of increased design complexity for the unit cells.

We should conceive the reconfigurable unit cell after we define the radiating element and AMC arrangement. Inverting this order can lead to a waste of resources because the design of the unit cell depends on the AMC arrangement. We should plan at least two reflection schemes without overlapping blind zones according to the same number of sets of radars that we can alternately address. Then, we should use tunable elements in the unit cells to change the AMC response accordingly. We can use not only varactors but also PIN diodes to change between different operating states.

Finally, it's time to put a finite-array AMC composed of the reconfigurable unit cells and the radiating element together. At this point, some new phenomena can take place. A close look at Figs. 6.11b and 6.13b reveals that the ranges where condition (6.22) holds in the former do not really match the bands B1–B3 in the latter. The same happens for Figs. 6.16b and 6.17b. There are three main reasons for this point:

i) The first is obvious to those who read Section 3.4.2; we simulated the unit cell of Fig. 6.11b considering a normal incidence of TEM waves; in the complete device, on the other hand, each infinitesimal part of the radiating element illuminates each infinitesimal part of the AMC with an infinite combination of polarization and incident angles; hence, in the complete device, the effective AMC response slightly differs from the one of the standalone unit cell;
ii) The second reason becomes evident after the understanding of Section 2.1.7; the illumination of the unit cell from different incident angles associated with the fact that $\mu_{eff}\varepsilon_{eff} > \mu_0\varepsilon_0$ gives rise to surfaces waves over the metasurface; here, μ_{eff} and ε_{eff} are the effective magnetic permeability and electric permittivity for waves propagating within the AMC substrate; in the finite-array AMC, these surface waves reflect on the edges creating extra resonant phenomena [100];
iii) Finally, putting AMC and radiating element close to each other may create extra coupling phenomena not accounted for in the respective standalone simulations.

This divergence between responses is not a problem for the antenna engineer. All they have to do is to optimize the complete device after putting all the pieces together. This will be as easy as deep is the understanding of how the parameters of each piece affects its own performance.

6.4.6 Composite Surfaces and Inhomogeneous AMCs

Even though the multiband-AMC reflection scheme may provide valuable advantages in terms of ES detection range and accuracy of classification of LPI signals, such strategy sacrifice the POI in frequency whether the device is tunable or not. For applications where we desire full, instantaneous spectral coverage, inhomogeneous AMCs providing unidirectional radiation patterns with a moderate enhancement of the antenna gain are more appropriate.

We saw in Section 6.2 that antennas such as the equiangular/Archimedean spirals and log-periodic present active regions that vary per the operating frequency. The idea behind inhomogeneous AMCs is to set different unit cells in different regions of the metasurface. In this way, energy radiated from different active regions of these elements, that is, at different frequencies, will interact mostly with the unit cells that provide constructive interferences at the broadside direction.

Figure 6.19a,b, illustrates an Archimedean spiral and a log-periodic antenna presenting two different active regions according to two bands. To better understand the concept of inhomogeneous AMC reflectors, let us start with an example of a simpler composite surface. Figure 6.19c shows a surface made of two regions, one encircling the other, composed of different materials. Suppose that we back one of the radiating elements in Fig. 6.19a,b with this surface. If material 1 is an AMC and material 2 is a PEC, one can choose the proper spacing h_{refl} to satisfy condition

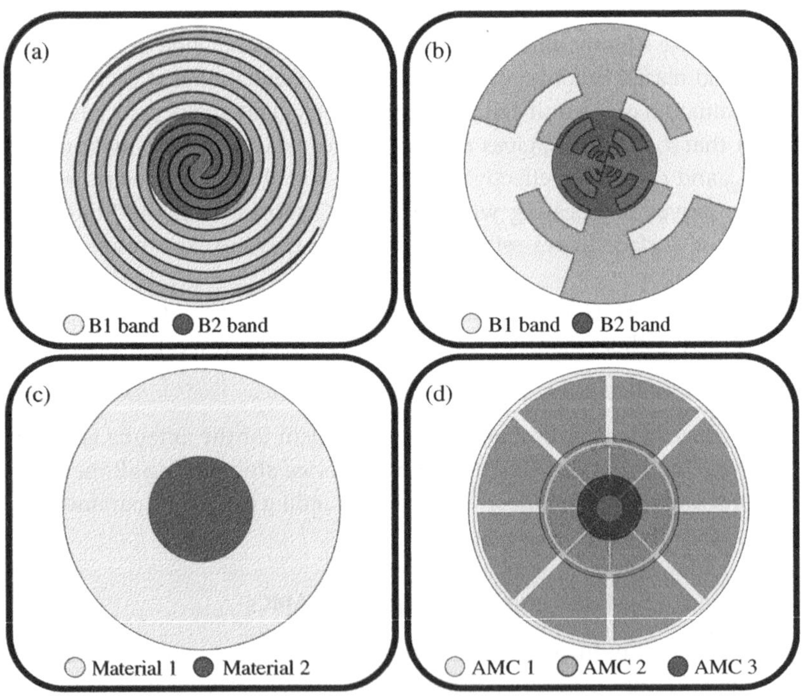

Fig. 6.19 Inhomogeneous reflector concept: (a and b) the active region of the spiral and log-periodic antennas varies for the B1 and B2 bands; (c) a hybrid surface may provide different properties for each band; (d) an inhomogeneous AMC provides proper phases for each band

(6.22) for the B1 and B2 bands, similar to the explained in Section 6.4.3. If one of the materials is an absorber, h_{refl} will be chosen according to the other.

The literature proposes some combinations of composite surfaces such as PEC with absorber [83, 101, 102], or AMC plus absorber [96, 103, 104], or PEC and AMC together [105–107]. In general, these combinations provide in each band the benefits of the respective structures. However, it turns out that the transition between two kinds of material can create some gain instabilities when the frequency varies and the active region moves,. For this reason, we are interested here in the case where diverse types of AMCs make up the composite structure[108–112]. This configuration provides a unique control of reflected waves. We optimize the different AMC unit cells to provide the proper reflection coefficient phases φ_{Γ} to each of the bands, and also to minimize the gain instabilities in the transition between them.

Fig. 6.19d illustrates this case. Here, we optimize AMCs 1 and 3 to cover adjacent bands in the spectrum. On the other hand, AMC 2 is an intermediate version thereof, covering some parts of both bands (even if not optimally) and providing whereby smoothness to the transition.

H. Nakano et al. [109] associate an Archimedean spiral with an inhomogeneous reflector composed of two regions of different squared-patch AMCs without vias, one encircling the other. The AMC of the central region provides a reflection coefficient phase $\varphi_{\Gamma} = 0°$ at 6.5 GHz while the other does so at 4.8 GHz. In the end, the antenna provides unidirectional radiation patterns with good axial ratio from 4.0 to 10.0 GHz, that is, the BW is 2.5 : 1. The gain is around 8.0 dBiC which represents an enhancement of 3.0 dB in relation to the standalone version (see Section 6.2.2). The total thickness is 7.0 mm or 0.09 λ at 4.0 GHz.

S. Mohamad et al. [110] propose another combination of Archimedean spiral and inhomogeneous reflector composed of two regions. This time, the AMCs provide reflection coefficient phases of $\varphi_{\Gamma} = 0°$ at 6.0 GHz and 3.0 GHz. Again, the strategy provides unidirectional radiation patterns with good axial, now from 3.0 to 9.0 GHz (BW of 3 : 1). A comparison between the final antenna and the standalone spiral shows improvements around 3.0 dB at 3.0, 6.0, and 9.0 GHz. The total thickness is 8.6 mm or 0.09 λ at 3.0 GHz.

D. Zeppettella et al. [111] present an equiangular spiral backed by a radial AMC composed of seven AMC rings, each one with different patch parameters to address different frequencies in the UHF range. After optimization, the device provides unidirectional patterns from 400 to 800 MHz (BW of 2 : 1) with a remarkable thickness of only 7.8 mm, or 0.01 λ at 400 MHz.

D. Kim et al. [112] propose a log-periodic antenna with four crossed radiating elements, akin to the one of Fig. 6.5, backed by an inhomogeneous reflector composed of concentric AMC rings. In this way, the antenna can provide polarization advantages in ES systems [61]. One of the rings is made of I-shaped metal patches connected through vias to the ground plane. The others are squared-patch AMCs

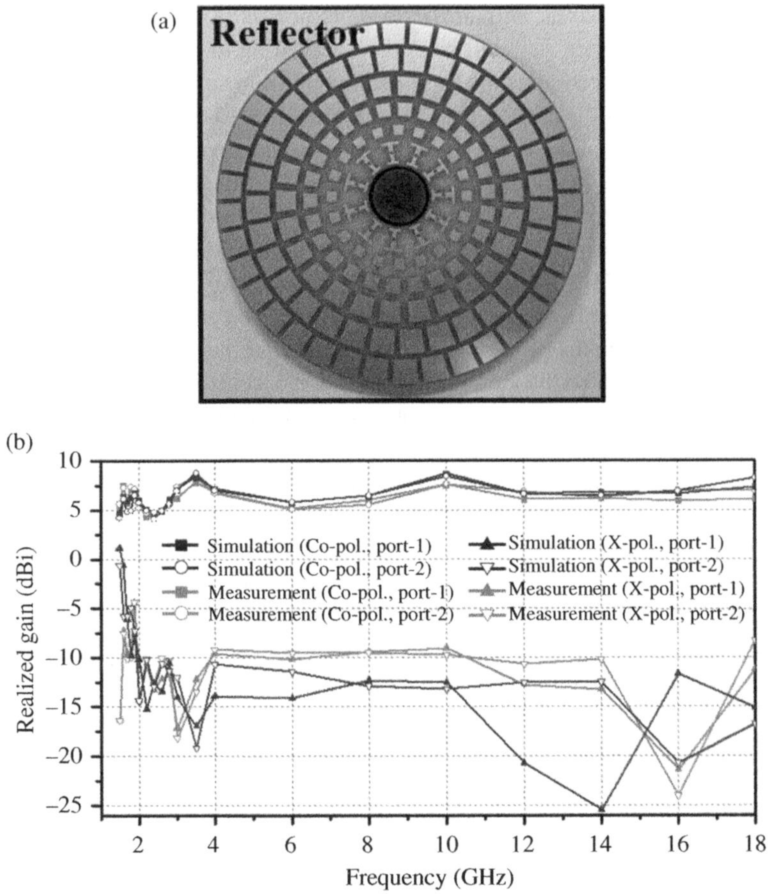

Fig. 6.20 Inhomogeneous reflector of [112]: (a) fabricated metasurface; (b) simulated and measured broadside realized gain. "Reproduced from Kim et al. [112]/IEEE/CC BY 4.0."

with different parameters and without vias. Fig. 6.20a shows the fabricated reflector. We can see unidirectional radiation patterns from 1.5 to 18.0 GHz. Hence, the BW is 12 : 1. Fig. 6.20b sketches a broadside gain around 7.5 dBi, which represents an enhancement of 2.5 dB with respect to typical log-periodic antennas (see Section 6.2.1). The total thickness is 15 mm, or 0.08 λ at 1.5 GHz.

In general, the multiband-AMC reflection scheme of Section 6.4.3 and the inhomogeneous reflector differ mainly in terms of gain enhancement (of the order of 6.0 dB in the former and 3.0 dB in the latter) and the presence of blind zones. The reason is that we optimize the former for maximum gain in each band, whereas we pursue smoothness in the band transition in the latter. Hence, while the former doubles the maximum ES detection range r_{max}, the latter multiplies it by 1.4 according to (6.3). In the end, we can consider both modern alternatives for ES systems depending on whether the priority is performance against LPI radars or frequency coverage.

6.5 Simulation: Impact of Radiation Pattern in Collected Signals

So far in this chapter, we have centered our discussions regarding the benefits of metasurface-based antennas mainly on some figures of merit. Specifically, such discussions considered the maximum ES detection range r_{max} calculated with (6.3) and some statements about the accuracy of prominent AI-based classification methods [9–13]. However, we can obtain a much deeper visualization and subsequent understanding of these benefits by checking the performance of these antennas in our EW scenario simulator.

In this chapter, the reader will conduct most of the simulations. Consider a scenario akin to the one presented in Section 5.6. The same surveillance radar is again at the position $P(0,0,0)$ operating at a carrier frequency $f_0 = 5.8$ GHz. Besides the DRFM, the aircraft now has an ES system composed of six channels: A, B, C, D, E, and F. The channels A–F have identical antennas spaced by 60° in the azimuthal domain like in the disposition shown in Fig. 6.1. For simplicity, we consider that the mutual coupling between the antennas is negligible. Therefore, their respective receivers A–F can perfectly discriminate the signals they intercept. We will focus here only on the azimuthal DOA, but we can obtain the elevation DOA similarly.

We can choose two antenna types for the ES system. One is the grooved bow-tie backed by a lossy cavity, which represents the traditional method of Section 6.3.1. The other is the grooved bow-tie backed by a multiband AMC, which illustrates the case of Section 6.4.3. We compute their simulated features [22, 57]. The broadside realized gain of the lossy-cavity version is 3.8 dBi at 5.8 GHz, while the gain of the AMC-backed version is 10.8 dBi. We compute the radiation patterns using the *phased.CustomAntennaElement* object. For noise purposes, we consider that the ES system instantaneously covers from 2.0 to 18.0 GHz, that is, $BW = 16$ GHz. We consider the listed features of the CA218-4112 amplifier (gain of 29 dB and noise figure of 5.0 dB) from Ciao Wireless, Inc. [113] for the LNA in the Rx front ends.

In the first simulation, the aircraft is not really flying toward the radar. Rather, its orientation is tilted 10° to the left of the radar in the azimuthal domain. Hence, in such domain, the broadside direction of antenna F is the closest one to the transmitter (observe the disposition of the antennas in Fig. 6.1 for better visualization).

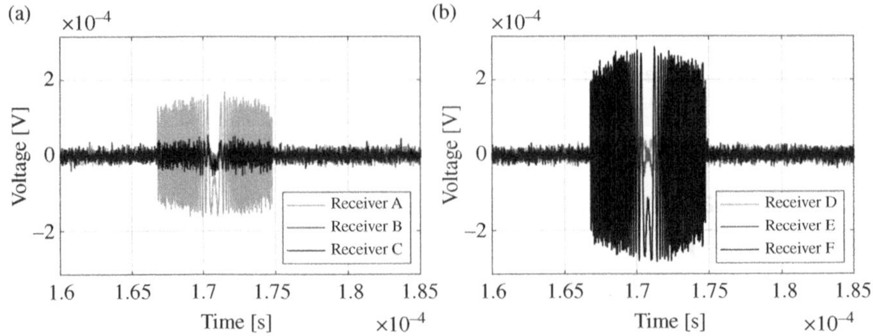

Fig. 6.21 In-phase component of intercepted signals in (a) receivers A–C; (b) receivers D–F

Figure 6.21a,b shows the in-phase component of the baseband signals in the receivers A–F of the ES system. Indeed, the intercepted pulse presents a higher amplitude in channel F.

6.6 Exercises

It is time for the reader to check how some ES functions respond to changes in the antenna gain/radiation pattern. The reader can access the simulator with the ES system by referring to the folder entitled *Ch06_ES* in the companion website and running the *main.m* file. They should choose their preferred methods and embed them in the correct places as indicated next. To achieve reasonable simulation times, you will probably want to divide the sampling rate by eight by setting the *prms.Fs* variable accordingly in the *def_params* function. The algorithm will automatically adjust the radar BW in this case.

A first task is to see the radiation patterns of all the ES antennas in the final azimuthal disposition. You can do it by defining the *vis.show_ES_ant_pat* variable as *true* in the *def_visualization* function. Another essential task is to vary the orientation of the aircraft and see how the intercepted signals change in the receivers A–F. You can control the aircraft orientation through the *shift_act_bearing* variable in the *build_acf_DRFM_ES* function. You can plot the signals by defining the *vis.plot_ES_signals* variable as *true* in the *def_visualization* function.

6.6.1 Simulating the Impact of Gain in Direction Finding

Difficult level: low.

After getting used to the directional nature of the antennas we use in the simulated ES system, you should estimate the DOA of the intercepted signals. In traditional amplitude-based methods [14–20], the system first finds the two largest amplitudes of a same signal measured in adjacent channels. Then, it evaluates the azimuthal DOA using the inverse function of expressions similar to [7]:

$$DF(\Phi) = \frac{|V_{left}(\Phi)| - |V_{right}(\Phi)|}{|V_{left}(\Phi)| + |V_{right}(\Phi)|} \tag{6.23}$$

where

- $|V_{left}|$ and $|V_{right}|$ are the signal amplitudes in adjacent channels [V];
- DF is the azimuthal direction-finding function [unitless].

All the variables above depend on the azimuth Φ domain referred to the platform orientation axes.

From the above, you can see that your first task is to find the amplitudes of the intercepted pulses in each receiver A–F. The appropriate place for that is in the *process_ES_PDW* function. This task is easy when a single pulse is intercepted. In this case, you can simply set an amplitude threshold above the noise floor to trigger a pulse detection. This threshold will define the minimum SNR required for ES detection, $SNR_{\min}$, which in turn leads to the maximum range $r_{\max}$ in (6.3).

When a detection happens in a channel, get the maximum signal amplitude, and then output it somehow in a pulse descriptor word (PDW) along with its PW and TOA. When multiple radars are present, many pulses may appear superimposed to one another in a same stream of data. The system should create a PDW for each one of them. If the reader is interested in implementing one such scenario, they will need pulse feature extraction [114]. An $SNR_{\min}$ will also exist in this case for proper ES detection.

Once you generate a PDW containing the pulse amplitude in each channel, you can write the DOA estimation algorithm in the *process_ES_DF* function. Plot the estimation of the azimuthal DOA through time together with the actual angular position of the radar in relation to the ES platform. You should also compute the deviation of DOA estimations.

Now, check that the AMC-backed bow-tie antenna is set by defining the *acf_DRFM_ES.pcs{2}.ant* variable in the *build_acf_DRFM_ES* function as *AMC_antenna*. Next, run the simulation with the aircraft at different places of the scenario. Use different values for the radius r_{acf} by changing the *acf_DRFM_ES.pcs{1}.rho* variable in that same function. First, observe for which values of r_{acf} your pulse detection algorithm works well or not. In practice, you are somehow trying to discover which is the maximum ES detection range r_{max} for the current scenario. After that, observe how the deviations in the DOA estimations change when you vary r_{acf}. In particular, observe how is the behavior when r_{acf} is close to $r_{\max}$.

Finally, to see the impact of the antenna gain in the estimation of DOA, repeat this procedure replacing the AMC-backed bow-tie antenna by the lossy-cavity one. You can do that by defining *acf_DRFM_ES.pcs{2}.ant* as *abs_antenna*. Repeat the procedure of the previous paragraph, observing the influence of the antenna on the pulse detection and deviation of DOA estimations.

6.6.2 Simulating a DOA-Driven DRFM

Difficult level: medium.

In Section 5.7.2, we proposed the exercise of implementing a DRFM taking into account the digitization of intercepted signals and realistic RF components. There, we set the DRFM to always point its antenna beam over the target radar assuming that an ES system correctly provides the angular position of the latter through the DOA estimation. If you did the exercise of Section 6.6.1, you now have one such ES system. Also, as you may have realized by looking at the *build_acf_DRFM_ES* function, we kept the DRFM embedded in the simulator of this chapter.

The first task of this exercise is to simply allocate the estimated DOAs in a variable that the DRFM can access in the next processing loop. After that, look for the comment *% load your DOA below* and do so by updating the *antenna_steer(1)* variable accordingly.

Now, when you turn on the DRFM (through changing the *pltfms{2}.pcs{1}.DRFM_state* variable from 0 to 1 in the main code), you can see the DRFM antenna beam in the scenario depicted by the *phased.ScenarioViewer* object. Find a way to update the *pltfms{2}.pcs{1}.beamSteering* variable per the DOA estimations. Note that, in this variable, the azimuth information is referred to the global coordinate system, not to the aircraft axes.

When you complete this task, you will be able to see oscillations in the antenna beam in the illustrated scenario when variations in the DOA estimations exist. Once again, run the simulation with the aircraft at different places, specially by using values for the radius r_{acf} close to the value of r_{max} you found in the previous exercise. Likewise, replace the AMC-backed bow-tie antenna with the lossy-cavity one and see the impact on the oscillations of the DRFM antenna beam.

6.6.3 Simulating the Impact of DOA Estimation in Active Stealth

Difficult level: low.

We studied in Section 5.3.1 how inaccuracies in the DOA estimation can affect the active-cancellation stealth application. If you did the exercises of Sections 5.7.3 and 6.6.2, we could say that you have everything needed to simulate the impact of such inaccuracies.

To conduct these simulations, all you must do is adjust the files mentioned in Section 6.6.2 with the parameters we suggested in Section 5.7.3. Again, we recommend you run the simulation with the aircraft at multiple positions and different

ES antennas. As such, their impact on the final active-cancellation stealth performance will become evident.

6.6.4 Simulating the Impact of Gain in LPI Signal Classification

Difficult level: medium.

As already stated, one valuable method to deinterleave and identify LPI radar signals is the classification of their coded pulses using AI. However, prominent techniques typically demand high SNR levels [9–13]. The current exercise consists in implementing one such a method to check performance with different gains for the ES antennas.

We have generated the radar waveform we used so far with the *phased.LinearFMWaveform* object and a specific BW, PW, and PRF. We can also model other waveforms appropriate for LPI radars by using the following objects in Matlab: *phased.SteppedFMWaveform*, *phased.PhaseCodedWaveform*, *phased.NonlinearFMWaveform*, and *phased.CustomFMWaveform*.

For this exercise, create a set of waveforms with the aforementioned objects and different parameters (let us say, five different values for BW, PW, and PRF for each waveform). When in possession of this dataset, directly use it to train the AI model of your choice. Next, embed it within the *process_ES_PDW* function in order to assign a label to each pulse the ES system detects. Like in the previous exercises in this chapter, run the simulation with different ES antennas and the aircraft at different places specially by using values for the radius r_{acf} close to $r_{\max}$. Observe how the accuracy of your classification method varies when you do that.

As a last step, create a new dataset that does not directly come from the waveforms you obtained with the Matlab objects. Do so by using these waveforms in the surveillance radar and by intercepting signals with the ES system in the EW scenario simulator. Register these signals at different conditions of intensity, noise, and multipath (you can set new platforms between the aircraft and the radar that will lead to extra scattering and signal distortion in the ES receivers). Retrain your model using this new, realistic dataset and test it again at different conditions.

6.7 Chapter Final Remarks

ES systems benefit from a one-way propagation when facing radars. However, LPI radars have the advantage of a signal processing gain due to their coded signals. In this battle, every dB counts when it comes to detect opponents in advance. In this chapter, we proposed metasurface-based methods for the design of low-profile, high-performing ES antennas.

Multiband AMCs provide unidirectional radiation patterns in multiple, wide frequency ranges when associated with wideband radiating elements. We can obtain improvements in the antenna gain around 6 dB in relation to the absorber-backed antenna versions. This advantage doubles the maximum ES detection range. Also, 6 dB in the SNR can have a significant impact on the accuracy of the classification of LPI signals using prominent AI-based methods [9–13].

The proposed method creates some blind zones between the covered bands. The proper design of the AMC can avoid leaving radars of interest uncovered. Besides, reconfigurable AMCs allow the dynamic shift of these zones in the frequency domain increasing the dynamism of the scenario. We also studied inhomogeneous metasurface reflectors improving the antenna gain by about 3 dB and without spectral blind zones. These features push even further the unfolding cat-and-mouse EW game.

References

1 Adamy, D. (2001). *EW 101: a first course in electronic warfare*. Norwood, MA: Artech House, Chapter 1, pp. 4.

2 Adamy, D. (2001). *EW 101: a first course in electronic warfare*. Norwood, MA: Artech House, Chapter 4, pp. 49–71.

3 Adamy, D. (2001). *EW 101: a first course in electronic warfare*. Norwood, MA: Artech House, Section 3.2.1, pp. 36–37.

4 Adamy, D. (2001). *EW 101: a first course in electronic warfare*. Norwood, MA: Artech House, Section 5.2.4, pp. 80.

5 Mardia, H.K. (1988). Adaptive multidimensional clustering for ESM. *Proceedings IEE Colloquium on Signal Processing for ESM Systems*, London, UK (26 April 1988), New York, US: IEEE.

6 Mardia, H.K. (1989). New techniques for the deinterleaving of repetitive sequences. *Proceedings of the Institution of Electrical Engineers* 136 (4): 149–154. https://doi.org/10.1049/ip-f-2.1989.0025

7 Gonçalves Licursi de Mello, R., De Sousa, F.R. (2019). Non-analytical direction-finding method as a key step in pursuing low size, weight, costs, and computational power in the deinterleaving of radar pulses. *IET Radar, Sonar & Navigation* 13 (11): 1876–1882. https://doi.org/10.1049/iet-rsn.2018.5011

8 Gonçalves Licursi de Mello, R. (2018). Modern techniques to locate and identify radars with low volume, weight, costs and processing capabilities. Master Thesis. University of Federal de Santa Catarina. https://repositorio.ufsc.br/handle/123456789/198162 (accessed 03 October 2023).

9 Xiao, Z., Yan, Z. (2021). Radar emitter identification based on novel time-frequency spectrum and convolutional neural network. *IEEE Communications Letters* 25 (8): 2634–2638. https://doi.org/10.1109/LCOMM.2021.3084043

10 Bhatti, S.G., Bhatti, A.I. (2022). Radar signals intrapulse modulation recognition using phase-based STFT and BiLSTM. *IEEE Access* 10: 80184–80194. https://doi.org/10.1109/ACCESS.2022.3195273

11 Qu, Z., Hou, C., Hou, C. et al. (2020). Radar signal intra-pulse modulation recognition based on convolutional neural network and deep Q-learning network. *IEEE Access* 8: 49125–49136. https://doi.org/10.1109/ACCESS.2020.2980363

12 Alrubeaan, T., Albagami, K., Ragheb, A. et al. (2019). An investigation of LPI radar waveforms classification in RoF channels. *IEEE Access* 7: 124844–124853. https://doi.org/10.1109/ACCESS.2019.2938317

13 Meng, X., Shang, C., Dong, J. et al. (2021). Automatic modulation classification of noise-like radar intrapulse signals using cascade classifier. *ETRI Journal* 43 (6): 991–1003. https://doi.org/10.4218/etrij.2020-0338

14 Adamy, D. (2001). *EW 101: a first course in electronic warfare*. Norwood, MA: Artech House, Section 8.4.3, pp. 156–158.

15 Al-Tarifi, M.A., Filipovic, D.S. (2018). On the assessment of antenna patterns for wideband amplitude-only direction finding. *IEEE Antennas and Wireless Propagation Letters* 17 (3): 385–388. https://doi.org/10.1109/LAWP.2018.2791199

16 Sakomura, E.S., Ferreira, D.B., Bianchi, I. et al. (2018). Compact planar two-arm compound spiral antenna for L–/X-band direction finding applications. *Proceedings of IEEE International Symposium Antennas and Propagation & USNC/URSI National Radio Science Meeting*, Boston, MA, USA (09–14 July 2017), New York, US: IEEE. https://doi.org/10.1109/APUSNCURSINRSM.2018.8608579

17 Mitrović, N.M., Ponjavić, M.M. (2015). Multichannel 2-D direction finding based on differential amplitude detection. *IEEE Sensors Journal* 15 (9): 5064–5070. https://doi.org/10.1109/JSEN.2015.2434876

18 Cazden, J., Filipovic, D. (2017). Wideband amplitude-only direction finding subsystem with conical spirals. *Proceedings of IEEE International Symposium on Antennas and Propagation and USNC-URSI Radio Science Meeting*, San Diego, CA, USA (09–14 July 2017), New York, US: IEEE. https://doi.org/10.1109/APUSNCURSINRSM.2017.8072594

19 Lasser, G., Filipovic, D. (2016). A wide-band spiral based amplitude-only azimuth direction finding system. *Proceedings of IEEE International Symposium Antennas and Propagation (APSURSI)*, Fajardo, PR, USA (26 June–01 July 2016), New York, US: IEEE. https://doi.org/10.1109/APS.2016.7696618

20 Lopez, D.G., Filipovic, D.S. (2017). On the design of milimeter-wave antennas for amplitude-only direction finding. *Proceedings of the IEEE International Symposium Phased Array Systems and Tech. (PAST)*, Waltham, MA, USA (18–21 October 2016), New York, US: IEEE. https://doi.org/10.1109/ARRAY.2016.7832575

21 Rumsey, V.H. (1966). *Frequency independent antennas*. New York, NY: Academic Press.

22 Gonçalves Licursi de Mello, R., Lepage, A.C., Begaud X. (2023). A low-profile, triple-band, and wideband antenna using dual-band AMC. *MDPI Sensors* 23 (4): 1920. https://doi.org/10.3390/s23041920

23 Gonçalves Licursi de Mello, R., Lepage, A.C., Begaud X. (2022). Réflecteur multibande à métamatériaux. *Speech presented at the Journées Scientifiques URSI France 2022* in Palaiseau, France (22–23 March 2022).

24 Gonçalves Licursi de Mello, R., Lepage, A.C., Begaud X. (2023). A Compact, quad-band, and wideband antenna using triple-band AMC. *Proceedings of 13th International Conference on Metamaterials, Photonic Crystals and Plasmonics (Meta2023)*, Paris, France (18–21 July 2023), Paris, France: Metaconferences. https://metaconferences.org/META23/files/meta23_proceedings.pdf

25 Kern, D.J., Werner, D.H., Monorchio, A. et al. (2005). The design synthesis of multiband artificial magnetic conductors using high impedance frequency selective surfaces. *IEEE Transactions on Antennas and Propagation* 53 (1): 8–17. https://doi.org/10.1109/TAP.2004.840540

26 Sarrazin, J., Lepage, A.C., Begaud, X. (2012). Dual-band artificial magnetic conductor. *Applied Physics A* 109: 1075–1080. https://doi.org/10.1007/s00339-012-7409-1

27 Mantash, M., Tarot, A. (2016). On the bandwidth and geometry of dual-band AMC structures. *10th European Conference on Antennas and Propagation (EuCAP)*, Davos, Switzerland (10–15 April 2016), New York, US: IEEE. https://doi.org/10.1109/EuCAP.2016.7481589

28 Costa, F., Monorchio, A., Talarico, S. et al (2008). An active high-impedance surface for low-profile tunable and steerable antennas. *IEEE Antennas and Wireless Propagation Letters* 7: 676–680. https://doi.org/10.1109/LAWP.2008.2006070

29 Yang, S., Chen, Y., Yu, C. et al. (2020). Design of a low-profile, frequency-reconfigurable, and high gain antenna using a varactor-loaded AMC ground. *IEEE Access* 8: 158635–158646. https://doi.org/10.1109/ACCESS.2020.3020853

30 Sanz-Izquierdo, B., Liang, B., Parker, E.A. et al. (2016). An application of active frequency selective surface to reconfigurable antenna technology. *Proceedings of the Active and Passive RF Devices Seminar*, London, UK (17 February 2016), New York, US: IEEE. https://doi.org/10.1049/ic.2016.0006

31 Chen, D., Yang, W., Che, W. et al. (2019). Polarization-reconfigurable and frequency-tunable dipole antenna using active AMC structures. *IEEE Access* 7: 77792–77803. https://doi.org/10.1109/ACCESS.2019.2919518

32 Maulana, M.F., Ismail, N., Zaki Hamidi, E.A. et al. (2020). Frequency reconfigurable multiband microstrip antenna composed of artificial magnetic conductor. *International Conference on Radar, Antenna, Microwave, Electronics and Telecommunications (ICRAMET)*, Tangerang, Indonesia (18–20 November 2020), New York, US: IEEE. https://doi.org/10.1109/ICRAMET51080.2020.9298610

33 Dixon, R. (2020). Azerbaijan's drones owned the battlefield in Nagorno-Karabakh — and showed future of warfare. *The Washington Post* (11 November 2020). https://www.washingtonpost.com/world/europe/nagorno-karabkah-drones-azerbaijan-aremenia/2020/11/11/441bcbd2-193d-11eb-8bda-814ca56e138b_story.html (accessed 08 August 2023).

34 Hecht, E. (2022). Drones in the Nagorno-Karabakh war: analyzing the data. *Military Strategy Magazine* 7 (4): 31–37. https://www.militarystrategymagazine.com/article/drones-in-the-nagorno-karabakh-war-analyzing-the-data/ (accessed 08 August 2023).

35 Pietralunga, C. (2023). Russia and Ukraine take drone warfare to unprecedented scale. *Le Monde* (18 June 2023). https://www.lemonde.fr/en/international/article/2023/06/18/russia-and-ukraine-take-drone-warfare-to-unprecedented-scale_6033281_4.html (accessed 08 August 2023).

36 Cook, E. (2024). Ukraine drones learning to dodge Russian EW signal jammers. *Newsweek* (04 March 2024). https://www.newsweek.com/ukraine-fpv-drones-russia-electronic-warfare-1875577 (accessed 11 March 2024).

37 Tegler, E. (2023). A new micro kind of electronic warfare may be unfolding in Gaza. *Forbes* (03 November 2023). https://www.forbes.com/sites/erictegler/2023/11/03/a-new-micro-kind-of-electronic-warfare-may-be-unfolding-in-gaza/ (accessed 03 November 2023).

38 Watling, J., Reynolds, N. (2023). Stormbreak: fighting through Russian defences in Ukraine's 2023 offensive. *Special Report*. Royal United Services Institute. https://www.rusi.org/explore-our-research/publications/special-resources/stormbreak-fighting-through-russian-defences-ukraines-2023-offensive (accessed 10 October 2023).

39 Balanis, C.A. (2016). *Antenna theory: analysis and design*, 4th ed. New York, NY: Wiley, Section 2.17.1, pp. 88–90.

40 Adamy, D. (2001). *EW 101: a first course in electronic warfare*. Norwood, MA: Artech House, Section 4.1, pp. 51–52.

41 Adamy, D. (2001). *EW 101: a first course in electronic warfare*. Norwood, MA: Artech House, Section 4.4, pp. 54–55.

42 Gao, G., Liang, Q., Liu, Z. et al. (2021). A photonics-based superheterodyne RF reception approach. *Advances in Astronautics Science and Technology* 4: 121–131. https://doi.org/10.1007/s42423-021-00089-y

43 Ghelfi, P., Scotti, F., Onori, D. et al. (2019). Photonics for ultrawideband RF spectral analysis in electronic warfare applications. *IEEE Journal of Selected Topics in Quantum Electronics* 25 (4): 1–9. https://doi.org/10.1109/JSTQE.2019.2902917

44 Ghelfi, P., Onori, D., Laghezza, F. et al. (2015). An RF scanning receiver based on photonics for Electronic Warfare applications. *European Radar Conference (EuRAD)*, Paris, France (09–11 September 2015). New York, US: IEEE. https://doi.org/10.1109/EuRAD.2015.7346271

45 Neri, F. (2018). *Introduction to electronic defense systems*, 3rd ed. Boston, MA: Artech House, Sections 4.2.5 and 4.2.6, pp. 304–347.

46 Kandoian, A.G. (1943). Broad band antenna. US 2,368,663A, filed 15 May 1943 and issued 06 February 1945.

47 Chatterjee, J.S. (1953). Radiation field of a conical helix. *Journal of Applied Physics* 24 (5): 550–559. https://doi.org/10.1063/1.1721328

48 Turner, E.M. (1955). Spiral slot antenna. US 2,863,145A, filed 19 October 1955 and issued 02 December 1958.

49 DuHamel, R., Ore, F. (2003). Logarithmically periodic antenna designs. *IRE International Convention Record*, New York, NY, USA (21–25 March 1966), New York, US: IEEE. https://doi.org/10.1109/IRECON.1958.1150707

50 DuHamel, R., Isbell, D. (2003). Broadband logarithmically periodic antenna structures. *IRE International Convention Record*, New York, NY, USA (21–25 March 1966), New York, US: IEEE. https://doi.org/10.1109/IRECON.1957.1150566

51 Love, A.E.H. (1901). The integration of the equations of propagation of electric waves. *Philosophical Transactions of the Royal Society, Series A* 197: 1–45, 1901.

52 Rumsey, V.H. (1966). *Frequency independent antennas*. New York, NY: Academic Press, Section 1.1, pp. 1–3.

53 Schelkunoff, S.A. (1943). *Electromagnetic waves*. New York, NY: D. Van Nostrand Inc., Section 8.12, pp. 286–290.

54 Rutledge, D., Muha, M. (1982). Imaging antenna arrays. *IEEE Transactions on Antennas and Propagation* 30 (4): 535–540. https://doi.org/10.1109/TAP.1982.1142856

55 Rumsey, V.H. (1966). *Frequency independent antennas*. New York, NY: Academic Press, Section 2.1, pp. 13–16.

56 Compton, R., McPhedran, R., Popovic, Z. et al. (1987). Bow-tie antennas on a dielectric half-space: theory and experiment. *IEEE Transactions on Antennas and Propagation* 35 (6): 622–631. https://doi.org/10.1109/TAP.1987.1144162

57 Gonçalves Licursi de Mello, R., Lepage, A.C., Begaud X. (2022). The bow-tie antenna: performance limitations and improvements. *IET Microwaves, Antennas & Propagation* 16 (5): 283–294. https://doi.org/10.1049/mia2.12242

58 Rumsey, V.H. (1966). *Frequency independent antennas*. New York, NY: Academic Press, Section 2.4, pp. 19–21.

59 Dyson, J. (1959). The equiangular spiral antenna. *IRE Transactions on Antennas and Propagation* 7 (2): 181–187. https://doi.org/10.1109/TAP.1959.1144653

60 McFadden, M., Scott, W.R. (2007). Analysis of the equiangular spiral antenna on a dielectric substrate. *IEEE Transactions on Antennas and Propagation* 55 (11): 3163–3171. https://doi.org/10.1109/TAP.2007.908838

61 Gonçalves Licursi de Mello, R., Junqueira, C. (2017). Polarization diversity on ESM systems. *Journal of Microwaves, Optoelectronics and Electromagnetic Applications* 16 (1): 273–283. https://doi.org/10.1590/2179-10742017v16i1888

62 Sammeta, R., Filipovic, D.S. (2018). Reduced size planar dual-polarized log-periodic antenna for bidirectional high power transmit and receive applications. *IEEE Transactions on Antennas and Propagation* 62 (11): 5453–5461. https://doi.org/10.1109/TAP.2014.2352645

63 Tran, H.H., Park, I. (2016). Wideband circularly polarized cavity-backed asymmetric crossed bowtie dipole antenna. *IEEE Antennas and Wireless Propagation Letters* 15: 358–361. https://doi.org/10.1109/LAWP.2015.2445939

64 Tatomirescu, A., Badescu, A. (2020). Wideband dual-polarized VHF antenna for space observation applications. *MDPI Sensors* 20 (4351): 1–10. https://doi.org/10.3390/s20154351

65 Vahdani, M., Begaud, X. (2008). A directive ultra wideband sinuous slot antenna. *Proceedings of the 1st European Conference on Antennas and Propagation*, Nice, France (06–10 November 2006), New York, US: IEEE. https://doi.org/10.1109/EUCAP.2006.4584965

66 Kim, D., Park, C.Y., Kim, Y. et al. (2022). Four-arm sinuous antenna with low input impedance for wide gain bandwidth. *IEEE Access* 10: 35265–35272. https://doi.org/10.1109/ACCESS.2022.3163821

67 Dorostkar, M.A., Azim, R., Islam, M.T. (2013,). A novel Γ-shape fractal antenna for wideband communications. *Procedia Technology* 11: 1285–1291. https://doi.org/10.1016/j.protcy.2013.12.326

68 Choukiker, Y.K., Behera, S.K. (2017), Wideband frequency reconfigurable Koch snowflake fractal antenna. *IET Microwaves, Antennas & Propagation* 11: 203–208. https://doi.org/10.1049/iet-map.2016.0238

69 Balanis, C.A. (2012). *Advanced engineering electromagnetics*, 2nd ed. New York, NY: Wiley, Section 1.2.2, pp. 3–5.

70 Hayt, W.H., Buck, J.A. (2012). *Engineering electromagnetics*, 8th ed. New York, NY: McGraw-Hill, Section 4.3, Eq. 9, pp. 82.

71 Balanis, C.A. (2012). *Advanced engineering electromagnetics*, 2nd ed. New York, NY: Wiley, Section 7.2, pp. 311–312.

72 Hayt, W.H., Buck, J.A. (2012). *Engineering electromagnetics*, 8th ed. New York, NY: McGraw-Hill, Section 7.6, Eq. 45, pp. 212.

73 Booker, H.G. (1946). Slot aerials and their relation to complementary wire aerials (Babinet's principle)'. *Journal of the Institution of Electrical Engineers - Part IIIA: Radiolocation* 93 (4): 620–626. https://doi.org/10.1049/ji-3a-1.1946.0150

74 Rumsey, V.H. (1966). *Frequency independent antennas*. New York, NY: Academic Press, Section 3.2, pp. 25–28.

75 Balanis, C.A. (2016). *Antenna theory: analysis and design*, 4th ed. New York, NY: Wiley, Section 11.3.1, pp. 594–598.

76 Balanis, C.A. (2016). *Antenna theory: analysis and design*, 4th ed. New York, NY: Wiley, Section 11.4.1, pp. 599–602.

77 Hamza, M., Khan, W.T. (2018). Hybrid utilization of loading techniques and cavity groove for performance enhancement of the UWB (2–18 GHz) spiral antenna. *International Journal of Antennas and Propagation* 2018 (9167154): 1–8. https://doi.org/10.1155/2018/9167154

78 Rozanov, K.N. (2000). Ultimate thickness to bandwidth ratio of radar absorbers. *IEEE Transactions on Antennas and Propagation* 48 (8): 1230–1234. https://doi.org/10.1109/8.884491

79 Balanis, C.A. (2012). *Advanced engineering electromagnetics*, 2nd ed. New York, NY: Wiley, Section 7.4.2, pp. 321–323.

80 Djoma, C., Jousset, M., Lepage, A.C. et al. (2014). Maximal bandwidth of an Archimedean spiral antenna above a reflector. *IEEE Antennas and Wireless Propagation Letters* 13: 333–336. https://doi.org/10.1109/LAWP.2014.2303858

81 Rahman, N.E.A., Sharma, A., Afsar, M.N. et al. (2011). Dielectric characterization and optimization of wide-band, cavity-backed spiral antennas. *The Applied Computational Electromagnetics Society Journal (ACES)* 26 (2): 123–130.

82 Nakano, H., Nogami, K., Arai, S. (1986). A spiral antenna backed by a conducting plane reflector. *IEEE Transactions on Antennas and Propagation* 34 (6): 791–796. https://doi.org/10.1109/TAP.1986.1143893

83 Yang, J.O., Bo, S., Zhang, J. et al. (2011). A low-profile unidirectional cavity-backed log-periodic slot antenna. *Progress in Electromagnetics Research* 119: 423–433. https://doi.org/10.2528/PIER11070503

84 Varault, S., Soiron, M., Barka, A. et al. (2017). RCS reduction with a dual polarized self-complementary connected array antenna. *IEEE Transactions on Antennas and Propagation* 65 (2): 567–575. https://doi.org/10.1109/TAP.2016.2637860

85 Sievenpiper, D.F., Zhang, L., Broas, R.F.J. et al. (1999). High-impedance electromagnetic surfaces with a forbidden frequency band. *IEEE Transactions on Microwave Theory and Techniques* 47 (11): 2059–2074. https://doi.org/10.1109/22.798001

86 Thior, A., Lepage, A.C., Begaud, X. (2009). Low profile, directive and ultra wideband antenna on a high impedance surface. *Proceedings of the 3rd European Conference on Antennas and Propagation*, Berlin, Germany (23–27 March 2009), New York, US: IEEE.

87 Damaj, L., Lepage, A.C., Begaud, X. (2014). Compact wideband antenna above a wideband non-uniform artificial magnetic conductor. *Applied Physics A* 117: 705–711. https://doi.org/10.1007/s00339-014-8726-3

88 Bell, J.M., Iskander, F. (2004). A low-profile Archimedean spiral antenna using an EBG ground plane. *IEEE Antennas and Wireless Propagation Letters* 3: 223–226. https://doi.org/10.1109/LAWP.2004.835753

89 Malvino, A., Bates, D. (2016). *Electronic principles*, 8th ed. New York, NY: McGraw-Hill Ed., Section 2-14, pp. 47.

90 Abbasi, N.A., Langley, R.J. (2011). Multiband-integrated antenna/artificial magnetic conductor. *IET Microwaves Antennas & Propagation* 5 (6): 711–717. https://doi.org/10.1049/iet-map.2010.0200

91 Mouhouche, F., Azrar, A., Dehmas, M. et al. (2018). Gain enhancement of monopole antenna using AMC surface. *Advanced Electromagnetics* 7 (3): 69–74. http://doi.org/10.7716/aem.v7i3.747

92 Paracha, K.N., Abdul Rahim, S.K., Soh, P. et al. (2018). A dual band stub-loaded AMC design for the gain enhancement of a planar monopole antenna. *Microwave* *and Optical Technology Letters* 60 (9): 2108–2112. https://doi.org/10.1002/mop.31308

93 Liu, Q., Liu, H., He, W. et al. (2020). A low-profile dual-band dual-polarized antenna with an AMC reflector for 5G communications. *IEEE Access* 8: 24072–24080. https://doi.org/10.1109/ACCESS.2020.2970473

94 Zhai, H., Zhang, K., Yang, S. et al. (2017). A low-profile dual-band dual-polarized antenna with an AMC surface for WLAN applications. *IEEE Antennas and Wireless Propagation Letters* 16: 2692–2695. https://doi.org/10.1109/LAWP.2017.2741465

95 Tran, H.H., Park, I. (2016). A dual-wideband circularly polarized antenna using an artificial magnetic conductor. *IEEE Antennas and Wireless Propagation Letters* 15: 950–953. https://doi.org/10.1109/LAWP.2015.2483589

96 Grelier, M., Djoma, C., Jousset, M. et al. (2012). Axial ratio improvement of an Archimedean spiral antenna over a radial AMC reflector. *Applied Physics A* 109: 1081–1086. https://doi.org/10.1007/s00339-012-7401-9

97 Tanabe, M., Nakano, H. (2020). Low-profile wideband spiral antenna with a circular HIS reflector composed of homogenous fan-shaped patch elements. *IEEE Transactions on Antennas and Propagation* 68 (10): 7219–7222. https://doi.org/10.1109/TAP.2020.2980365

98 Yang, T., Ren, J., Xi, R.Q. et al. (2021). A low-profile circularly polarized antenna based on a circularly symmetric high impedance surface. *International Journal of RF and Microwave Computer-Aided Engineering* 31 (11): 1–9. https://doi.org/10.1002/mmce.22821

99 Sarrazin, J., Lepage, A.-C., Begaud, X. (2012). Circular high-impedance surfaces characterization. *IEEE Antennas and Wireless Propagation Letters* 11: 260–263. https://doi.org/10.1109/LAWP.2012.2189349

100 Costa, F., Luukkonen, O., Simovski et al. (2011). TE surface wave resonances on high-impedance surface based antennas: analysis and modeling. *IEEE Transactions on Antennas and Propagation* 59 (10): 3588–3596. https://doi.org/10.1109/TAP.2011.2163750

101 Nakano, H., Kikkawa, K., Iitsuka, Y. et al. (2008). Equiangular spiral antenna backed by a shallow cavity with absorbing strips. *IEEE Transactions on Antennas and Propagation* 56 (8): 2742–2747. https://doi.org/10.1109/TAP.2008.927573

102 Nakano, H., Oyanagi, H., Yamauchi, J. (2009). Extremely low-profile, wideband spiral antenna with absorbing material. *Proceedings of 3rd European Conference on Antennas and Propagation*, Berlin, Germany (23–27 March 2009), New York, US: IEEE.

103 Nakano, H., Kikkawa, K., Kondo, N. et al. (2009). Low-profile equiangular spiral antenna backed by an EBG reflector. *IEEE Transactions on Antennas and Propagation* 57 (5): 1309–1318. https://doi.org/10.1109/TAP.2009.2016697

104 Grelier, M., Jousset, M., Mallégol, S. et al. (2010). Wideband QAMC reflector's antenna for low profile applications. *Proceedings of 2nd International Conference Series on Metamaterials, Photonic Crystals and Plasmonics (Meta2010)*, Cairo, Egypt (22–25 February 2010), Paris, France: Metaconferences. https://metaconferences.org/META/files/meta10_proceedings.pdf

105 Liu, C., Lu, Y., Du, C. et al. (2010). The broadband spiral antenna design based on hybrid backed-cavity. *IEEE Transactions on Antennas and Propagation* 58 (6): 1876–1882. https://doi.org/10.1109/TAP.2010.2041147

106 Wang, Z., Karathanasis, K., Volakis, J.L. (2015). Axial ratio reduced ultra wideband slot spiral on hybrid impedance surfaces. *Journal of Electromagnetic Waves and Applications* 29 (2): 143–153. https://doi.org/10.1080/09205071.2014.981035

107 Joshi, C., Lepage, A.C., Sarrazin, J. et al. (2016). Enhanced broadside gain of an ultrawideband diamond dipole antenna using a hybrid reflector. *IEEE Transactions on Antennas and Propagation* 64 (7): 3269–3274. https://doi.org/10.1109/TAP.2016.2565695

108 Tanabe, M., Oyama, M., Oishi, Y. et al. (2015). A spiral antenna over a high-impedance surface consisting of fan-shaped patch cells. *Proceedings of IEEE-APS Topical Conference on Antennas and Propagation in Wireless Communications (APWC)*, Turin, Italy (07–11 September 2015), New York, US: IEEE. https://doi.org/10.1109/APWC.2015.7300221

109 Nakano, H., Oyanagi, H., Yamauchi, J. (2010). Spiral antenna above a composite HIS reflector. *Proceedings of IEEE Antennas and Propagation Society International Symposium*, Toronto, ON, Canada (11–17 July 2010), New York, US: IEEE. https://doi.org/10.1109/APS.2010.5561163

110 Mohamad, S., Cahill, R., Fusco, V. (2014). Selective high impedance surface active region loading of Archimedean spiral antenna. *IEEE Antennas and Wireless Propagation Letters* 13: 810–813. https://doi.org/10.1109/LAWP.2014.2314860

111 Zeppettella, D., Ali, M. (2020). A broadband directional circularly polarized spiral antenna on EBG structure. *Journal of Electromagnetic Waves and Applications* 34 (11): 1563–1585. https://doi.org/10.1080/09205071.2020.1774927

112 Kim, D., Park, C.Y., Lee, C.-H. et al. (2023). Four-arm log-periodic toothed antenna with hybrid reflector composed of inhomogeneous metamaterial elements for electronic support measures. *IEEE Access* 11: 104735–104744. https://doi.org/10.1109/ACCESS.2023.3318871

113 Ciao Wireless, Inc. (2023). Ciao Wireless, Solid State Hybrid RF & Microwave Amplifiers and Sub-Assemblies for every application. https://www.ciaowireless.com/ (accessed 08 August 2023).

114 Gonçalves Licursi de Mello, R., De Sousa, F.R. (2018). Precise techniques to detect superimposed radar pulses on ESM systems. *IET Radar, Sonar & Navigation* 12 (7): 735–741. https://doi.org/10.1049/iet-rsn.2017.0563

7

Metasurface-Driven Electronic Attack

Since the first extensive use of their modern version in WWII, radars have evolved from primitive detection systems to advanced tracking [1], cognitive [2–5], and multifunctional AESA radars [6–8] in a fast pace. Interestingly, the many forms this technology can adopt differ not only in functionality but also in resilience to jamming, setting up the need for a whole plethora of EA techniques instead of a generalized approach. For example, a vulgar application of noise could neutralize the first detection systems [9]. On the other hand, noise jammers could become a beacon even to legacy tracking systems, such as conical-scan and sequential-lobe radars, putting their own platforms at risk in this case [10]. These radars use information from multiple time-sequenced skin returns to resolve the angular position of targets [11]. Against them, the use of angle deception jamming techniques mimicking false sequences would be more suitable [12].

In the cat-and-mouse EW game, monopulse radars have represented a big challenge for EW engineers since their extensive deployment in the theater of operation back in the 1960s [13]. Figure 7.1 illustrates this kind of radar, which uses four feed antennas A, B, C, and D sharing a common reflector. In the receiving, the monopulse radar can reasonably discriminate the skin returns in four receivers A–D, allowing for an amplitude-comparison method [14]. In this way, this radar can resolve for the angular position of targets with a single skin return, differently from its predecessors, which makes it robust to jamming. If one tries to interfere over the monopulse radar with angle deception techniques, they also risk becoming a beacon [15].

In the last decades, EW engineers envisioned many EA techniques to tackle the monopulse radar. Among them, formation jamming, blinking jamming, terrain bounce, skirt jamming, cross-eye jamming, and so forth. These techniques present different levels of implementation complexity and effectiveness [16, 17]. Together with all the possible EP measures, they build up a huge scenario diversification. If we also consider cognitive and multifunctional AESA radars, the possibilities become countless, and the existence of a generalized EA measure seems less and less reasonable.

Metasurface-Driven Electronic Warfare, First Edition. Rafael Gonçalves Licursi de Mello.

Published 2025 by John Wiley & Sons, Inc.
Companion website: www.wiley.com/go/licursidemello/metasurfacedrivenelectronicwarfare1

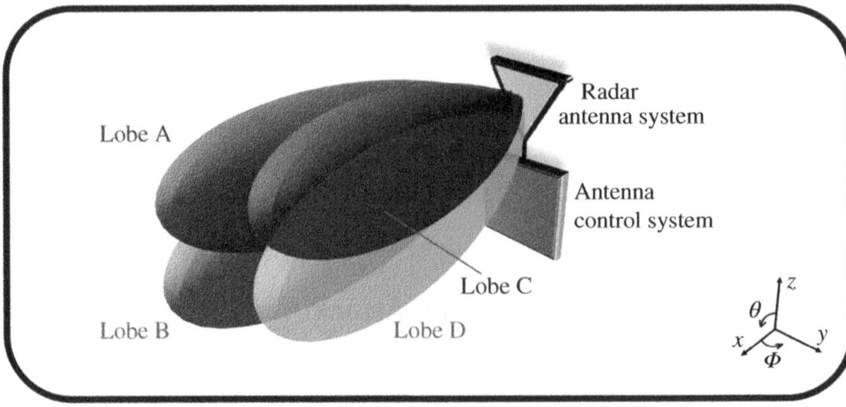

Fig. 7.1 Monopulse radar principle: four independent channels are associated with four feed antennas sharing the same reflector. "Gonçalves Licursi de Mello [14]/MDPI/CC BY 4.0."

Another particularity in EA, as traditionally performed, concerns a propagation disadvantage in relation to the radar, as it will become obvious when we study the jamming-to-signal ratio (JSR) equation.

Metasurfaces can change the game not only by suppressing this disadvantage but also by unlocking the so-needed generalized EA measure. To illustrate that, we will use the monopulse as the target radar. It is true that multifunctional and cognitive radars will predominate in the scenario of the future. However, the particular logic of tracking systems usually is only known by their designers [18], and examples based on these radars would provide a poor generality. The monopulse radar, on the other hand:

- Has a well-known working principle which our EW scenario simulator can model with better generality;
- Has a well-known performance and can serve as a reference to assess the illustrated techniques;
- It is a hard opponent from the EA standpoint, as already stated.

In this chapter, we will demonstrate how to carry out two classical EA measures with metasurfaces. The first is the cross-polarization jamming. Conventionally, cross-polarization jamming is an active, self-screening technique that exploits one vulnerability existing in some radar antennas, like the parabolic dish [19–24]. However, such kind of measure do not affect every radar. F. Neri [25] reports that cross-polarization jamming has not shown any effectiveness against radars with phased-array antennas. To address this case, we will take blinking jamming as a second measure. Traditionally, blinking jamming is also defined as an active, self-screening technique. It employs a formation of two or more aircraft that alternately transmit jamming signals over a target radar [11, 14, 26–30], consisting of a good example of multi-platform jamming.

Naturally, we carry out nowadays these techniques with DRFMs, introduced in Section 5.2. As such, we avoid a huge disadvantage due to the signal processing gain G_{match} of the radar (see Section 5.1). However, we will demonstrate in this chapter that active and even passive metasurfaces can perform the aforesaid techniques, providing the following benefits:

- Metasurfaces do not transmit a jamming signal on their own, and they rather act over the original radar signals; in this case, radars cannot obtain advantages by increasing their power to counter the measure;
- Due to propagation reasons, a burn-through range is inexistent for techniques metasurfaces perform;
- Metasurfaces circumvent the inconveniences of DRFMs earlier seen in Sections 5.2.1–5.2.3 (complexity, power consumption, and need for a dedicated compartment and an antenna); ultimately, we anticipate the use of many lightweight, cheap drones for EA purposes.

Cheap drones are key for the implementation of massive, autonomous swarms. With the illustrated concepts, swarms enable the glimpse of a generalized EA measure to make any tracking impossible, whether from multifunctional, cognitive, monopulse, or any other radar. Further, swarms also allow the envisioning of massive EW networks. Per the latest research, EW networks provide noticeable benefits to perceiving pertinent information for cognitive jamming [31–33]. We will study the latter in Chapter 9, whereas we will introduce the former here. Before that, this chapter will provide a brief introduction to monopulse radars. Then, it will discuss the theory of cross-polarization and blinking jamming, as well as their respective implementations with metasurfaces. We will consider the performance of already fabricated metasurfaces to assess these techniques in our EW scenario simulator.

7.1 The Monopulse Radar

The monopulse radar of Fig. 7.1 has four channels A–D, each one using a single feed antenna symmetrically squinted by θ_{sqt} and Φ_{sqt} in relation to the broadside direction θ_{ant} and Φ_{ant} of a shared reflector [14]. Together, the four reflector-backed antennas A–D constitute the radar antenna system. For simplicity, we consider that the mutual coupling between the feed antennas is negligible. In this way, we identify four identical, isolate co-polarized radiation patterns with main lobes A–D squinted in relation to the broadside direction θ_{ant} and Φ_{ant} in Fig. 7.1. Consequently, channels A–D transmit identical signals independently, and can reasonably discriminate the skin returns of potential targets.

At this point, it is worth recalling the notation we defined earlier in Section 4.1, where $m = h$ or v refers to the polarization of waves in the radar Tx path and $n = h$ or v in the Rx path. For each polarization, we can denote the isolate radiation patterns $G_{A-D,m/n}(\theta, \Phi)$ associated with the respective channels A–D in terms of a reference radiation pattern $G_{0,m/n}(\theta, \Phi)$ squinted in relation to the broadside direction, as follows:

$$G_{A,m/n}(\theta,\Phi) = G_{0,m/n}(\theta - \theta_{sqt} - \theta_{ant}, \Phi + \Phi_{sqt} - \Phi_{ant}) \tag{7.1}$$

$$G_{B,m/n}(\theta,\Phi) = G_{0,m/n}(\theta + \theta_{sqt} - \theta_{ant}, \Phi + \Phi_{sqt} - \Phi_{ant}) \tag{7.2}$$

$$G_{C,m/n}(\theta,\Phi) = G_{0,m/n}(\theta - \theta_{sqt} - \theta_{ant}, \Phi - \Phi_{sqt} - \Phi_{ant}) \tag{7.3}$$

$$G_{D,m/n}(\theta,\Phi) = G_{0,m/n}(\theta + \theta_{sqt} - \theta_{ant}, \Phi - \Phi_{sqt} - \Phi_{ant}) \tag{7.4}$$

Now, consider an aircraft flying at velocity $\boldsymbol{v}_1$ illuminated by the monopulse radar. The radar is at the origin of the cartesian coordinate system, that is, $P(0,0,0)$. The aircraft is in a radius r_{acf} at the elevation angle θ_{acf} and azimuthal angle Φ_{acf}, that is, in the position $P(r_{acf}, \theta_{acf}, \Phi_{acf})$ of the spherical coordinate system indicated in Fig. 7.1. Figure 7.2a illustrates this scenario in the azimuth plane, whereas Fig. 7.2b does so in the elevation plane [14].

We can calculate the skin returns S_{A-D} of the aircraft within each channel A–D by summing all the polarization components as in (4.5). If the radar is pointing the broadside of its antenna system to the aircraft, that is, $\theta_{ant} = \theta_{acf}$ and $\Phi_{ant} = \Phi_{acf}$, we can find such components by using (7.1)–(7.4) in the monostatic radar equation (4.7) as:

$$S_{A,nm} = \frac{P_t \lambda^2}{(4\pi)^3 r_{acf}^4} G_{0,m}(-\theta_{sqt}, +\Phi_{sqt}) G_{0,n}(-\theta_{sqt}, +\Phi_{sqt}) RCS_{nm}(\theta'_{rdr}, \Phi'_{rdr}) \tag{7.5}$$

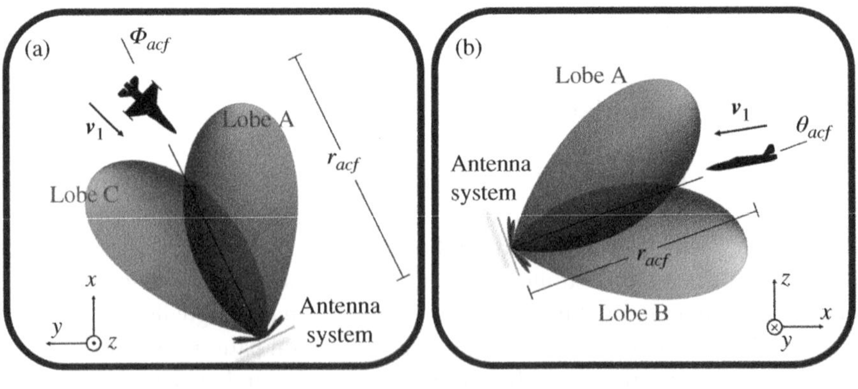

Fig. 7.2 Aircraft illuminated by a monopulse radar: (a) azimuth plane; (b) elevation plane. "Gonçalves Licursi de Mello [14]/MDPI/CC BY 4.0."

$$S_{B,nm} = \frac{P_t\lambda^2}{(4\pi)^3 r_{acf}^4} G_{0,m}(+\theta_{sqt}, +\Phi_{sqt}) G_{0,n}(+\theta_{sqt}, +\Phi_{sqt}) RCS_{nm}(\theta'_{rdr}, \Phi'_{rdr}) \tag{7.6}$$

$$S_{C,nm} = \frac{P_t\lambda^2}{(4\pi)^3 r_{acf}^4} G_{0,m}(-\theta_{sqt}, -\Phi_{sqt}) G_{0,n}(-\theta_{sqt}, -\Phi_{sqt}) RCS_{nm}(\theta'_{rdr}, \Phi'_{rdr}) \tag{7.7}$$

$$S_{D,nm} = \frac{P_t\lambda^2}{(4\pi)^3 r_{acf}^4} G_{0,m}(+\theta_{sqt}, -\Phi_{sqt}) G_{0,n}(+\theta_{sqt}, -\Phi_{sqt}) RCS_{nm}(\theta'_{rdr}, \Phi'_{rdr}) \tag{7.8}$$

Hence, the skin returns S_{A-D} in each channel A–D depend on the squinted radiation patterns $G_{A-D,m/n}(\theta, \Phi)$. In this manner, we can carry out an amplitude comparison to estimate the angles θ_{acf} and Φ_{acf} of the aircraft whenever they diverge from θ_{ant} and Φ_{ant}. To this end, the monopulse radar uses the actual channels A–D to create three virtual channels as:

$$S_{\Sigma} = S_A + S_B + S_C + S_D \tag{7.9}$$

$$S_{\Delta\Phi} = (S_A + S_B) - (S_C + S_D) \tag{7.10}$$

$$S_{\Delta\theta} = (S_A + S_C) - (S_B + S_D) \tag{7.11}$$

where S_{Σ} is known as the sum channel, and $S_{\Delta\Phi}$ and $S_{\Delta\theta}$ are, respectively, the azimuthal and elevation difference channels. Note that, due to the summation (4.5), each term in (7.9) through (7.11) is made up of all the polarization components of $S_{A-D,nm}$. In expanded form, we can thus express (7.9)–(7.11) as:

$$\begin{aligned} S_{\Sigma} = {} & (S_{A,hh} + S_{A,hv} + S_{A,vh} + S_{A,vv}) + (S_{B,hh} + S_{B,hv} + S_{B,vh} + S_{B,vv}) \\ & + (S_{C,hh} + S_{C,hv} + S_{C,vh} + S_{C,vv}) + (S_{D,hh} + S_{D,hv} + S_{D,vh} + S_{D,vv}) \end{aligned} \tag{7.12}$$

$$\begin{aligned} S_{\Delta\Phi} = {} & [(S_{A,hh} + S_{A,hv} + S_{A,vh} + S_{A,vv}) + (S_{B,hh} + S_{B,hv} + S_{B,vh} + S_{B,vv})] \\ & - [(S_{C,hh} + S_{C,hv} + S_{C,vh} + S_{C,vv}) + (S_{D,hh} + S_{D,hv} + S_{D,vh} + S_{D,vv})] \end{aligned} \tag{7.13}$$

$$\begin{aligned} S_{\Delta\theta} = {} & [(S_{A,hh} + S_{A,hv} + S_{A,vh} + S_{A,vv}) + (S_{C,hh} + S_{C,hv} + S_{C,vh} + S_{C,vv})] \\ & - [(S_{B,hh} + S_{B,hv} + S_{B,vh} + S_{B,vv}) + (S_{D,hh} + S_{D,hv} + S_{D,vh} + S_{D,vv})] \end{aligned} \tag{7.14}$$

We can next calculate the azimuth and elevation errors Φ_ϵ and θ_ϵ of the target in relation to the broadside direction of the radar antenna system as:

$$\Phi_\epsilon = k_\Phi \frac{S_{\Delta\Phi}}{S_{\Sigma}} \sim \Phi_{acf} - \Phi_{ant} \tag{7.15}$$

$$\theta_\epsilon = k_\theta \frac{S_{\Delta\theta}}{S_\Sigma} \sim \theta_{acf} - \theta_{ant} \tag{7.16}$$

where k_Φ and k_θ are proportionality constants mapping the sensitiveness of the normalized amplitudes $S_{\Delta\Phi}/S_\Sigma$ and $S_{\Delta\theta}/S_\Sigma$ to such errors.

We use the azimuth and elevation errors Φ_ϵ and θ_ϵ to feed the radar antenna base control system, which steers the antenna system to keep the aircraft at its broadside direction. In this book, we model the monopulse antenna base control system as two independent servos whose behaviors are similar to a simple mass-spring-damper system. The servos make the orientation of the antenna system Φ_{ant} and θ_{ant} converge to the aircraft angular position θ_{acf} and Φ_{acf}, whereby minimizing Φ_ϵ and θ_ϵ:

$$\underbrace{\Phi_\epsilon}_{\substack{\textit{driving}\\\textit{force}}} = \underbrace{m_{ant}\frac{\partial^2 \Phi_{ant}}{\partial t^2}}_{\textit{inertia}} + \underbrace{m_{ant} d_\Phi \frac{\partial \Phi_{ant}}{\partial t}}_{\textit{friction}} - \underbrace{m_{ant}\omega_\Phi^2 \Phi_\epsilon}_{\substack{\textit{restoring}\\\textit{force}}} \tag{7.17}$$

$$\underbrace{\theta_\epsilon}_{\substack{\textit{driving}\\\textit{force}}} = \underbrace{m_{ant}\frac{\partial^2 \theta_{ant}}{\partial t^2}}_{\textit{inertia}} + \underbrace{m_{ant} d_\theta \frac{\partial \theta_{ant}}{\partial t}}_{\textit{friction}} - \underbrace{m_{ant}\omega_\theta^2 \theta_\epsilon}_{\substack{\textit{restoring}\\\textit{force}}} \tag{7.18}$$

where m_{ant} is an equivalent mass of the antenna system, d_Φ and d_θ are damping coefficients, and $\omega_\Phi = 2\pi f_\Phi$ and $\omega_\theta = 2\pi f_\theta$ are the natural angular frequencies of the system in the azimuth and elevation dimensions, respectively.

Fig. 7.3 illustrates the behavior of the antenna control system. The initial orientation is $\Phi_{ant} = 0°$ and $\theta_{ant} = 90°$ but, hereafter in similar graphs, we express the

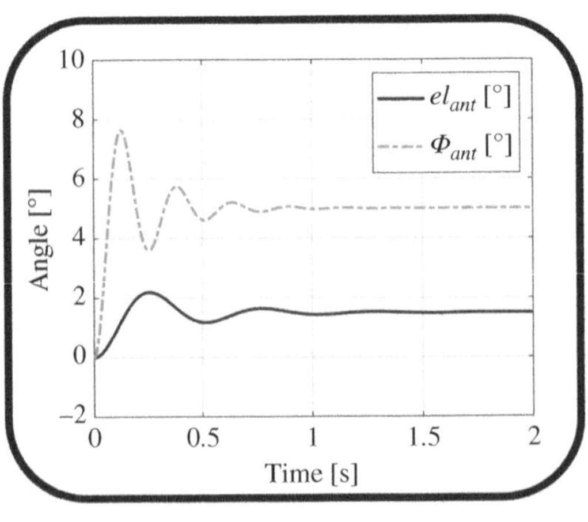

Fig. 7.3 Response of the antenna control system to an aircraft suddenly detected at the angular position el_{acf} = 1.5° and Φ_{acf} = 5°. "Gonçalves Licursi de Mello [14]/MDPI/CC BY 4.0."

elevation angle as $el_{ant} = 90° - \theta_{ant}$. The antenna equivalent mass is $m_{ant} = 10$ kg, the damping coefficients in the azimuthal and elevation dimensions are $d_{\Phi} = 10/\text{s}$ and $d_{\theta} = 6/\text{s}$, and the natural frequencies are $f_{\Phi} = 4$ Hz and $f_{\theta} = 2$ Hz. We run a Matlab loop where angle errors of $\Phi_{\epsilon} = \Phi_{acf} - \Phi_{ant}$ and $\theta_{\epsilon} = \theta_{acf} - \theta_{ant}$ are input to (7.17) and (7.18) at each microsecond. Here, $\Phi_{acf} = 5°$ and $\theta_{acf} = 88.5°$, which simulates that an aircraft is approaching the radar with an elevation angle $el_{acf} = 1.5°$ from the ground.

Observe in Fig. 7.3 how the antenna system converges to the target angular position. Initially, the system oscillates per the natural frequencies f_{Φ} and f_{θ}. As time goes by, the damping coefficients d_{Φ} and d_{θ} attenuate the respective oscillations.

7.2 The Jamming Equation and the Burn-Through Range

Consider a monopulse radar with four channels A–D located at $P(0,0,0)$. A jamming aircraft emits electromagnetic waves with vertical and horizontal polarization, that is, for $n = h$ and $n = v$ from the position $P(r_{acf}, \theta_{acf}, \Phi_{acf})$. The waves impinge over the radar with a density of power [W/m^2]:

$$W_{j,n} = \frac{P_j G_{j,n}\left(\theta'_{rdr}, \Phi'_{rdr}\right)}{4\pi r_{acf}^2} \tag{7.19}$$

where

- P_j is the power transmitted by the jammer [W];
- $G_{j,n}$ is the jammer antenna realized gain in the horizontal or vertical polarization ($n = h$ or v) [dBi];
- θ'_{rdr} and Φ'_{rdr} denote to the angular position of the radar receiver referred to the jammer platform orientation axes [rad].

The effective areas of the four antennas A–D of the monopulse radar collect the power density (7.19) leading, in each channel, to a total power of:

$$J_{A-D} = J_{A-D,h} + J_{A-D,v} \tag{7.20}$$

where

$$\begin{aligned} J_{A-D,n} &= W_{j,n} \frac{\lambda^2}{4\pi} G_{A-D,n}\left(\theta_{acf} - \theta_{ant}, \Phi_{act} - \Phi_{ant}\right) \\ &= \frac{P_j \lambda^2}{\left(4\pi r_{acf}\right)^2} G_{j,n}\left(\theta'_{rdr}, \Phi'_{rdr}\right) G_{A-D,n}\left(\theta_{acf} - \theta_{ant}, \Phi_{act} - \Phi_{ant}\right) \end{aligned} \tag{7.21}$$

We can associate the effectiveness of the EA measure with the ratio J_{A-D}/S_{A-D} between the jamming signals and the *true* skin returns within each channel A–D. For example, consider the specific case where both the jammer and the radar are vertically polarized with perfect polarization purity. By using (7.21) and (4.7), we can express such ratio as:

$$\frac{J_{A-D,v}}{S_{A-D,vv}} = \frac{4\pi r_{acf}^2 P_j G_{j,v}\left(\theta'_{rdr}, \Phi'_{rdr}\right)}{P_t G_{A-D,v}\left(\theta_{acf} - \theta_{ant}, \Phi_{act} - \Phi_{ant}\right) RCS_{vv}\left(\theta'_{rdr}, \Phi'_{rdr}\right)} \tag{7.22}$$

Equation (7.22) is known as the JSR equation (represented here for the specified polarization conditions) [34]. We should retain two important points from (7.22):

- First, the jamming effectiveness depends not only on the jammer itself (through P_j and $G_{j,v}$) but also on the radar performance (through P_t and $G_{A-D,v}$);
- Also, the jamming is more effective for long ranges r_{acf} between the target radar and the jamming platform; this happens because (7.21) is a one-way propagation equation, whereas (4.7) is a round-trip propagation equation.

To put the first point above more explicitly, the radar has a chance to react to the jammer if it detects that it is under attack. About the second point, the relationship between (4.7) and (7.21) is similar to (4.7) and (6.2), used for the ES detection. However, if the ES system needs to detect radars from the farthest possible regions, the goal of the EA system is to jam them from as close as possible so that friendly forces could penetrate the radar zone. Hence, it is fair to consider that it is the radar that has the propagation advantage in the jammer case. Analogous to the ES detection case in Section 6.1, advantage here does not mean that the radar will always overcome the jammer because the other terms in (7.22) can make all the difference.

For analysis purposes, we often consider that an EA technique is successful when a specific effect occurs at each channel A–D. To this end, we consider a minimum JSR in each channel, JSR_{min}, for each technique. The fact that the ratio J_{A-D}/S_{A-D} diminishes with decreasing r_{acf} implies thus a minimum range below which the technique will not succeed. For the specific polarization conditions considered in (7.22), we can express this range as:

$$r_{burnA-D} = \sqrt{\frac{JSR_{\min} P_t G_{A-D,v}\left(\theta_{acf} - \theta_{ant}, \Phi_{act} - \Phi_{ant}\right) RCS_{vv}\left(\theta'_{rdr}, \Phi'_{rdr}\right)}{4\pi P_j G_{j,v}\left(\theta'_{rdr}, \Phi'_{rdr}\right)}} \tag{7.23}$$

The range r_{burn} is known in the literature as the burn-through range [35]. It establishes a zone within which the channels A–D correctly process the skin returns despite the jamming signals. Within this zone, we often consider that the aircraft is under the same risks as it would be if the EA measure did not exist.

Although (7.22) and (7.23) provide valuable insight into the behavior of EA measures, for the purposes of this book, we can consider them overly simplified representations of what happens in the real world. In a dynamic scenario, platforms and antennas are subjected to movements and rotations. Antennas can also perform scans. According to (7.1) through (7.4), the gain $G_{A-D,m/n}(\theta, \Phi)$ of the monopulse radar antennas can make both skin returns S_{A-D} and jamming signals J_{A-D} vary dramatically according to the angular disposition between antennas and platforms in the scenario. Further, it turns out that the jamming signals may considerably affect one channel but not others due to the fact that the antennas A–D are squinted. Besides that, the radar antenna control system presents inertial and oscillatory features, as represented by (7.17) and (7.18). All that said, the radar tracking may behave differently depending not only on *where* but also on *when* the jammer is activated. Fortunately, the EW scenario simulator available on this book's companion website takes all this into account. In this way, we can study next two classical EA measures with metasurfaces, arriving later at a generalized approach.

7.3 Cross-Polarization Jamming with Metasurfaces

Cross-polarization jamming exploits the fact that some radar antennas present lobes in the cross-polarized radiation pattern squinted in relation to the main lobe of the co-polarized pattern.

As an example, consider an antenna consisting of a vertically polarized dipole illuminating a parabolic dish. We generate such an antenna with the Matlab Antenna Toolbox [36]. Specifically, we use the *reflectorParabolic* object with a radius of 5λ and a focal length of 2.5λ at 10 GHz. Fig. 7.4a,b respectively depicts the vertically and horizontally polarized radiation patterns using the *pattern* function of the Antenna Toolbox, both normalized per the maximum value of Fig. 7.4a.

In Fig. 7.4a, a single, vertically polarized, main lobe exists pointing at the broadside direction of the parabolic dish. On the other hand, Fig. 7.4b shows four significant, horizontally polarized lobes squinted in relation to the broadside direction. It is worth mentioning that, even though the horizontally polarized, squinted lobes exist, their maximum gain is lower than the one of the vertically polarized, main lobe (approximately 16 dB lower here).

Many works exist on cross-polarization jamming [19–24]. However, to the best of the author's knowledge, none provides a reasonably complete measurement analysis. A significant reason for this gap is because of the strictly defense nature of the application, involving classified information. Other reasons are related to the complexity of measurement campaigns and the constant updates required by new technologies, as we discussed earlier in Section 5.5.

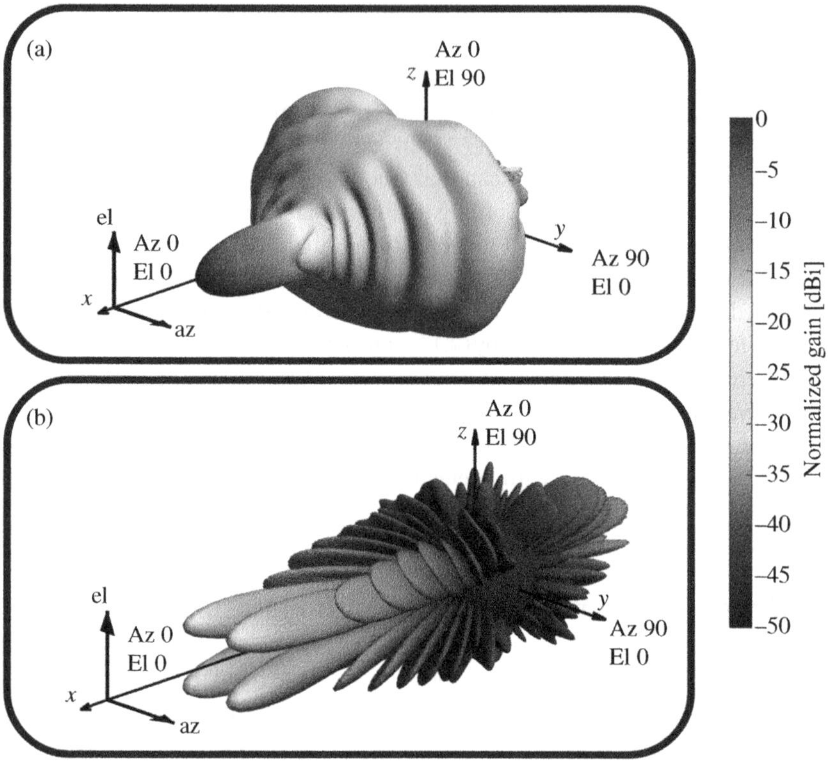

Fig. 7.4 Typical radiation patterns of a parabolic dish antenna: (a) co-polarized component (vertical polarization); (b) cross-polarized component (horizontal polarization)

7.3.1 Working Principle with Conventional EA Systems

To illustrate the technique, consider a monopulse radar with four feed antennas A–D sharing a common parabolic dish. Consider that the radiation patterns of this antenna system are equivalent to four isolate co- and cross-polarized radiation patterns as those in Fig. 7.4, each one squinted by θ_{sqt} and Φ_{sqt} in relation to the broadside direction θ_{ant} and Φ_{ant} of the parabolic dish. We designed this particular antenna system to be vertically polarized, that is, usually the $S_{A-D,vv}$ components will be much larger than all the others in (7.12) through (7.14). However, a jammer platform equipped with a DRFM intercepts the radar signals, retransmitting them with a high power toward the radar in the cross-polarization, that is, with horizontal polarization.

Consider that the jamming signals suffer the same attenuations and amplifications as those of the jammer platform skin returns in the radar processing. This assumption is as reasonable as the signals reproduced by the DRFM have a waveform similar to the original skin returns to benefit from a similar signal processing gain G_{match} in the target radar, as explained in Section 5.2. Thus, considering the jamming signals expressed by (7.21), the virtual channels earlier represented by (7.12)–(7.14) become:

$$S_{\Sigma} = (S_{A,vv} + J_{A,h}) + (S_{B,vv} + J_{B,h}) + (S_{C,vv} + J_{C,h}) + (S_{D,vv} + J_{D,h}) \tag{7.24}$$

$$S_{\Delta\Phi} = [(S_{A,vv} + J_{A,h}) + (S_{B,vv} + J_{B,h})] - [(S_{C,vv} + J_{C,h}) + (S_{D,vv} + J_{D,h})] \tag{7.25}$$

$$S_{\Delta\theta} = [(S_{A,vv} + J_{A,h}) + (S_{C,vv} + J_{C,h})] - [(S_{B,vv} + J_{B,h}) + (S_{D,vv} + J_{D,h})] \tag{7.26}$$

When the jamming signals $J_{A-D,h}$ are significantly stronger than the skin returns $S_{A-D,vv}$ in (7.24) through (7.26), the virtual channels will be mainly weighted through the cross-polarization patterns $G_{A-D,h}(\theta, \Phi)$ of the radar antennas A–D, each one represented by a squinted version of Fig. 7.4b. However, in the radar design, we defined the co-polarized radiation patterns $G_{A-D,v}(\theta, \Phi)$ of the radar antennas A–D, each one represented by a squinted version of Fig. 7.4a, as references for the calculation of the angular errors Φ_ϵ and θ_ϵ with (7.15) and (7.16). Hence, in the illustrated situation, the shape of the cross-polarized patterns will produce erroneous calculations. Consequently, feeding Φ_ϵ and θ_ϵ into the radar antenna base control system represented by (7.17) and (7.18) will make the radar orient its antenna system per the wrong shape, missing the actual angular position of the jammer platform.

In terms of effectiveness, we can derive the JSR within each channel A–D for the illustrated cross-polarization jamming from (7.21) and (4.7) as:

$$\frac{J_{A-D,h}}{S_{A-D,vv}} = \frac{4\pi r_{acf}^2 P_j G_{j,h}(\theta'_{rdr}, \Phi'_{rdr}) G_{A-D,h}(\theta_{acf} - \theta_{ant}, \Phi_{act} - \Phi_{ant})}{P_t G_{A-D,v}^2(\theta_{acf} - \theta_{ant}, \Phi_{act} - \Phi_{ant}) RCS_{vv}(\theta'_{rdr}, \Phi'_{rdr})} \tag{7.27}$$

and we can calculate the burn-through range for each channel A–D as:

$$r_{burnA-D} = \sqrt{\frac{JSR_{\min} P_t G_{A-D,v}^2(\theta_{acf} - \theta_{ant}, \Phi_{act} - \Phi_{ant}) RCS_{vv}(\theta'_{rdr}, \Phi'_{rdr})}{4\pi P_j G_{j,h}(\theta'_{rdr}, \Phi'_{rdr}) G_{A-D,h}(\theta_{acf} - \theta_{ant}, \Phi_{act} - \Phi_{ant})}} \tag{7.28}$$

Observe in (7.27) and (7.28) that the radar can react by simply increasing the transmitted power P_t if it somehow detects that it is under cross-polarization jamming (a metasurface-based technique with this purpose will be presented in Chapter 8). Moreover, even though parabolic dishes present cross-polarized lobes, the gain of the co-polarized patterns at most of the directions is typically much higher, as already stated. Hence, the presence of $G_{A-D,h}(\theta, \Phi)/G_{A-D,v}(\theta, \Phi)$ in (7.27) and (7.28) may imply in a significant cross-polarization disadvantage for the jammer at most of the directions. We define here such disadvantage as $G_{x-pol}(\theta, \Phi)$, which may require the jamming signals to be much stronger than the skin returns. F. Neri [25] states that the jamming signal impinging over the radar should typically be 25–30 dB stronger than the skin returns for effectiveness. Although newsworthy, we consider this statement too simplistic for the reasons stated at the end of Section 7.2.

The cross-polarization disadvantage $G_{x-pol}(\theta, \Phi)$ is heavily dependent on the rotations and scans of the radar antennas and the platform movements in a dynamic scenario. This dependence appears in (7.1)–(7.4) but becomes even clearer by comparing the co- and cross-polarized patterns in Fig. 7.4. Hence, we could hardly consider (7.27) and (7.28) in the analysis of a dynamic scenario. Nonetheless, our EW scenario simulator can take not only such angular dependence into account but also the complexity of the radar antenna control system, represented by (7.17) and (7.18).

7.3.2 Working Principle with Metasurfaces

We can also carry out cross-polarization jamming with passive, power-amplifying, and reconfigurable polarization-conversion metasurfaces, the first two introduced in Sections 4.4.3 and 5.4.3, and the last one briefly commented in Section 6.4.2. From a radar standpoint, these metasurfaces act over the terms RCS_{nm} of the $\widehat{RCS}$ tensor expressed by (4.2), ultimately impacting the skin returns obtained with (4.6) or (4.7).

For the sake of exemplification, consider again the same vertically polarized monopulse radar from Section 7.3.1. This time, however, the platform the radar illuminates does not have a DRFM, being rather coated by polarization-conversion metasurfaces. Consider that these metasurfaces do not impact the waveform of the reflected waves, changing only the polarization of a certain amount of energy from vertical to horizontal. As such, the signals reflected by the metasurfaces will again be subjected to the same attenuations and amplifications as those of the original skin returns in the radar processing. In addition, the $\widehat{RCS}$ of the platform will be mainly made by the terms RCS_{hv} and RCS_{vv}, with $RCS_{hh} = RCS_{vh} \cong 0$. Using these terms in (7.5)–(7.8), we can express the virtual channels (7.12)–(7.14) as:

$$S_{\sum} = (S_{A,hv} + S_{A,vv}) + (S_{B,hv} + S_{B,vv}) + (S_{C,hv} + S_{C,vv}) + (S_{D,hv} + S_{D,vv}) \tag{7.29}$$

$$S_{\Delta\Phi} = [(S_{A,hv} + S_{A,vv}) + (S_{B,hv} + S_{B,vv})] - [(S_{C,hv} + S_{C,vv}) + (S_{D,hv} + S_{D,vv})] \tag{7.30}$$

$$S_{\Delta\theta} = [(S_{A,hv} + S_{A,vv}) + (S_{C,hv} + S_{C,vv})] - [(S_{B,hv} + S_{B,vv}) + (S_{D,hv} + S_{D,vv})] \tag{7.31}$$

In (7.29) through (7.31), it is the ratio between the respective terms $S_{A-D,hv}$ and $S_{A-D,vv}$ that determines whether the shape of the co- or the cross-polarized patterns of the radar antennas A–D will prevail in the virtual channels. In other words, now, it is the ratio $S_{A-D,hv}/S_{A-D,vv}$ that will disturb the calculation of the angular errors Φ_ϵ and θ_ϵ with (7.15) and (7.16) causing a wrong orientation of the radar antenna system using (7.17) and (7.18). We can calculate such ratio using (7.5)–(7.8) as:

$$\frac{S_{A-D,hv}}{S_{A-D,vv}} = \frac{\dfrac{P_t\lambda^2}{(4\pi)^3 r_{acf}^4} G_{A-D,v}(\theta,\Phi)G_{A-D,h}(\theta,\Phi)RCS_{hv}\left(\theta'_{rdr},\Phi'_{rdr}\right)}{\dfrac{P_t\lambda^2}{(4\pi)^3 r_{acf}^4} G_{A-D,v}(\theta,\Phi)G_{A-D,v}(\theta,\Phi)RCS_{vv}\left(\theta'_{rdr},\Phi'_{rdr}\right)} = \frac{G_{A-D,h}(\theta,\Phi)RCS_{hv}\left(\theta'_{rdr},\Phi'_{rdr}\right)}{G_{A-D,v}(\theta,\Phi)RCS_{vv}\left(\theta'_{rdr},\Phi'_{rdr}\right)} \tag{7.32}$$

Eq. (7.32) makes evident three main points:

a) To start with, a burn-through range does not exist for the metasurface case simply because all signals in (7.29)–(7.31) perform a round-trip propagation between the radar and the platform;
b) Second, the radar cannot react to the metasurface-driven cross-polarization jamming by increasing the transmitted power because P_t does not appear in the equation (this condition holds as long as the metasurface response remains linear);
c) However, the angularly dependent disadvantage $G_{x-pol}(\theta, \Phi) = G_{A-D,h}/G_{A-D,v}$ that occurs for the cross-polarization jamming with DRFMs is also present in the metasurface case.

To obtain a large ratio $S_{A-D,hv}/S_{A-D,vv}$ in (7.32), the ratio $RCS_{x-pol} = RCS_{hv}\left(\theta'_{rdr},\Phi'_{rdr}\right)/RCS_{vv}\left(\theta'_{rdr},\Phi'_{rdr}\right)$ should exceed $G_{x-pol}(\theta, \Phi)$ accordingly. The RCS_{x-pol} ratio will primarily depend on two factors:

d) On the PCR of the metasurface, defined by (4.15);
e) On the deployment rate previously studied in Section 4.6.2.

Hence, in terms of effectiveness, double attention should be paid to coating the entireness of the platform if we use passive metasurfaces [37–39]. We saw in Section 4.4.3 that the working principle of this kind of metasurface is just to impose different reflection coefficient phases φ_r to orthogonal polarization components of impinging waves. Thus, if this mechanism is perfect in a particular metasurface, the PCR equals 1. In this case, all the energy of waves impinging over the metasurface has its polarization converted with no power amplification in the reflection. If in the platform in question the deployment rate is also perfect, the terms $S_{A-D,vv}$ will vanish. In this case, the RCS_{x-pol} ratio tends to infinite, making the cross-polarization jamming unavoidably succeed. However, all it takes is a part of the aircraft not to be coated and a certain amount of reflected energy will keep the vertical polarization. In this case, a poor RCS_{x-pol} ratio may be insufficient for a successful jamming, particularly when a significant gain disadvantage $G_{x-pol}(\theta, \Phi)$ exists at the direction the radar antenna system is pointing.

A last point to note for passive metasurfaces is that they can only carry out static cross-polarization jamming. Static here means that the radar will stably point in the wrong direction if the technique succeeds. The jammer platform, in turn, can do nothing to control the radar in a dynamic manner.

On the other hand, reconfigurable metasurfaces [40–43] can change the RCS_{x-pol} ratio in a dynamic manner. This is possible because their unit cells employ reconfigurable components, such as varactor or PIN diodes, to switch the polarization conversion on and off, as we commented in Section 6.4.2. As such, the platform can interchangeably "attract" the co- and cross-polarized lobes of the radar antennas, playing with the inertial and oscillatory features of the control system inherently present in (7.17) and (7.18). This advantage comes at the expense of higher complexity and costs in relation to passive metasurfaces. Also, we give birth to power consumption that was inexistent before because the reconfigurable components require feeding lines for the biasing thereof.

Nothing changes from the passive to the reconfigurable metasurface case with regard to the severe deployment rate requirement. However, power-amplifying metasurfaces [44–47] have considerable advantages in this matter. We saw in Section 5.4.3 that these metasurfaces typically employ amplifiers in a right-angle transmission line between two orthogonal slots or metal patches. As such, the metasurface couples the energy of impinging waves and then reradiates it in an orthogonal polarization with a power amplification. Consequently, even if we do not coat the platform entirely, a case where significant terms $S_{A-D,vv}$ may exist in (7.29)–(7.31), the power amplification can still provide terms $S_{A-D,hv}$ large enough for satisfactory ratios in (7.32).

The power amplification is also key for the following two factors:

- The amplification gain can provide RCS_{x-pol} ratios strong enough to overcome significant disadvantages $G_{x-pol}(\theta, \Phi)$ at any direction in relation to the radar, ensuring an unconditionally successful jamming;
- The amplification can also act in a dynamic manner, allowing once again the jammer to control the RCS_{x-pol} ratio and play with the inertial and oscillatory features of the control system present in (7.17) and (7.18).

Of course, these advantages do not come for free again. We expect increased complexity, costs, and power consumption in relation to the passive and reconfigurable metasurfaces discussed above. Now, the power consumption relates not only to a simple biasing of components but also to the power amplification itself. Still, the power consumption related to a digital processing, such as in DRFMs, does not exist.

7.4 Simulation: Comparative Study on Cross-Polarization Jamming

In the sequel, we aim to make it evident the differences between carrying out cross-polarization jamming with conventional EA systems or metasurfaces. In this context, our EW scenario simulator provides sufficiently high levels of realism when handling all the dynamism related not only to moving platforms but also to the rotations and scans of antennas.

Consider a scenario where a monopulse radar composed of four channels A–D, like the one described in Section 7.1, is located at the position $P(0,0,0)$ and operates at f_0 = 10.0 GHz. Fig. 7.1 indicates the coordinate system. For convenience, we model each of the four feeding antennas A–D as a *phased.CustomAntennaElement* object. The antennas are identical. Each one is made of a vertically polarized dipole illuminating a parabolic dish, as described in Section 7.3, and presents the radiation patterns earlier depicted in Fig. 7.4. Their gain is 26.4 dBi, and the main-plane 3-dB beamwidths are $\theta_{3dB} = 8.40°$ and $\Phi_{3dB} = 5.92°$. The squint angles in relation to the broadside direction θ_{ant} and Φ_{ant} of the antenna system are $\theta_{sqt} = \theta_{3dB}/4$ and $\Phi_{3dB} = \Phi_{sqt}/4$. Figure 7.5 depicts all the co-polarized radiation patterns together in

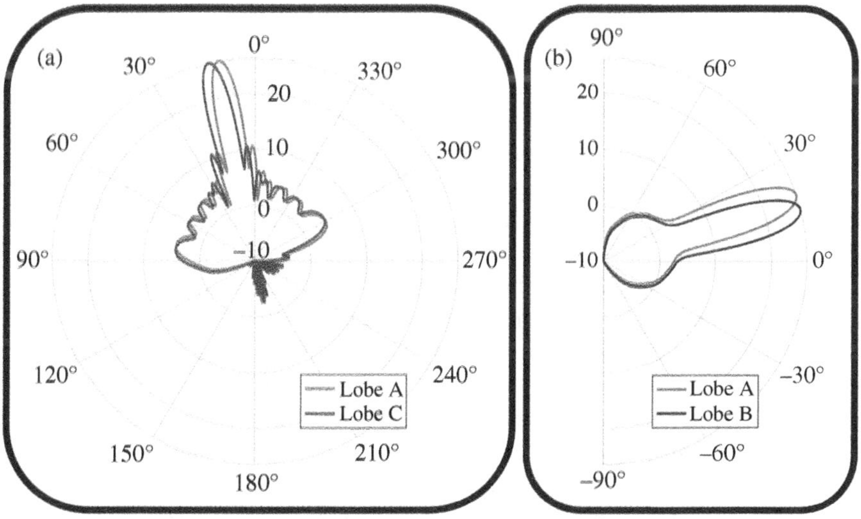

Fig. 7.5 Radiation patterns of the antennas A–D in (a) azimuth; (b) elevation (90° − θ)

the azimuth and elevation planes considering the antenna system in its initial position, which in turn is $el_{ant} = 90° - \theta_{ant} = 17°$ and $\Phi_{ant} = 12°$.

In the simulations, we consider that the mutual coupling between the antennas A–D is negligible. Therefore, the respective receivers A–D can perfectly discriminate the signals they intercept. An aircraft is flying toward this radar at Mach 2 ($v_1 = 686$ m/s) at a dynamic position $P(r_{acf}, \theta_{acf} = 65°, \Phi_{acf} = 20°)$. Note that the elevation of the aircraft, from the ground, is $el_{acf} = 90° - \theta_{acf} = 25°$. This angular disposition allows us to consider, for simplicity, a constant aircraft RCS which initially is $RCS_{hh} = RCS_{vv} = 1$ m^2 and $RCS_{hv} = RCS_{vh} = 0$.

We design a radar waveform with chirp pulses with $PRF = 12.5$ kHz, $PW = 1$ μs, and $BW = 50$ MHz. We take the CA812-3111 LNA from Ciao Wireless, Inc. [48] (gain of 27 dB and noise figure of 1.6 dB) as a reference for realistic features in the Rx front end. In the Tx front end, on the other hand, we consider the CA812-6116 PA, also from Ciao Wireless, Inc. (gain of 30 dB). The number of non-coherently integrated pulses is 20.

We model the monopulse tracker through the implementation of the antenna control system as in (7.17) and (7.18). The antenna equivalent mass is $m_{ant} = 10$ kg, and the damping coefficients are $d_\theta = 9$/s and $d_\Phi = 18$/s. The system's natural frequencies are $f_\theta = 2$ Hz and $f_\Phi = 4$ Hz. Lastly, we model a missile guided by the monopulse radar and flying at Mach 10 ($\mathbf{v}_1 = 3430$ m/s) with a *phased.Platform* object for visualization purposes.

7.4.1 Initial Results for Conventional EA Systems and Metasurfaces

In the first simulation, we check the performance of a conventional EA system carrying out cross-polarization jamming with a DRFM. We model the DRFM like in Section 5.7.2 except for two points. First, the features of the Tx and Rx front ends follow those we use in the monopulse radar (LNA with 27 dB of gain and 1.6 dB of noise figure and PA with 30 dB). Second, its antenna elements are horizontally polarized. At this stage, we consider the DRFM ideal; that is, the signals it transmits are a perfect copy of the signals in its receiver. Further, the DRFM receives information about the radar location from an ES system and points the antenna beams accordingly.

We consider that at $t = 0$ the monopulse radar receives information from an acquisition radar about an aircraft located at $r_{acf} = 10$ km. We use the *phased.ScenarioViewer* object to depict this scenario in Fig. 7.6. This is the beginning of the simulation, whose results Fig. 7.7a shows in terms of angular position of the monopulse radar antenna system. First, observe the convergence from the initial position, $el_{ant} = 17°$ and $\Phi_{ant} = 12°$, to the aircraft's angular position, $el_{acf} = 25°$ and $\Phi_{acf} = 20°$. As expected, some oscillations occur. The DRFM is turned on at 1.5 s when $r_{acf} = 9$ km, thereby introducing energy into the radar receivers through

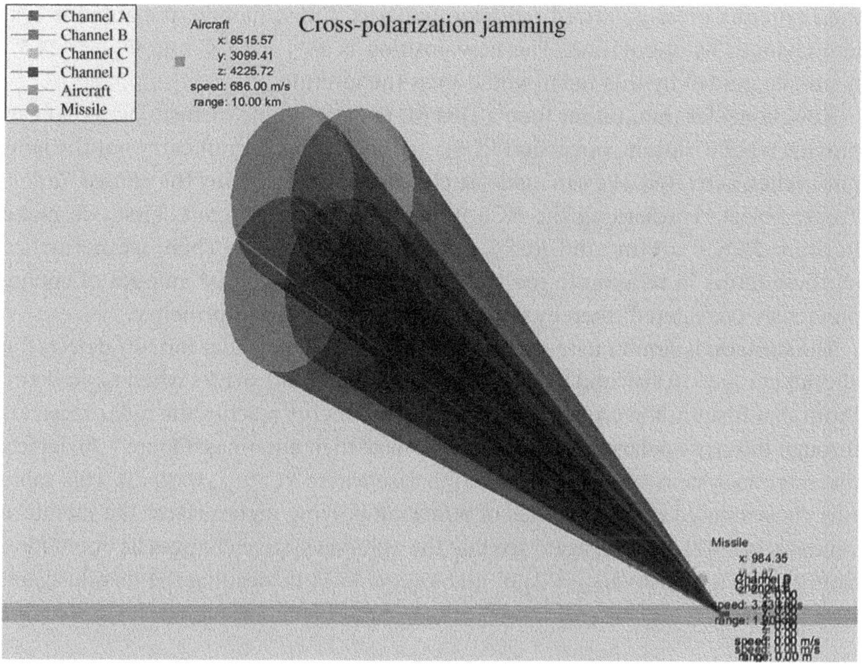

Fig. 7.6 EW scenario represented with the *phased.ScenarioViewer* object

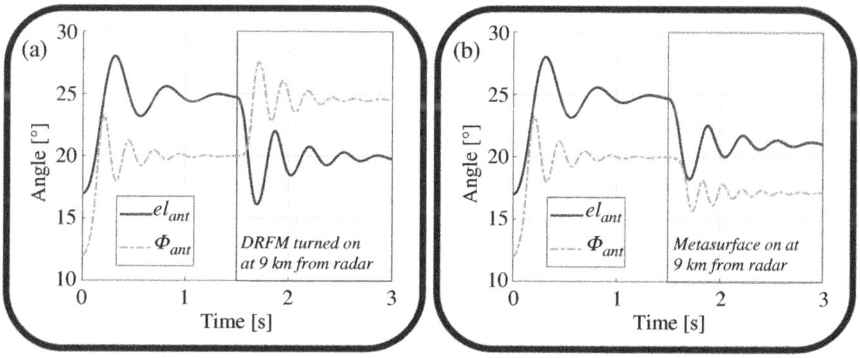

Fig. 7.7 Response of the antenna system to cross-polarization jamming carried out from r_{acf} = 9 km: (a) conventional EA system case; (b) reconfigurable metasurface case

their antenna cross-polarized radiation patterns. Consequently, the antenna system changes its orientation. The new position is $el_{ant} = 19.8°$ and $\Phi_{ant} = 24.5°$. A missile guided by this radar would miss the aircraft.

Now, consider that, rather than a DRFM, reconfigurable metasurfaces that can provide a polarization-conversion RCS_{x-pol} ratio to the aircraft carry out the jamming when activated. We can model such scenario by changing the *phased.RadarTarget* object representing the RCS tensor (4.2) of the aircraft. First, we make $RCS_{hv} = RCS_{vh} = 1$ m^2 and $RCS_{hh} = RCS_{vv} = 1/RCS_{x-pol}$. Then, we normalize all these terms in relation to the total power to keep the total amount of energy previously considered, thereby respecting the conservation principle.

The scenario is similar to the one related to Fig. 7.7a: the radar initially detects the aircraft at $r_{acf} = 10$ km, and the jamming is turned on at $t = 1.5$ s when $r_{acf} = 9$ km. From that instant, the cross-polarized, reflected energy reaches the radar receivers through the cross-polarized radiation patterns of their antennas. Figure 7.7b depicts the antenna system angular position for a case where $RCS_{x-pol} = 30$ dB. This value was chosen based on the features of power-amplifying metasurfaces the literature validates [44–47]. We can again see that the antenna system changes its orientation from $t = 1.5$ s, now to $el_{ant} = 21.0°$ and $\Phi_{ant} = 17.1°$ misleading the missile again.

7.4.2 Burn-Through Range Mitigation

Now, we are interested in making it evident the statement (a) of Section 7.3.2, namely, that a burn-through range that does exist for conventional EA approaches indeed does not exist for the metasurface case.

First, we examine what happens when the radar detects the aircraft at a range as close as $r_{acf} = 2.5$ km from it. Figure 7.8a illustrates the results for the DRFM case. The jamming is turned on at $t = 1.5$ s when $r_{acf} = 1.5$ km. However, this time, nothing happens to the antenna system. The main reason is that the jamming signals are not strong enough at this range to overwhelm the skin returns within the radar system. Translating the discussion to the terminology of (7.27), the JSR within each channel A–D, $J_{A-D,h}/S_{A-D,vv}$, does not reach $JSR_{\min}$ because r_{acf} is too low. In other words, the aircraft is within the burn-through range.

Following, we study the same scenario for the reconfigurable metasurface case. Fig. 7.8b presents the results. They are practically identical to those of Fig. 7.7b. They confirm our analytical prediction based on (7.32) concerning the inexistence of a burn-through range for the metasurface case.

At this point, the reader may wonder what happens if the aircraft activates cross-polarization jamming at $r_{acf} = 9$ km (like in Section 7.4.1) instead of activating it at $r_{acf} = 1.5$ km and indefinitely approaches the radar right after. Fig. 7.9a sketches the simulated results for this situation. We can see that the measure initially has an

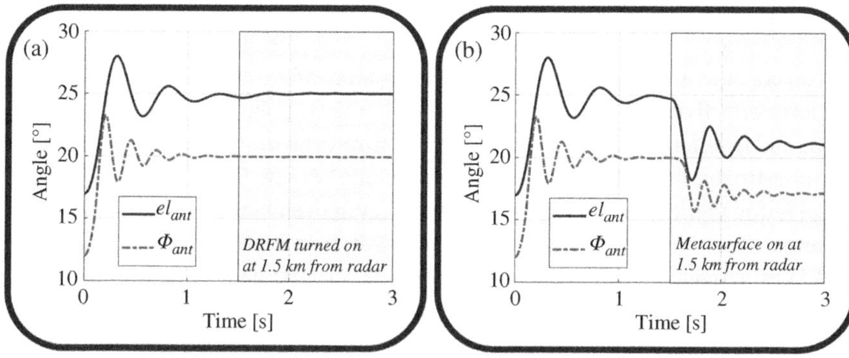

Fig. 7.8 Response of the antenna system to cross-polarization jamming carried out from r_{acf} = 1.5 km: (a) conventional EA system case; (b) reconfigurable metasurface case

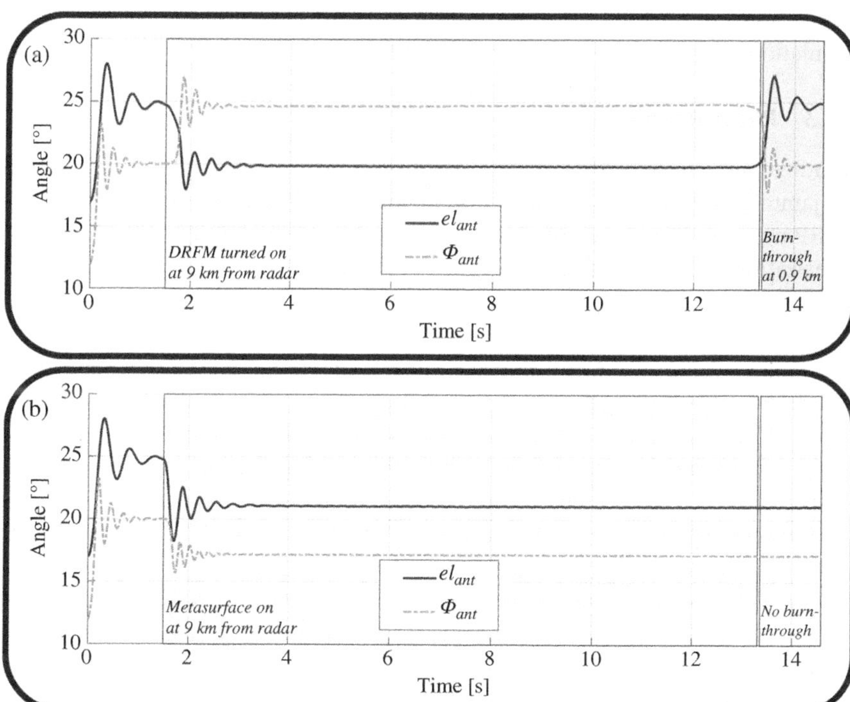

Fig. 7.9 Results for cross-polarization jamming carried out from r_{acf} = 9 km to r_{acf} = 0: (a) conventional EA system case; (b) reconfigurable metasurface case

impact on the radar for the DRFM case. However the range becomes too low at $t = 13.3$ s and $J_{A-D,h}/S_{A-D,vv}$ gets below $JSR_{\min}$ (specifically, $r_{acf} = 0.9$ km, which we can consider the burn-through range for such configuration). Hence, the measure loses effectiveness. From that moment on, the co-polarized radiation patterns of the radar antennas prevail, and the system returns to its previous position.

Interestingly, this range is shorter than the one we saw in Fig. 7.8a. The shorter range is due to the fact that the radar antenna system stables the cross-polarization patterns over the aircraft and the co-polarization patterns elsewhere after the DRFM is activated. In this steady situation, a specific gain disadvantage G_{x-pol} takes place. On the other hand, in the previous situation of Fig. 7.8a, the jamming was activated when the system was still oscillating, allowing a larger receiving of energy through the co-polarization patterns at some moments.

This analysis demonstrates the usefulness of the EW scenario simulator to the detriment of simpler approaches strictly based on the JSR equation, as we deliberated at the end of Section 7.2. Finally, Fig. 7.9b shows the results for the metasurface case where, of course, the impact over the radar lasts until the very end of the simulation.

7.4.3 Radar Reaction

Another configuration that deserves attention is the one where the radar reacts to the jamming by increasing its transmitted power P_t. We can see such a reaction as an EP measure. Here, we are interested in making it evident the fact that cross-polarization jamming with metasurfaces is invulnerable to this measure, as item (b) of Section 7.3.2 previously stated.

We start by simulating again a scenario with the aircraft activating the jamming at $t = 1.5$ s when $r_{acf} = 9$ km. However, now we assume that the radar somehow detects that it is under jamming at $t = 3.0$ s, increasing of 30 dB its transmitted power P_t as a reaction. Figure 7.10a shows the results for the DRFM case. We can see that the radar succeeds in pointing its antenna system back over the aircraft after the reaction, indicating that the jamming lost effectiveness. Referring to the terminology of (7.27) again, increasing P_t yields a larger $S_{A-D,vv}$ avoiding the JSR within the channels A–D to reach $JSR_{\min}$.

Now, Fig. 7.10b illustrates the results for the case when metasurfaces carry out the jamming. As expected, nothing changes after the radar reaction because P_t does not appear in (7.32). This confirms that the metasurface implementation is indeed invulnerable to such kind of EP.

7.4.4 Impact of Polarization-Conversion Ratio/Deployment Rate

A last point we are interested in is the performance of the metasurface-driven cross-polarization jamming as a function of the polarization-conversion RCS_{x-pol} ratio the metasurfaces provide to the aircraft. This is a matter of first importance

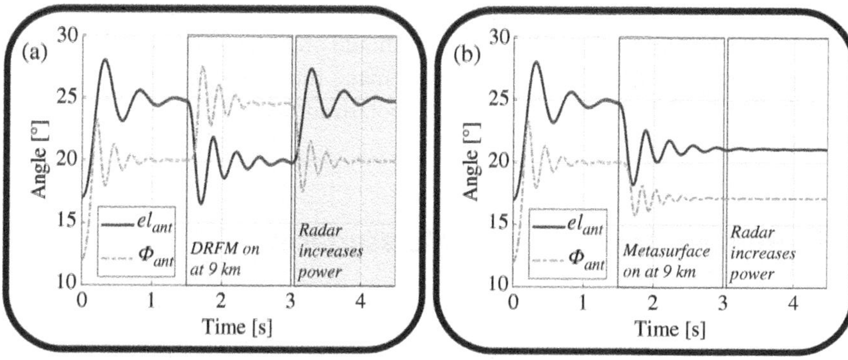

Fig. 7.10 Response of the antenna system to cross-polarization jamming when the radar reacts: (a) conventional EA system case; (b) reconfigurable metasurface case

because, even though RCS_{x-pol} naturally depends on the performance of the metasurface, a poor deployment rate can significantly degrade such figure of merit, as discussed in items (d) and (e) of Section 7.3.2.

Simulating the impact of the polarization-conversion ratio is straightforward by repeating the simulation whose results Fig. 7.7b represents with other values for RCS_{x-pol}. By gradually degrading RCS_{x-pol} in our simulations, we find that the jamming succeeds in the current scenario as long as $RCS_{x-pol} = 19$ dB. Figure 7.11a depicts the results for this case where we see that the antenna system still deviates from the original aircraft angular position after we activate the jamming. Contrarywise, Fig. 7.11b shows results for RCS_{x-pol} = 15 dB where we see that the radar indeed does not lose tracking of the aircraft.

These results corroborate with the fact that the deployment rate is critical in the application in question. Hence, we should pay special attention to the coating of

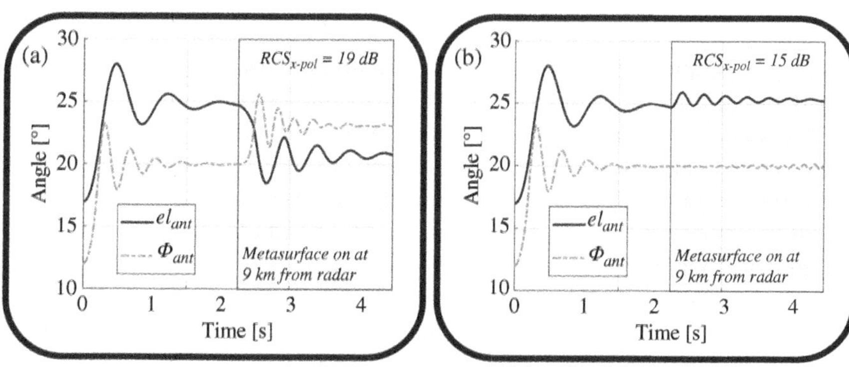

Fig. 7.11 Response of the antenna system to metasurface-driven cross-polarization jamming with: (a) RCS_{x-pol} = 19 dB; (b) RCS_{x-pol} = 15 dB

the entireness of the platform as well as to reduce the effects of the embedded antennas. Such results also indicate that we should prefer power-amplifying metasurfaces among the different types so as to make the cross-polarized reflections over the metasurface stronger than the co-polarized ones throughout the rest of the platform.

7.5 Blinking Jamming with Metasurfaces

Not every radar antenna has radiation patterns of the same kind of those in Fig. 7.4. Phased-array antennas, in particular, may present much lower levels of cross-polarization. As it will become clearer in Chapter 8, because the reason is that the radiation patterns of phased arrays are derived from their antenna elements – which, in turn, the designer can select among those with the highest polarization purities.

To exemplify, consider a phased-array antenna with 10×10 elements spaced $\lambda/2$ from one another at 5.8 GHz. The antenna element geometry is a vertically polarized metal patch. We generate such phased array, first, by creating the antenna element using the *patchMicrostrip* object from the Matlab Antenna Toolbox [36]. Then, by applying the element in a uniform rectangular array using the *phased.URA* object from the Phased-Array System Toolbox. Fig. 7.12a,b depicts the vertically and horizontally polarized radiation patterns, both normalized per the maximum value of Fig. 7.12a. The vertically polarized pattern in Fig. 7.12a is similar to the one in Fig. 7.4a, with a strong main lobe pointing at the broadside direction of the array. However, the horizontally polarized pattern in Fig. 7.12b significantly differs from that of Fig. 7.4b. Now, the strongest cross-polarized lobe is much lower but, more importantly, it is pointing at the broadside direction. This configuration makes cross-polarization jamming fail even if the jammer succeeds in overcoming the significant cross-polarization gain advantage.

One method to tackle monopulse radars equipped with phased arrays is blinking jamming, a technique whose potential EW engineers envisioned back in the 1950s [26]. Blinking jamming exploits the fact that signals of two or more platforms cannot be reasonably distinguished by the target radar when they lie in the same radar resolution cell. Like cross-polarization jamming, the literature on the technique is not scarce [11, 14, 26–30]. However, it lacks reasonable measurement analyses for the same reasons we listed in the last paragraph of Section 7.3.

Consider a phased-array monopulse radar illuminating a dual-aircraft formation. The formation is tight enough to keep both aircraft 1 and 2 in an azimuthal angular range $\Delta\Phi < \Phi_{3dB}$ as Fig. 7.13a illustrates. Similarly, aircraft 1 and 2 lie in the elevation angular range $\Delta\theta < \theta_{3dB}$ as Fig. 7.13b depicts. Here, Φ_{3dB} and θ_{3dB} are respectively the 3-dB beamwidths in the azimuthal and elevation planes of the

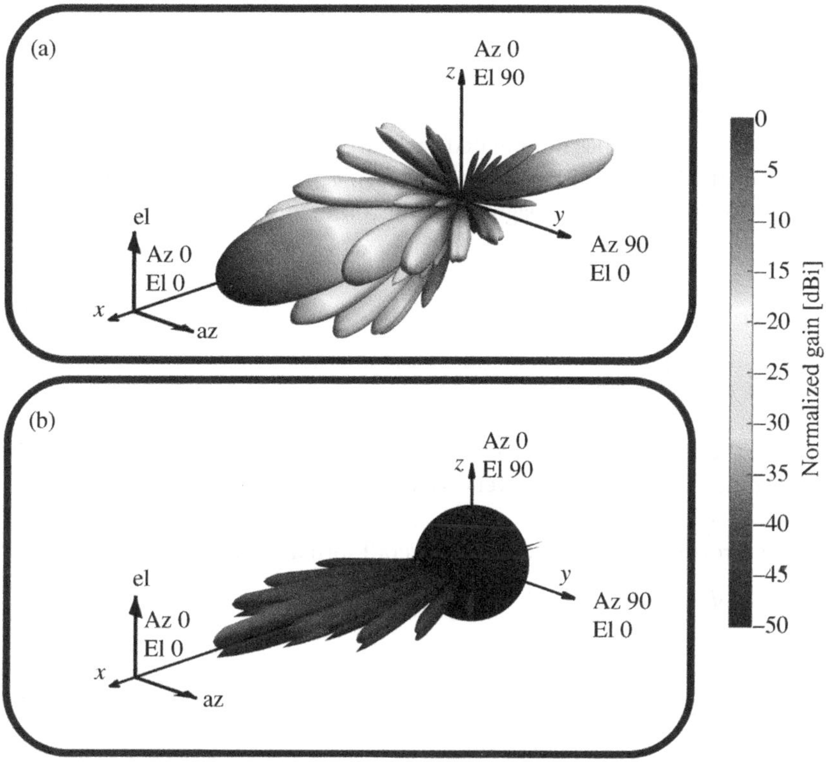

Fig. 7.12 Typical radiation patterns of a 10 × 10 patch-element phased array: (a) co-polarized component (vertical polarization); (b) cross-polarized component (horizontal polarization)

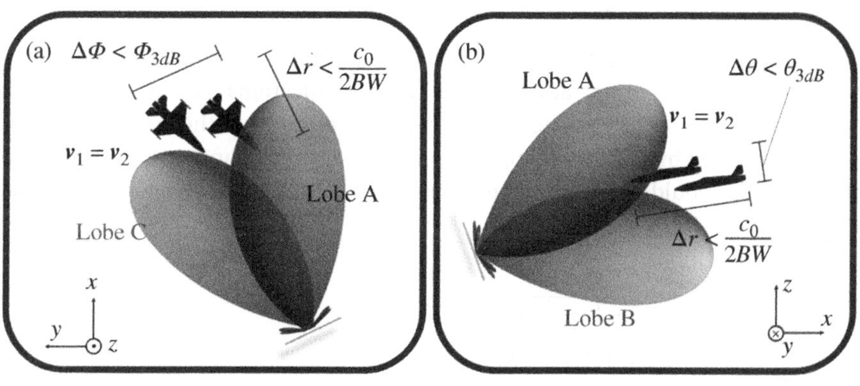

Fig. 7.13 Dual-aircraft formation lying in the same resolution cell of a monopulse radar: (a) azimuth plane; (b) elevation plane. "Gonçalves Licursi de Mello [14]/MDPI/CC BY 4.0."

monopulse virtual lobe in the sum channel $S_{\sum}$ expressed by (7.9). In addition, the range difference of aircraft 1 and 2 is $\Delta r < c_0/2BW$, where c_0 is the speed of the light and BW is the radar bandwidth. Their velocities are $\boldsymbol{v}_1 = \boldsymbol{v}_2$. In other words, both aircraft lie in the same radar resolution cell [14].

If the four antenna lobes illuminated each aircraft 1 and 2 separately, the radar would estimate different angular error sets $[\theta_{\epsilon 1}, \Phi_{\epsilon 1}]$ and $[\theta_{\epsilon 2}, \Phi_{\epsilon 2}]$, respectively. Using (7.15) and (7.16), the signals in the difference channels $S_{\Delta \Phi}$ and $S_{\Delta \theta}$ would be:

$$S_{\Delta \Phi 1} = \frac{\Phi_{\epsilon 1} S_{\sum 1}}{k_{\Phi}}, \quad S_{\Delta \Phi 2} = \frac{\Phi_{\epsilon 2} S_{\sum 2}}{k_{\Phi}} \tag{7.33}$$

$$S_{\Delta \theta 1} = \frac{\theta_{\epsilon 1} S_{\sum 1}}{k_{\theta}}, \quad S_{\Delta \theta 2} = \frac{\theta_{\epsilon 2} S_{\sum 2}}{k_{\theta}} \tag{7.34}$$

Because the aircraft lie in the same resolution cell, their skin returns will be superimposed in the radar receivers. Hence, in (7.15) and (7.16), $S_{\sum} = S_{\sum 1} + S_{\sum 2}$, $S_{\Delta \Phi} = S_{\Delta \Phi 1} + S_{\Delta \Phi 2}$, and $S_{\Delta \theta} = S_{\Delta \theta 1} + S_{\Delta \theta 2}$. Then, using (7.33) and (7.34), the angle errors estimated for the dual-aircraft formation are:

$$\Phi_{\epsilon} = \frac{\Phi_{\epsilon 1} S_{\sum 1} + \Phi_{\epsilon 2} S_{\sum 2}}{S_{\sum 1} + S_{\sum 2}} \tag{7.35}$$

$$\theta_{\epsilon} = \frac{\theta_{\epsilon 1} S_{\sum 1} + \theta_{\epsilon 2} S_{\sum 2}}{S_{\sum 1} + S_{\sum 2}} \tag{7.36}$$

What (7.35) and (7.36) are showing us is that, in the particular case without jamming, whether aircraft 1 or 2 will attract Φ_{ϵ} and θ_{ϵ} will depend on the intensity of their skin returns [11].

7.5.1 Working Principle with Conventional EA Systems

Now suppose that aircraft 1 has a DRFM, being capable of transmitting jamming signals that suffer similar attenuations and amplifications as those of the skin returns in the radar processing. Therefore, we consider that the jamming signals $J_{A-D,n}$ expressed by (7.21) will appear in the sum virtual channel as a signal J_1. In this case, we should now express the angle errors represented above by (7.35) and (7.36) as:

$$\Phi_{\epsilon} = \frac{\Phi_{\epsilon 1} \left(S_{\sum 1} + J_1\right) + \Phi_{\epsilon 2} S_{\sum 2}}{S_{\sum 1} + S_{\sum 2} + J_1} \tag{7.37}$$

$$\theta_{\epsilon} = \frac{\theta_{\epsilon 1} \left(S_{\sum 1} + J_1\right) + \theta_{\epsilon 2} S_{\sum 2}}{S_{\sum 1} + S_{\sum 2} + J_1} \tag{7.38}$$

Here, it is clear that $\Phi_\epsilon \rightarrow \Phi_{\epsilon 1}$ and $\theta_\epsilon \rightarrow \theta_{\epsilon 1}$ when $J_1 \gg S_{\sum 2}$. Analogously, if aircraft 2 also has a DRFM, then its jamming signals J_2 can make $\Phi_\epsilon \rightarrow \Phi_{\epsilon 2}$ and $\theta_\epsilon \rightarrow \theta_{\epsilon 2}$ when $J_2 \gg S_{\sum 1}$ in particular when aircraft 1 is not transmitting any jamming signals, that is, $J_1 = 0$.

From the above development, we can see that DRFMs embedded in both aircraft can make $\Phi_\epsilon \rightarrow \Phi_{\epsilon 1}, \Phi_{\epsilon 2}$ and $\theta_\epsilon \rightarrow \theta_{\epsilon 1}, \theta_{\epsilon 2}$ alternately. In this way, aircraft 1 and 2 can interchangeably "attract" the monopulse radar, playing with the inertial and oscillatory features of its control system seen in (7.17) and (7.18). By alternately transmitting jamming signals J_1 and J_2 over the target radar with a blinking frequency f_{bl}, we can exploit the natural frequencies f_Φ and f_θ of the control system. As such, the oscillations can even exceed $\Phi_{\epsilon 1}$ and $\Phi_{\epsilon 2}$, and/or $\theta_{\epsilon 1}$ and $\theta_{\epsilon 2}$, making the guidance of weapons by the target radar bounce around the jamming platforms and become ineffective [14].

To highlight how blinking jamming can exploit the oscillatory features of the radar antenna control system, we depict in Fig. 7.14 the response of (7.17) and (7.18) considering an initial orientation of $\Phi_{ant} = 7°$ and $el_{ant} = 90° - \theta_{ant} = 12°$.

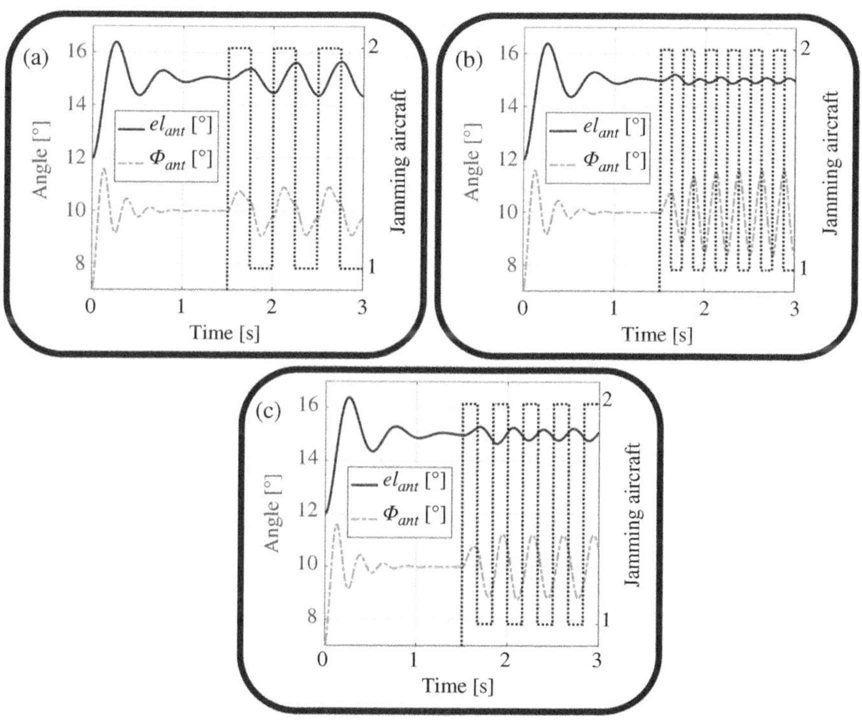

Fig. 7.14 Response of the antenna control system to diverse blinking frequencies: (a) $f_{bl} = f_\theta$; (b) $f_{bl} = f_\Phi$; (c) $f_{bl} = (f_\Phi + f_\theta)/2$. "Gonçalves Licursi de Mello [14]/MDPI/CC BY 4.0."

All system parameters are identical to those used in Fig. 7.3 (antenna equivalent mass $m_{ant} = 10$ kg, damping coefficients $d_{\Phi} = 10$/s and $d_{\theta} = 6$/s, and natural frequencies $f_{\Phi} = 4$ Hz and $f_{\theta} = 2$ Hz). From $t = 0$, we run a Matlab loop inputting to (7.17) and (7.18) angle errors $\Phi_{\epsilon} = \Phi_{acf} - \Phi_{ant}$ at each microsecond, where $\Phi_{acf} = 10°$, and $\theta_{\epsilon} = \theta_{acf} - \theta_{ant}$. Further, $el_{acf} = 90° - \theta_{acf} = 15°$, which simulates a target at this angular position. Then, from the instant $t = 1.5$ s, we alternate new values $\Phi_{acf} = 10 \pm 1°$ and $el_{acf} = 15 \pm 0.5°$ with different blinking frequencies f_{bl}. This last configuration simulates that a blinking jamming is performed from $t = 1.5$ s by a dual-aircraft formation with azimuthal and elevation separations $\Delta\Phi = 1°$ and $\Delta\theta = 0.5°$.

Figure 7.14a shows the radar antenna angular position el_{ant} (continuous line) and Φ_{ant} (dashed line) when the blinking frequency f_{bl} matches the natural frequency f_{θ} of the radar antenna control system in the elevation dimension ($f_{bl} = f_{\theta} = 2$ Hz). It also indicates the aircraft that is effectively transmitting the jamming signals at each instant (dotted line). We can see that, from $t = 1.5$ s, the radar cannot be stable toward the formation anymore. The matching between f_{bl} and f_{θ} causes significant oscillations in this dimension, ranging from 14.3° to 15.7° and exceeding the angular position of both aircraft. In azimuth, the oscillations range from 9.1° to 10.9°, also exceeding the aircraft separation. A weapon guided by such radar would probably miss both aircraft. Figure 7.14b illustrates the case where $f_{bl} = f_{\Phi} = 4$ Hz. Significant oscillations occur at the azimuthal dimension, ranging from 8.4° to 11.6°. In the elevation dimension, the system oscillations decrease to the range 14.9°–15.1°. Nevertheless, a weapon guided by such radar would probably miss both aircraft since the error in one dimension is sufficient to this end. Finally, Fig. 7.14c represents the case where the blinking frequency is the average of the natural frequencies in both azimuthal and elevation dimensions, that is, $f_{bl} = (f_{\Phi} + f_{\theta})/2 = 3$ Hz. We see a good compromise, with oscillations in the azimuthal dimension ranging from 8.8° to 11.2° while in the elevation dimension they range from 14.8° to 15.2° [14].

The above simulations show the potential of blinking jamming over a monopulse radar using a tracking system with oscillatory features. As already mentioned, conventional EA systems will achieve such response by using DRFMs. As such, we should account for all inconveniences thereof (complexity, power consumption, need for a dedicated compartment and antenna, and presence of spurs in jamming signals).

In the matter of effectiveness, for both radar and jammer vertically polarized with perfect purity, we can derive the JSR for each channel A–D using (7.21) for the jamming in aircraft 1 and (4.7) for the true skin return of aircraft 2:

$$\frac{J_{A-D,v1}}{S_{A-D,vv2}} = \frac{4\pi r_{acf}^2 P_j G_{j,v}(\theta'_{rdr}, \Phi'_{rdr})}{P_t G_{A-D,v}(\theta_{acf} - \theta_{ant}, \Phi_{act} - \Phi_{ant}) RCS_{vv2}(\theta'_{rdr}, \Phi'_{rdr})} \tag{7.39}$$

Additionally, we can calculate the burn-through range for each channel A–D as:

$$r_{burn} = \sqrt{\frac{JSR_{\min} P_t G_{A-D,v}\left(\theta_{acf} - \theta_{ant}, \Phi_{act} - \Phi_{ant}\right) RCS_{vv2}\left(\theta'_{rdr}, \Phi'_{rdr}\right)}{4\pi P_j G_{j,v}\left(\theta'_{rdr}, \Phi'_{rdr}\right)}} \tag{7.40}$$

In comparison to (7.27) and (7.28), we can see in (7.39) and (7.40) that a cross-polarization disadvantage of the type $G_{x-pol}(\theta, \Phi) = G_{A-D,h}(\theta, \Phi)/G_{A-D,v}(\theta, \Phi)$ is not present now. Nonetheless, it is evident that the radar can also react to blinking jamming by simply increasing its transmitted power P_t.

7.5.2 Working Principle with Metasurfaces

A close look at (7.35) and (7.36) reveals that another way to obtain $\Phi_\epsilon \rightarrow \Phi_{\epsilon 1}$ and $\theta_\epsilon \rightarrow \theta_{\epsilon 1}$ is by making $S_{\sum 2} \ll S_{\sum 1}$. Likewise, $\Phi_\epsilon \rightarrow \Phi_{\epsilon 2}$ and $\theta_\epsilon \rightarrow \theta_{\epsilon 2}$ when $S_{\sum 1} \ll S_{\sum 2}$.

Strictly speaking, it is evident in (7.5) through (7.8) that from the standpoint of a platform illuminated by the radar, there are two ways of acting over $S_{\sum 1}$ and $S_{\sum 2}$: the most obvious (and unreasonable) one is to fly toward or away from the radar, varying the range r_{acf}; however, varying the platform RCS is also an option. Specifically, the latter is the only way to vary $S_{\sum 1}$ and/or $S_{\sum 2}$ without losing any degree of freedom in the spatial domain.

There are three ways of making $S_{\sum 2} \ll S_{\sum 1}$ with metasurfaces: by decreasing the RCS of aircraft 2 with RCS reduction metasurfaces; by increasing that of aircraft 1 with power-amplifying metasurfaces; or finally by combining both methods. In any case, to perform an alternate RCS variation and make the aircraft "blink" from the radar perspective, a dynamic control is necessary, for which reconfigurable metasurfaces are essential.

We can derive the effectiveness of blinking jamming with metasurfaces for a vertically polarized radar with perfect purity by simply using (4.7) for the couple of aircraft as:

$$\frac{S_{A-D,vv1}}{S_{A-D,vv2}} = \frac{RCS_{vv1}\left(\theta'_{rdr}, \Phi'_{rdr}\right)}{RCS_{vv2}\left(\theta'_{rdr}, \Phi'_{rdr}\right)} = RCS_{blink} \tag{7.41}$$

As expected, (7.41) confirms that effectiveness depends only on the jammer formation through the RCS_{blink} ratio in the metasurface-driven blinking jamming. Therefore:

a) Increasing the power P_t transmitted by the radar will not provide any impact because the metasurfaces will reflect/absorb energy proportionally (this again holds as long as the metasurface response remains linear);
b) A burn-through range does not exist.

The RCS_{blink} ratio, in turn, will primarily depend on two factors:

c) On the amount of energy that the metasurfaces deployed on the aircraft can either absorb or power-amplify;
d) Once again, on the deployment rate studied in Section 4.6.2.

7.5.3 Implementation with Reconfigurable Metasurface Absorbers

Section 4.2 introduced the fundamental mechanisms behind the RCS reduction. Sections 4.4–4.6 presented a detailed study of appropriate, static metasurfaces for this task based on different mechanisms. In the context of passive stealth, the performance we pursue was mainly expressed in terms of a good absorption coefficient in a wide BW and incidence angle range. The problematic in the blinking jamming application is different. Here, the absorption BW is of minor importance. By exploiting reconfigurability, one can shift back and forth the operational frequency range of the absorber, making it "blink" from the viewpoint of a radar tuned at a specific frequency. Hence, it is the tuning ratio, defined by the limit tuning frequencies, that matters to tackle a wide number of radars. Of course, the absorption BW should be enough to cover the isolate BW of the target radars.

The literature reports many frequency-reconfigurable metasurface absorbers [49–52]. X. Xu et al. [49] use a PIN diode as a reconfigurable resistor between shaped metal elements over a ground plane in a 5-mm thick unit cell. In essence, this metasurface exploits a mechanism similar to the one Fig. 4.12 represented earlier, where an AMC with impedance Z_{AMC} is associated with a resistive element with impedance R_{sheet}. However, now we can vary R_{sheet} through the biasing of the PIN diode with different voltages. According to (3.15), Z_{AMC} varies with the angular frequency ω. Further, we can model the metasurface surface impedance Z_s as the parallel between Z_{AMC} and R_{sheet}. Thus, varying the latter allows a reasonable matching of Z_s to the free-space impedance η_0 at different frequencies. In this way, the metasurface can vary its RCS reduction from −5 dB to −20 dB in the range of 9.0 to 14.5 GHz approximately. This performance may provide an $RCS_{blink} = 15$ dB in a tuning ratio of $1.61 : 1$ if the deployment rate is 100%.

J. Zhao et al. [50] propose a unit cell composed of two orthogonal couples of capacitive ring resonators separated each by a varactor over a ground plane (see Fig. 7.15a). The working principle is also similar to the one of Fig. 4.12. Different bias voltages over the varactor carry out the control of the absorption. The thickness is 1.6 mm.The metasurface can tune the absorption coefficient between two states in the simulations: higher than 99%, which is equivalent to −17 dB of RCS reduction by using $|\tau_{TEM}| = 0$ in (4.23), and lower than 5% (equivalent to −0.01 dB of RCS reduction). Hence, we can consider that a RCS_{blink} ratio of 17 dB is achievable if the deployment rate is 100%. Figure 7.15b shows the measurement setup.

(a)

(b)

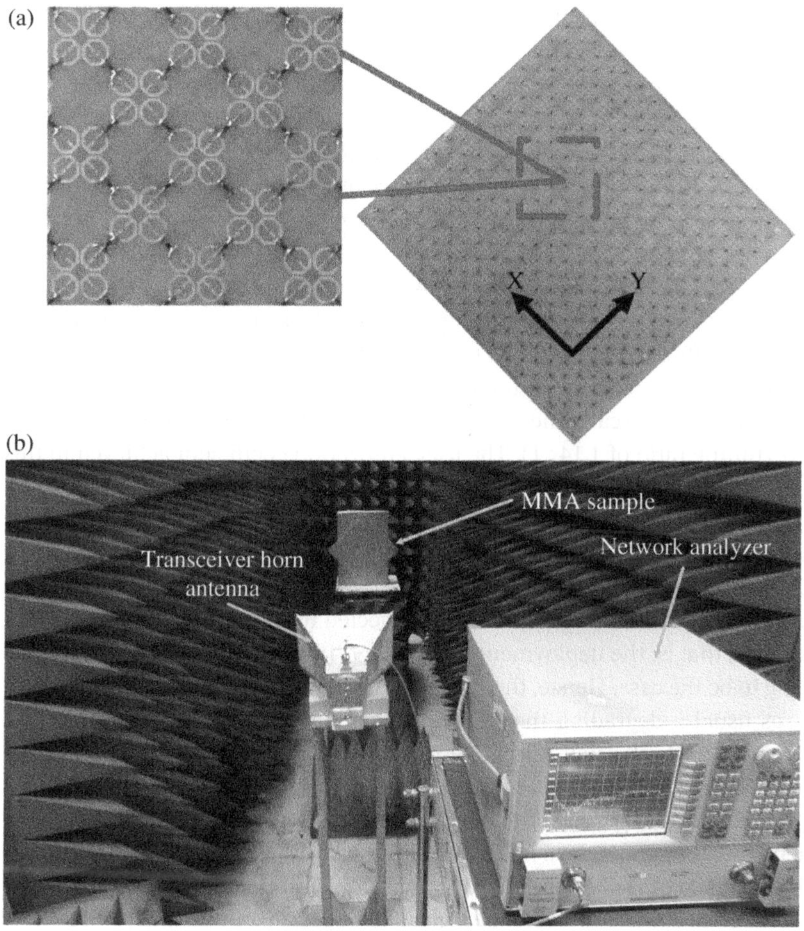

Fig. 7.15 Reconfigurable metasurface absorber: (a) fabricated metasurface; (b) measurement setup. "Reproduced from Zhao et al. [50]/IOP Publishing/CC BY 3.0."

We observe a good agreement with simulations. The fabricated metasurface can tune an absorption coefficient $\alpha \geq 90\%$ from 4.3 to 5.9 GHz, representing a tuning ratio of 1.37 : 1 (apparently, the measured absorption coefficient is $\alpha \geq 99\%$ when tuned from 4.60 to 5.45 GHz). J. Zhao et al. [50] also report the simulated results for diverse angles of incidence, and the absorption coefficient remains higher than 90% for up to 50° of angle of incidence in both transverse-electric and transverse-magnetic polarizations.

W. Wang et al. [51] present a unit cell consisting of a cuboid ferrite element over a ground plane. This metasurface provides a dual-band behavior with an absorptive coefficient measured as around 95% in the first band, which is equivalent to −10.1 dB of RCS reduction by using $|\tau_{TEM}| = 0$ in (4.23). Out of band, they measure the absorption coefficient as close to zero at some frequencies. Hence, this metasurface can provide an RCS_{blink} ratio equal to −10.1 dB. This time, we perform reconfiguration through an external magnetic field. By applying such field over the ferrite particles, we shift the first band from about 8.7 to 9.7 GHz, and we shift the zero-absorption frequencies similarly, representing a tuning ratio of 1.11 : 1. The thickness of the metasurface is 1.8 mm.

W. Wang et al. [52] also use a ferrite-based unit cell but now with U-shape elements. In this work, they propose a quad-band metasurface absorber with an absorption coefficient $\alpha \geq 98\%$ in one of the bands (which is equivalent to −14.0 dB of RCS reduction). With the application of a biasing magnetic field over the ferrite particles, we can tune the center frequency of this band from about 8.3 to 9.5 GHz (tuning ratio of 1.14 : 1). The least absorption coefficient achieved in this range is around 50% (equivalent to −1.2 dB of RCS reduction) which leads to an RCS_{blink} ratio of −12.8 dB if the deployment rate is 100%. The thickness of the metasurface is 1.6 mm.

As the reader may have realized, the RCS_{blink} ratios reported above consider that the skin returns of both aircraft are fully subjected to the effects of the employed metasurface, that is, the deployment rate is 100%. In the real world, this is usually not going to be the case. Hence, the platform is going to reflect a certain amount of energy as usual – degrading the blinking jamming performance. Therefore, we should pay special attention to coating the entireness of the platform to mitigate this degradation. Also, RCS-reconfigurable antennas [53, 54] are preferable.

Besides the advantages of simplicity, costs, and not requiring a dedicated compartment and antenna, here, a noticeable difference between using reconfigurable absorbers instead of DRFMs lies in the power consumption. While the latter consumes energy to digitally process the signals to be retransmitted and to effectively retransmit them, the former neither process nor transmits anything – it consumes only a neglectable amount of power in the biasing of the reconfigurable components deployed over the unit cells.

7.5.4 Implementation with Power-Amplifying Metasurfaces

With regards to increasing the RCS of the aircraft, power-amplifier metasurfaces [44–47, 55] can provide this capability, as Section 5.4 discussed earlier. In the context of active-cancellation stealth, we require these metasurfaces to reflect impinging waves with a specific amplitude and phase per illustrated in Fig. 5.6.

In the blinking jamming application, the job is much simpler. Here, metasurfaces can just use full power and ignore the phase in the power-amplified reflections because the goal is to alternately make the aircraft RCS larger than one another, as stated in the beginning of Section 7.5.2.

The natural reconfigurability of the amplifiers, achieved through the biasing thereof, allows the required “blink.” Besides, the frequency range where the technique will work corresponds to the metasurface BW.

In terms of effectiveness, we saw in Section 5.4 that gains of up to 25 dB in the reflection of waves are possible today with experimentally demonstrated power-amplifying metasurfaces. Such gain means that we can achieve a RCS_{blink} ratio of 25 dB with a deployment rate of 100%. Such ratio, higher than those provided nowadays by the reconfigurable metasurface absorbers studied in Section 7.5.3, allows a lower deployment rate, which can be a big advantage for many platforms.

In spite of this advantage, one should consider that power-amplifying metasurfaces are significantly more complex, expensive, and power-hungry than reconfigurable metasurface absorbers. Now, the design must employ active circuitry. Further, we spend energy not only with the biasing of reconfigurable components but also with the amplification of the reflected waves. Nevertheless, such issues are not comparable to the inconveniences of DRFMs.

In conclusion, we can also use the active-cancellation stealth technique to perform blinking jamming. However, at present, there is no apparent reason to go for this much more complex implementation.

7.6 Simulation: Comparative Study on Blinking Jamming

Now, we are interested in highlighting the differences between carrying out blinking jamming with conventional EA systems or metasurfaces in our EW scenario simulator. R.G.L. Mello [14] shows that, for the final performance, it does not matter whether we use power-amplifying or absorptive metasurfaces as long as a certain RCS_{blink} ratio is ensured. Further, R.G.L. Mello [14] also studies the behavior of the technique with various blinking frequencies f_{bl}. Thus, in what follows, we adopt $f_{bl} = f_{\Phi}$ and present new results concerning the radar reaction and the burn-through range for a single value of RCS_{blink} ratio.

Consider a scenario where a monopulse radar is located at the position $P(0,0,0)$ according to the coordinate system specified in Fig. 7.1. Such radar is identical to the one of Section 7.4 except for the specified hereafter. The radar operates at $f_0 = 5.8$ GHz. We model each of its four antennas A–D as a phased array composed of 30×30 vertically polarized square-patch elements. The array elements are spaced by $\lambda/2$ in a rectangular grid, which leads to a directivity of 34.6 dB to each of the four antennas. As such, the 3-dB beamwidths are $\Phi_{3dB} = \theta_{3dB} = 3.38°$. The squint angles in relation to the broadside direction θ_{ant} and Φ_{ant} of the antenna system are $\theta_{sqt} = \theta_{3dB}/4$ and $\Phi_{3dB} = \Phi_{sqt}/4$. Figure 7.16 depicts all the radiation patterns together in the azimuth and elevation planes, considering the antenna

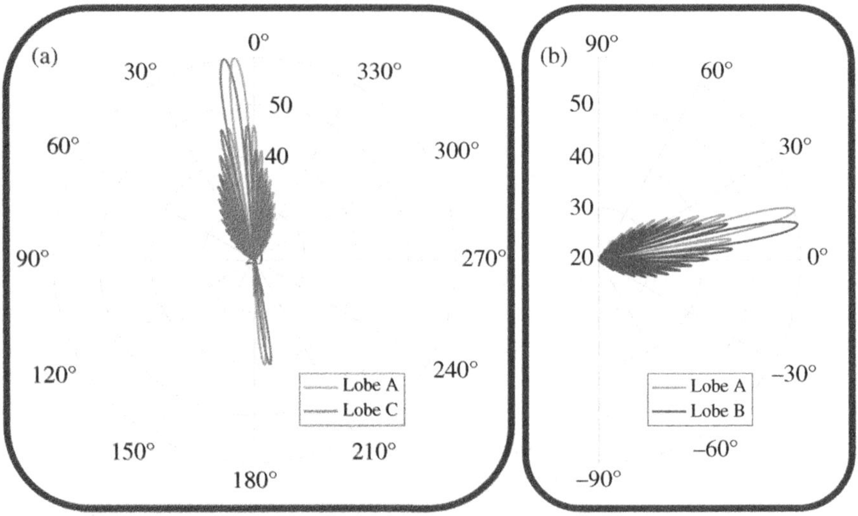

Fig. 7.16 Patterns of the antennas A–D in (a) azimuth and (b) elevation (90° − θ). "Gonçalves Licursi de Mello [14]/MDPI/CC BY 4.0."

system in its initial position $el_{ant} = 90° - \theta_{ant} = 12°$ and $\Phi_{ant} = 7°$. We take the CA26-4114 PA from Ciao Wireless, Inc. [48] (gain of 22 dB) as a reference in the Tx front end. In the Rx front end, we consider the CA48-2111 LNA, also from Ciao Wireless, Inc. (gain of 29 dB and noise figure of 1.3 dB).

A dual-aircraft formation, whose centroid is at a dynamic position $P(r_{acf}, \theta_{acf} = 75°, \Phi_{acf} = 10°)$, is flying toward this radar at Mach 2 ($v_1 = 686$ m/s). Note that the elevation of the formation, from the ground, is $el_{acf} = 90° - \theta_{acf} = 15°$. The aircraft are angularly separated by $\pm 0.25°$ from the centroid in the elevation plane and $\pm 0.5°$ in the azimuth. In this way, we can consider the RCS of each aircraft constant for the sake of simplicity. Initially, $RCS_{hh} = RCS_{vv} = 1\ \text{m}^2$ and $RCS_{hv} = RCS_{vh} = 0$. We model again a missile guided by the monopulse radar and flying at Mach 10 (3430 m/s) with a *phased.Platform* object for visualization purposes. Fig. 7.17 depicts the scenario at $t = 0$ using the *phased.ScenarioViewer* object.

7.6.1 Initial Results and Radar Reaction

First, we are interested in checking how a conventional EA system performs blinking jamming with DRFMs in the aircraft. We model the DRFMs rather as detailed in Section 5.7.2, except that we consider them ideal, that is, they transmit perfect copies of the signals in their receivers. Also, they receive information about the radar location from an ES system and point the antenna beams accordingly.

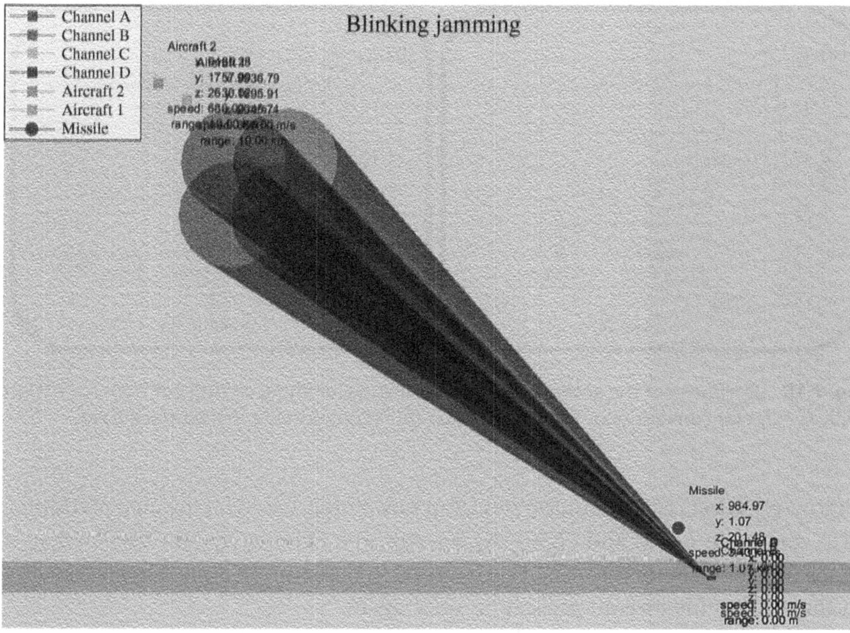

Fig. 7.17 EW scenario represented with the *phased.ScenarioViewer* object. "Gonçalves Licursi de Mello [14]/MDPI/CC BY 4.0."

We consider that, at $t = 0$, the monopulse radar receives information from an acquisition radar about what allegedly is an aircraft at $r_{acf} = 10$ km. Fig. 7.18a shows the results in terms of the angular position of the monopulse radar antenna system which, at first, converges to the formation centroid, $el_{acf} = 15°$ and $\Phi_{acf} = 10°$. From $t = 1.5$ s, when $r_{acf} = 9$ km, the aircraft DRFMs are alternately turned on, attracting the antenna system based on (7.37) and (7.38). We can see that from this moment on, the system starts to oscillate in a range of $\pm 1.5°$ around the formation centroid at the azimuth dimension, which exceeds three times the aircraft separation. This is more than enough to make a missile guided by this radar miss the aircraft.

The second case studied here concerns the use of reconfigurable metasurfaces that can dynamically change the aircraft RCS whether by absorbing or power-amplifying electromagnetic waves. By programmatically changing the *phased. RadarTarget* object representing the RCS tensor (4.2) of the aircraft, we model the RCS blinking. The scenario is similar to the one related to Fig. 7.18a (formation initially detected at $r_{acf} = 10$ km, jamming turned on at $t = 1.5$ s when $r_{acf} = 9$ km). Figure 7.18b depicts the antenna system response for $RCS_{blink} = 15$ dB. We chose this value because prominent reconfigurable metasurface absorbers [50] or power-amplifying metasurfaces [55] can reasonably provide it. Again, the formation

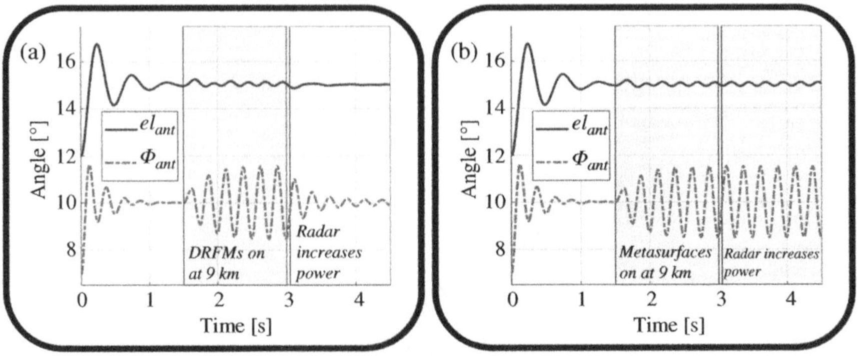

Fig. 7.18 Response of the antenna system to blinking jamming carried out from r_{acf} = 9 km with $f_{bl} = f_{\Phi}$: (a) conventional EA system case; (b) reconfigurable metasurface case.

alternately attract the antenna system, but now purely based on the own platforms' skin returns in (7.35) and (7.36). As a result, the antenna system oscillates in a range of $\pm 1.5°$ around the formation centroid in the azimuth dimension, exceeding in three times the aircraft separation once again.

In the same simulations, we also study what happens when the radar reacts to the blinking jamming by increasing its transmitted power P_t, which we can consider an EP measure. The goal is to make it clear that the implementation with metasurfaces is invulnerable to this measure, as deliberated in item (a) of Section 7.5.2. We assume that the radar somehow detects that it is under jamming at $t = 3.0$ s, increasing P_t of 30 dB as a reaction. In the DRFM case, represented in Fig. 7.18a, the jamming loses effectiveness after the reaction, and the radar points back the antenna system over the formation centroid. Referring to the terminology of (7.39), this happens because increasing P_t yields a larger $S_{A-D,vv2}$ avoiding the JSR within the channels A–D to reach JSR_{min}. On the other hand, nothing changes after the radar increases its power when metasurfaces carry out the jamming, as represented in Fig. 7.18b. The reason is that P_t simply does not appear in (7.41).

7.6.2 Burn-Through Range Mitigation

In the sequel, we study statement (b) of Section 7.5.2 concerning the inexistence of a burn-through range for the implementation of blinking jamming with metasurfaces.

First, we simulate the scenario with conventional EA systems, that is, DRFMs carrying out the measure from $r_{acf} = 9$ km to $r_{acf} = 0$. Figure 7.19a illustrates the result for this case. From $t = 1.5$ s, we alternately turn the DRFMs on. It is evident that, even though the measure initially works, the closer to the radar, the less

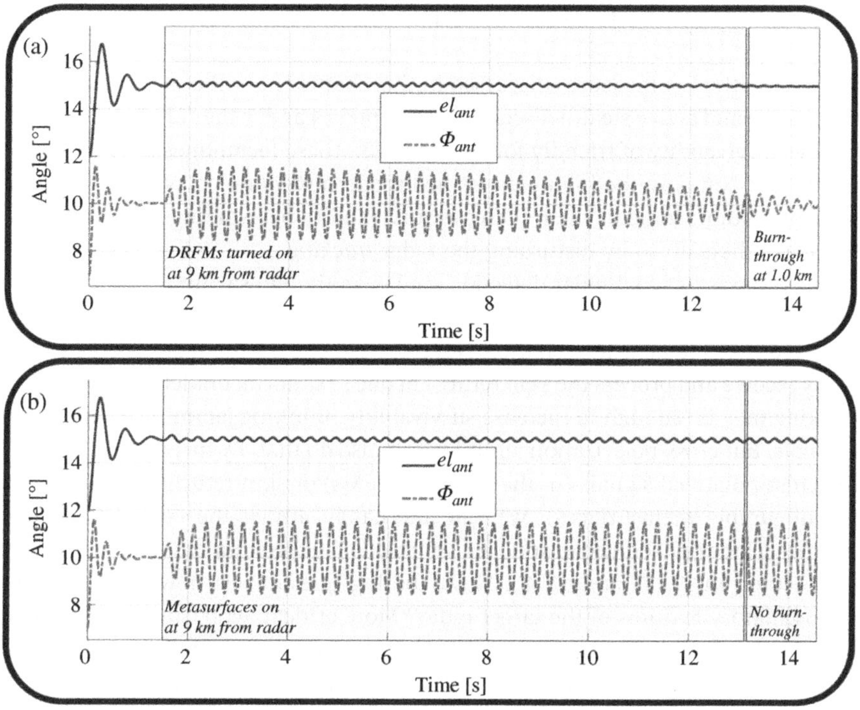

Fig. 7.19 Results for blinking jamming carried out from $r_{acf} = 9$ km to $r_{acf} = 0$: (a) conventional EA system case; (b) reconfigurable metasurface case

effective it becomes. The analysis here is straightforward if one refers to the terminology of (7.39): as r_{acf} decreases so does the JSR. Differently from what happens with the cross-polarization jamming in Fig. 7.9a, performance gradually degrades. Hence, we define the burn-through range as the range where the oscillations do not exceed the aircraft separation anymore – which is $r_{acf} = 1$ km in the case of Fig. 7.19a. Such behavior is particularly dangerous if an active radar-homing missile targets the aircraft formation. In this case, even though the EA measure will initially disturb the missile, sooner or later, the latter will get close to the formation, where the measure loses effectiveness, and hit it.

Following, we simulate the case where we perform blinking jamming with metasurfaces, as represented by the results in Fig. 7.19b. As expected once again, nothing happens as the time t runs and the range r_{acf} decreases simply because (7.41) does not depend on the latter. This result corroborates the inexistence of a burn-through range in the metasurface implementation. Moreover, among all the other advantages, it shows that even an active radar homing missile would not be able to hit a metasurface-coated formation.

7.7 Envision of Generalized EA with Metasurfaces

In essence, the EA measures discussed in this chapter have the same objective as the passive and active stealth discussed in Chapters 4 and 5: they all aim at increasing the survivability of friendly forces. However, these techniques (and the many others that compose the highly diversified modern scenario) may differ in a fundamental point: some of them increase survivability by avoiding radar *detection,* whereas others do so by disturbing the radar *tracking*.

One thing is to avoid radar detection, and even though stealth technologies are becoming more advanced each day, the required RCS reduction just cannot be achieved in some tactical scenarios. Nevertheless, even if the radar is able to correctly receive and process the skin returns of one's platform, hindering appropriate tracking may be enough to increase survivability. Blinking jamming is an explicit example, but cross-polarization jamming can also do that. By turning "on and off" the cross-polarized signals (in the case of DRFMs) or skin returns (in the case of reconfigurable metasurfaces), we can alternately attract the co- or the cross-polarized lobes of the radar antenna system, and exploit its oscillatory features, represented in (7.17) and (7.18). Now, the big question is: how can we know the oscillatory features of the target radar? More critical, who can guarantee that the tracking function of the target radar has oscillatory features?

Even though the monopulse radar with oscillatory features is widely adopted, this radar is still a particular case. Other radars may present tracking systems whose transfer functions substantially differ from (7.17) and (7.18). For example, the surveillance radar considered in the scenario of Section 5.6 has a target tracker that operates upon the final radar product, that is, the two-dimensional radar video, without changing its antenna scanning as a result. Also, cognitive radars can use advanced AI models to optimize the tracking rules themselves [2–5].

In the abovementioned cases, the tracking logic in the radar internal functioning may remain a total mystery to the EW system, and there is nothing new to the EW engineer in this case [18]. It is true that we can envision the use of reinforcement learning to tackle this problem if a radar reaction exists [56–68]. By applying a specific EA measure over a target radar, we may next check through ES means how its signals behave which can reveal how the system works. However, we leave this matter for Chapter 9 because, here, we are just going to work with the fact that radar tracking systems typically exploit the temporal behavior of targets nowadays. In other words, these systems typically rely on the fact that platforms are physically linked to reasonable velocities. For example, in the surveillance radar of Section 5.6, the tracker associates subsequent detections in the same region, considering that they can displace at a velocity around Mach 2. Besides, in cognitive radars, the target velocity is implicitly computed in the learning process.

This temporal dependence is a key point in terms of EA because one can make the radar somehow confused by creating numerous puzzling detections throughout the environment even if oscillatory features are not present.

Fig. 7.20 illustrates this concept with a massive, autonomous drone swarm surrounding a generic radar. The radar tries to detect and track potential targets. However, each drone is somehow able to create co-polarized and cross-polarized copies of the radar signals, in a chaotic, dynamic manner. In this way, the swarm alternately highlights each of the drones for the co- or the cross-polarized lobes of the radar antenna system. Further, the copied signals saturate the radar AGC circuits, forcing it to lower its receiving gain.

From the radar perspective, the detections will be meaningless (no aircraft would be able to fly from one position of detection to another with a reasonable velocity). If the tracker is similar to that of the surveillance radar of Section 5.6, failure will occur because it cannot associate the variation of the position of the detections with a velocity around Mach 2. If the target radar is cognitive, a weird behavior never seen before will contaminate its learning. Such behavior will disturb all the kinds of radar that compute the temporal domain in their tracking systems, making us believe that one such environment would be generally unpractical for radar operation. We can consider this scenario safe for the penetration of an aircraft formation to suppress the target radar, as Fig. 7.20 also illustrates. All that said, we can classify the creation of such an environment as a stand-off, active, generalized EA technique.

At this point, the reader may wonder what metasurfaces have to do with such EA measure. The point is that, by *massive* drone swarm here, we mean that we should use hundreds or even thousands of drones to cover the whole field of view of the radar. This massive swarm, in turn, implies the need for lightweight, cheap drones for practical, logistic, economic, and even political reasons in the war effort. As extensively discussed, it is not feasible to equip such platforms with DRFMs. On the other hand, we can deploy metasurfaces over lightweight, cheap drones and obtain the reconfigurability needed for such application. In Fig. 7.20, co-polarized power-amplifying metasurfaces coat each drone in part. Cross-polarized metasurfaces coat the rest of each drone. Hence, the drones can amplify reflected waves no matter their polarization in a dynamic manner, thereby allowing the swarm to play with the co- or the cross-polarized lobes of the radar antenna system.

It became evident in (7.32) and (7.41) that, for the metasurface-driven measures in question, first, radars cannot react by simply increasing their transmitted power P_t, and also that a burn-through range is inexistent. These are extra reasons for the use of metasurfaces in a drone swarm for EA purposes. The former provides an extra layer of certainty for friendly forces. The latter gives a huge advantage in keeping the radar confused until the aircraft formation has the radar within its weapon range.

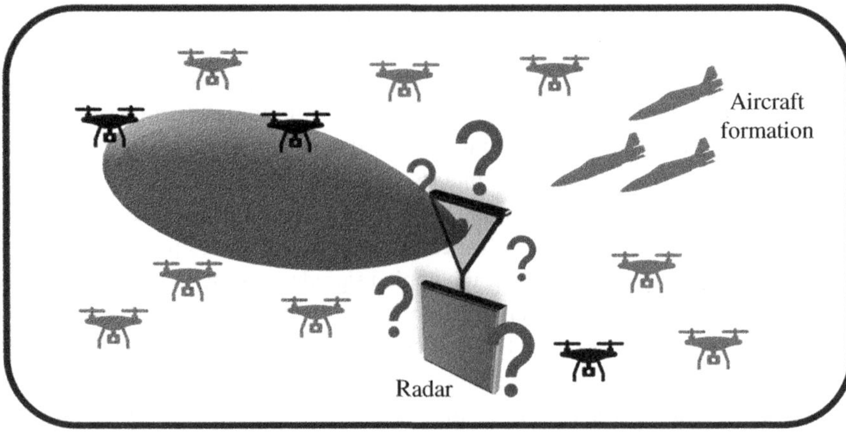

Fig. 7.20 Conceptual illustration of the chaotic EA environment

Of course, some challenges exist to bring the chaotic EA measure from the domain of ideas to the theater of operation, as the reader may realize. Most of them are not related to the metasurfaces themselves but to the autonomous drone swarm in general. Among them, we can cite the communication and synchronization of the drones in the swarm, the resilience of such communication to jamming, the definition of the best spatial arrangement for the swarm, and also the best blinking pattern to generate an impractical environment for radar operation. With regard to power-amplifying metasurface challenges, Section 5.4.5 is the place the reader should refer to.

7.8 Exercises

Now, it is your turn to practice the concepts developed in this chapter. You can access the simulator with the cross-polarization and blinking jamming functionalities by running the *main.m* file in the respective folders entitled *Ch07a_EA_cross_pol_jam* and *Ch07b_EA_blk_jam* on the companion website. After practicing with the exercises of Sections 5.7 and 6.6, the reader should be familiar enough with the codes to choose their preferred methods and embed them in the correct places. You can achieve more reasonable simulation times if you divide the sampling rate by eight by setting the *prms.Fs* variable in the *def_params* function. The algorithm will adjust the radar BW accordingly.

The first task here is to play with both the cross-polarization and the blinking jamming functionalities by changing the aircraft/formation position, RCS, polarization-conversion RCS_{x-pol} ratio, and RCS_{blink} ratio. Another essential task is to vary the aircraft/formation trajectory and see how the jamming performs.

7.8.1 Modeling the Missile Guidance

Difficult level: medium

After understanding how the studied jamming methods work, one good idea to further assess their performances is to implement realistic missile kinetics. In this way, we can directly use the aircraft survivability as one of the metrics.

To this end, we already placed the *mnp_guide_missile* function at the correct place in the code. When you open this function, you will see that the missile is originally set to follow the orientation of the radar antenna system (θ_{ant} and Φ_{ant}).

Make some research for realistic kinetics models [69] and edit such function. Then, run the simulation with different configurations for the cross-polarization and the blinking jamming. Before taking your own conclusions, note that the literature [20] provides good understanding of how we can do the survivability analysis.

7.8.2 Impact of DRFM on the Proposed EA Measures

Difficult level: low

Now that you have a complete simulator including a realistic missile, it is time to see how the DRFM nonidealities can impact jamming performance.

First, look for the *prms.realistic_DRFM* variable in the *def_params* function and set it as *true*. Then, repeat the pertinent steps of Section 5.7.2 and run the simulation, observing how the jamming performs. Finally, repeat the steps of Section 6.6.2 to provide your DRFM with DOA information coming from an ES system and run the simulation again.

With such an exercise, you will obtain insight into the complexity of the DRFM solution compared to metasurfaces.

7.8.3 Simulating the Generalized EA Measure

Difficult level: high

The importance of the previous exercises lies not only in the fact that you can reinforce the concepts studied in this chapter. It also consists in developing your own, complete simulator.

With such simulator in hand, you can programmatically set many platforms distributed around the radar to model a drone swarm. You can also programmatically assign a reconfigurable *RCS_reduction*, polarization-conversion RCS_{x-pol} ratio, and RCS_{blink} ratio to each one of them. In this way, you can simulate the generalized EA measure discussed in Section 7.7.

When your code is ready in what concerns the drone swarm, consider replacing the monopulse radar considered in the scenario of Section 7.6 with the surveillance radar of Section 5.6. Further, consider implementing your own cognitive radar as described in the literature [2–5].

Such an exercise will reveal the potential of the technique. By programmatically setting the reconfiguration of the drones, we emulate an ideal communication and synchronization between them. More audacious simulations can implement such features through an extra radio link that the adversaries can jam too. This also allows an analysis on the impact of the latency of such communication over the performance of the technique.

7.9 Chapter Final Remarks

The arrival of the DRFM in the last decades started a new era in EW. Far from ending, such era is characterized by high-performance EA systems capable of handling multiple radars at once. Notwithstanding their extensive use in the modern scenario, DRFMs are not convenient for all applications. Specifically, the criticality of size, weight, power consumption, and costs in lightweight drones constrain their use. Further, the DRFM resources in terms of processing, power, bandwidth, and antenna beams, naturally are not infinite.

In this chapter, we proposed metasurfaces as an alternative to DRFMs in the execution of EA measures. The simplicity, compactness, and power efficiency of metasurfaces are appropriate to operate in extremely constrained platforms, or to alleviate the workload of DRFMs in conventional applications. To illustrate concepts, we studied cross-polarization and blinking jamming. We derived an analytical formulation that showed that the implementation with metasurfaces is invulnerable to a radar reaction through the increase of power, differently from DRFMs. Moreover, a burn-through range does not exist, which is particularly interesting against active homing missiles. We demonstrated such features with our EW scenario simulator. The studied concepts allowed us to envision a new stand-off, active, generalized EA technique using drone swarms considering the lightweight and low-cost features of metasurfaces. Such technique has the potential to handle multifunctional, cognitive, monopulse, or any other kind of radar whose tracking depends on the temporal behavior of target detections.

The essence of this chapter is that the ability to do EA with metasurfaces may represent a huge impact, considering that numerous, cheap systems distributed in the scenario can make the use of the spectrum unfeasible. If the arrival of new radars represents new challenges for EW engineers in the cat-and-mouse EW game, the implementation of the generalize EA measure forces radar engineers to abandon velocity as a tracking parameter. In the matter of AI, this is a totally new behavior for which models will need retraining.

It is worthwhile to note that the implementation challenges related to the generalized EA measure are mostly at the drone swarm level – not the metasurfaces. But, most importantly, the key point in this chapter is the thought-provoking idea of discovering new ways to perform EA with metasurfaces. Doubtless, the adversaries will also innovate in protecting themselves and:

> All warfare is based on deception. Hence, when we are able to attack, we must seem unable; when using our forces, we must appear inactive; when we are near, we must make the enemy believe we are far away; when far away, we must make him believe we are near.
>
> —Sun Tzu, military theorist

References

1. Skolnik, M.I. (1990). *Radar handbook*, 2nd ed. New York, NY: McGraw-Hill Ed.
2. Haykin, S. (2006). Cognitive radar: a way of the future. *IEEE Signal Processing Magazine* 23 (1): 30–40. https://doi.org/10.1109/MSP.2006.1593335
3. Farina, A., De Maio, A., Haykin, S. (2017). *The impact of cognition on radar technology*. London, UK: Scitech Publishing.
4. Smith, G.E., Cammenga, Z., Mitchell, A. (2016). Experiments with cognitive radar. *IEEE Aerospace and Electronic Systems Magazine* 31 (12): 34–46. https://doi.org/10.1109/MAES.2016.150215
5. Gurbuz, S.Z., Griffiths, H.D., Charlish, A. et al. (2019). An overview of cognitive radar: past, present, and future. *IEEE Aerospace and Electronic Systems Magazine* 34 (12): 6–18. https://doi.org/10.1109/MAES.2019.2953762
6. Jeong, N.-H., Kim, M., Choi, J.-H. et al. (2023). Beam scheduling of maritime multifunctional radar based on binary integration. *IEEE Access* 11: 35796–35807. https://doi.org/10.1109/ACCESS.2023.3265801
7. Kellett, J., Dawber, W., Wallace, W. et al. (2022). Multifunction maritime radar and RF systems—technology challenges and areas of development. *IEEE Aerospace and Electronic Systems Magazine* 37 (4): 6–20. https://doi.org/10.1109/MAES.2021.3081168
8. Klemm, R., Nickel, U., Gierull, C. (2017). *Novel radar techniques and applications volume 1: real aperture array radar, imaging radar, and passive and multistatic radar*. Stevenage, UK: Scitech Publishing, Chapter 3, pp. 135–171. https://doi.org/10.1049/SBRA512F
9. Lothes, R.N., Szymanski, M.B., Wiley, R.G. (1990). *Radar vulnerability to jamming*. Norwood, MA: Artech House, Section 2.2, pp. 31–33.
10. Lothes, R.N., Szymanski, M.B., Wiley, R.G. (1990). *Radar vulnerability to jamming*. Norwood, MA: Artech House, Section 2.5, pp. 40.

11 Davidson, K., Bray, J. (2020). Theory and design of blink jamming. *MDPI Applied Sciences* 10 (1914): 1–20. https://doi.org/10.3390/app10061914

12 Lothes, R.N., Szymanski, M.B., Wiley, R.G. (1990). *Radar vulnerability to jamming*. Norwood, MA: Artech House, Section 4.2, pp. 86–96.

13 Barton, D., Sherman, S. (2011). *Monopulse Principles and Techniques*, 2nd ed. Norwood, MA: Artech House.

14 Gonçalves Licursi de Mello, R. (2023). Metasurfaces and blinking jamming: convergent study, comparative analysis, and challenges. *MDPI Micromachines* 14 (7): 1405. https://doi.org/10.3390/mi14071405

15 Adamy, D. (2001). *EW 101: a first course in electronic warfare*. Norwood, MA: Artech House, Section 9.9.1, pp. 210.

16 Adamy, D. (2001). *EW 101: a first course in electronic warfare*. Norwood, MA: Artech House, Sections 9.9.3–9.9.11, pp. 212–222.

17 Neri, F. (2002). Anti-monopulse jamming techniques. *Proceedings of SBMO/IEEE MTT-S International Microwave and Optoelectronics Conference (Cat. No.01TH8568)*, Belém, Brazil (06–10 August 2001), New York, US: IEEE. https://doi.org/10.1109/SBMOMO.2001.1007634

18 Neri, F. (2018). *Introduction to electronic defense systems*, 3rd ed. Boston, MA: Artech House, Section 5.4.20, pp. 449.

19 Compton, R. (1983). The performance of a tripole adaptive array against cross-polarized jamming. *IEEE Transactions on Antennas and Propagation* 31 (4): 682–685. https://doi.org/10.1109/TAP.1983.1143119

20 Global Business Media. (2011). Electronic countermeasures. *Special Report*. https://issuu.com/magazineproduction/docs/gbm_electronic_countermeasures_ezine/1 (accessed 17 November 2023).

21 Kalinbacak, İ., Pehlivan, M., Yeğin, K. (2017). Cross polarization monopulse jammer located on UAV. *8th International Conference on Recent Advances and Space Technology (RAST)*, Istanbul, Turkey (19–22 June 2017), New York, US: IEEE. https://doi.org/10.1109/RAST.2017.8002943

22 Han, H., Xu, X., Wang, H. et al. (2019). Analysis of cross-polarization jamming for phase comparison monopulse radars. *IEEE 2nd International Conference on Electronic Information and Communication Technology (ICEICT)*, Harbin, China (20–22 January 2019), New York, US: IEEE. https://doi.org/10.1109/ICEICT.2019.8846402

23 Markin, E. (2011). Stability criterion of radar angle tracking systems in real time under conditions of polarization jamming. *IEEE RadarCon (RADAR)*, Kansas City, MO, USA (23–27 May 2011), New York, US: IEEE. https://doi.org/10.1109/RADAR.2011.5960536

24 Du Plessis, W.P., Mosoma, K. (2022). Initial results for cross-polarisation jamming of monopulse radar. *International Conference Electromagnetcis Advanced Applications (ICEAA)*, Cape Town, South Africa (23–27 May 2011), New York, US: IEEE. https://doi.org/10.1109/ICEAA49419.2022.9900053

25 Neri, F. (2018). *Introduction to electronic defense systems*, 3rd ed. Boston, MA: Artech House, Section 5.4.15, pp. 438–441.

26 Kuck, J.H. (1957). Means for tracking multiple target formations by radar. US Patent 3,130,402, filed 12 February 1957 and issued 21 April 1964

27 Maksimov, M. (1979). *Radar anti-jamming techniques*. Norwood, MA: Artech House, Section 2.2.4, pp. 55–57.

28 Yan, X., Zhao, Y., Yang, Q. et al. (2013). Research on synchronized non-coherent blinking jamming to monopulse radar seeker. *Proceedings of the International Conference. on Automatic Control and Artificial Intelligence (ACAI 2012)*, Xiamen, China (03–05 Mar. 2012), New York, US: IEEE. https://doi.org/10.1049/cp.2012.1327

29 Li, P., Geng, X.-M., Zhang, Y. et al. (2009). The analysis on angle noise produced by blinking jamming. *Proceedings of International Asia Conference Informatics Control, Automation and Robotics*, Bangkok, Thailand (01–02 February 2009), New York, US: IEEE. https://doi.org/10.1109/CAR.2009.104

30 Bao, L., Li, F., Di, X. et al. (2020). Impact of cooperative jamming of dual stealth aircraft formation on detection performance of monopulse radar. *Aip Advances* 10 (125007): 1–12. https://doi.org/10.1063/5.0027273

31 Zhang, P., Huang, Y., Jin, Z. (2023). An electronic jamming method based on a distributed information sharing mechanism. *MDPI Electronics* 12 (2130): 1–18. https://doi.org/10.3390/electronics12092130

32 Zhang, Y., Chen, Z., Yan, Y. (2023). Research on single target cognitive electronic reconnaissance strategy for unmanned aerial vehicle. *IET Radar, Sonar & Navigation* 17 (11): 1711–1727. https://doi.org/10.1049/rsn2.12461

33 Ren, Y., Li, B., Wang, H. et al. (2019). A novel cognitive jamming architecture for heterogeneous cognitive electronic warfare networks. *Information Science and Applications* 621: 97–104. https://doi.org/10.1007/978-981-15-1465-4_11

34 Adamy, D. (2001). *EW 101: a first course in electronic warfare*. Norwood, MA: Artech House, Section 9.2.3, pp. 185–186.

35 Adamy, D. (2001). *EW 101: a first course in electronic warfare*. Norwood, MA: Artech House, Section 9.3, pp. 187–191.

36 The MathWorks Inc.(2023) MathWorks - Maker of MATLAB and Simulink. https://www.mathworks.com/ (accessed 04 June 2023).

37 Karamirad, M., Ghobadi, C., Nourinia, J. (2021). Metasurfaces for wideband and efficient polarization rotation. *IEEE Transactions on Antennas and Propagation* 69 (3): 1799–1804. https://doi.org/10.1109/TAP.2020.3012828

38 Nguyen, T.K.T., Nguyen, T.M., Nguyen, H.Q. et al. (2021). Simple design of efficient broadband multifunctional polarization converter for X-band applications. *Scientific Reports* 11 (2032): 1–12. https://doi.org/10.1038/s41598-021-81586-w

39 Khan, M.I., Khalid, Z., Tahir, F.A. (2019). Linear and circular-polarization conversion in X-band using anisotropic metasurface. *Scientific Reports* 9 (4552): 1–11. https://doi.org/10.1038/s41598-019-40793-2

40 Sanz-Izquierdo, B., Liang, B., Parker, E.A. et al. (2016). An application of active frequency selective surface to reconfigurable antenna technology. *Proceedings of the Active and Passive RF Devices Seminar*, London, UK (17 February 2016), New York, US: IEEE. https://doi.org/10.1049/ic.2016.0006

41 Chen, D., Yang, W., Che, W. et al. (2019). Polarization-reconfigurable and frequency-tunable dipole antenna using active AMC structures. *IEEE Access* 7: 77792–77803. https://doi.org/10.1109/ACCESS.2019.2919518

42 Zhang, N., Chen, K., Zhao, J. et al. (2022). A dual-polarized reconfigurable reflectarray antenna based on dual-channel programmable metasurface. *IEEE Transactions on Antennas and Propagation* 70 (9): 7403–7412. https://doi.org/10.1109/TAP.2022.3165872

43 Yu, H., Zhang, Z., Su, J. et al. (2023). Quad-polarization reconfigurable reflectarray with independent beam-scanning and polarization switching capabilities. *IEEE Transactions on Antennas and Propagation* 71 (9): 7285–7298. https://doi.org/10.1109/TAP.2023.3291460

44 Wang, X., Xia, D., Li, G. et al. (2023). Design and waveguide measurement of 2-bit reconfigurable amplification metasurface cell. *IEEE Antennas and Wireless Propagation Letters* 22 (9): 2090–2094. https://doi.org/10.1109/LAWP.2023.3275558

45 Yang, J., Yang, W., Qu, K. et al. (2023). Active polarization-converting metasurface with electrically controlled magnitude amplification. *Optic Express* 31 (18): 28979–28986. https://doi.org/10.1364/OE.499458

46 Dong, Y., Kang, W., Sima, B. et al. (2022). Retroreflector with polarization isolation based on nonreciprocal metasurface. *IEEE Antennas and Wireless Propagation Letters* 21 (10): 1940–1944. https://doi.org/10.1109/LAWP.2022.3186334

47 Lončar, J., Grbic, A., Hrabar, S. (2019). Ultrathin active polarization-selective metasurface at X-band frequencies. *Physical Review B: Condensed Matter* 100 (075131): 1–7. https://doi.org/10.1103/PhysRevB.100.075131

48 Ciao Wireless, Inc. (2024). Ciao Wireless, Solid State Hybrid RF & Microwave Amplifiers and Sub-Assemblies for every application. https://www.ciaowireless.com/ (accessed 08 August 2023).

49 Xu, X., Jiang, J., Miao, L. et al. (2012). Design of tunable metamaterial absorbers based on PIN diodes. *IEICE Electrocnics Express* 9 (17): 1408–1413. https://doi.org/10.1587/elex.9.1408

50 Zhao, J., Cheng, Q., Chen, J. et al. (2013). A tunable metamaterial absorber using varactor diodes. *New Journal of Physics* 15, (043049): 1–11. https://doi.org/10.1088/1367-2630/15/4/043049

51 Wang, W., Wang, J., Yan, M. et al. (2018). Dual band tunable metamaterial absorber based on cuboid ferrite particles. *Journal of Physics D: Applied Physics* 51 (31): 1–13. https://doi.org/10.1088/1361-6463/aacf09

52 Wang, W., Xu, C., Yan, M. et al. (2019). Broadband Tunable Metamaterial Absorber Based on U-shaped Ferrite Structure. *IEEE Access* 7: 150969–150975. https://doi.org/10.1109/ACCESS.2019.2947140

53 Zou, Y., Kong, X., Xing, L. et al. (2022). A slot antenna array with reconfigurable RCS using liquid absorber. *IEEE Transactions on Antennas and Propagation* 70 (7): 6095–6100. https://doi.org/10.1109/TAP.2021.3138525

54 Liu, Y., Zhang, W., Jia, Y. et al. (2021). Low RCS antenna array with reconfigurable scattering patterns based on digital antenna units. *IEEE Transactions on Antennas and Propagation* 69 (1): 572–577. https://doi.org/10.1109/TAP.2020.3004993

55 Taravati, S., Eleftheriades, G.V. (2021). Full-duplex reflective beamsteering metasurface featuring magnetless nonreciprocal amplification. *Nature Communications* 12 (4414): 1–11. https://doi.org/10.1038/s41467-021-24749-7

56 Pan, Z., Li, Y., Wang, S. et al. (2023). Joint optimization of jamming type selection and power control for countering multifunction radar based on deep reinforcement learning. *IEEE Transactions on Aerospace and Electronic Systems* 59 (4): 4651–4665. https://doi.org/10.1109/TAES.2023.3272307

57 Zhang, W., Ma, D., Zhao, Z. et al. (2023). Design of cognitive jamming decision-making system against MFR based on reinforcement learning. *IEEE Transactions on Vehicular Technology* 72 (8): 10048–10062. https://doi.org/10.1109/TVT.2023.3261318

58 Feng, L.-W., Liu, S.-T., Xu, H.-Z. (2022). Multifunctional radar cognitive jamming decision based on dueling double deep Q-network. *IEEE Access* 10: 112150–112157. https://doi.org/10.1109/ACCESS.2022.3214842

59 Wang, L., Song, F., Fang, G. et al. (2022). A multi-agent reinforcement learning-based collaborative jamming system: algorithm design and software-defined radio implementation. *China Communications* 19 (10): 38–54. https://doi.org/10.23919/JCC.2022.10.003

60 Aref, M.A., Jayaweera, S.K., Machuzak, S. (2017). Multi-agent reinforcement learning based cognitive anti-jamming. *IEEE Wireless Communications and Networking Conference (WCNC)*, San Francisco, CA, USA (19–22 March 2017), New York, US: IEEE. https://doi.org/10.1109/WCNC.2017.7925694

61 Zhang, C., Song, Y., Jiang, R. et al. (2023). A cognitive electronic jamming decision-making method based on Q-learning and ant colony fusion algorithm. *MDPI Remote Sensing* 15 (3108): 1–21. https://doi.org/10.3390/rs15123108

62 Liu, H., Zhang, H., He, Y. et al. (2022). Jamming strategy optimization through dual Q-learning model against adaptive radar. *MDPI Sensors* 22 (145): 1–17. https://doi.org/10.3390/s22010145

63 Xu, Y., Wang, C., Liang, J. et al. (2022). Deep reinforcement learning based decision making for complex jamming waveforms. *MDPI Entropy* 24 (1441): 1–22. https://doi.org/10.3390/e24101441

64 Zhang, S., Tian, H., Chen, X. et al. (2020). Design and implementation of reinforcement learning-based intelligent jamming system. *IET Communications* 14 (18): 3231–3238. https://doi.org/10.1049/iet-com.2020.0410

65 Slimeni, F., Chtourou, Z., Amor, A.B. (2021). Reinforcement learning based anti-jamming cognitive radio channel selection. *4th International Conference Advanced*

Systems and Emergent Technologies (ICASET), Hammamet, Tunisia (15–18 December 2020), New York, US: IEEE. https://doi.org/10.1109/IC_ASET49463.2020.9318287

66 Wang, L., Peng, J., Xie, Z. et al. (2020). Optimal jamming frequency selection for cognitive jammer based on reinforcement learning. *2nd International Conference on Information Communication and Signal Processing (ICICSP)*, Weihai, China (28–30 September 2019), New York, US: IEEE. https://doi.org/10.1109/ICICSP48821.2019.8958575

67 Haigh, K. Z., Andrusenko, J. (2021). *Cognitive electronic warfare: an artificial intelligence approach*, Boston, MA: Artech House, Chapter 7, pp. 139–158.

68 Haigh, K.Z., Mackay, A.M., Cook, M.R. et al. (2015). Parallel learning and decision making for a smart embedded communications platform. *BBN Technical Report*, BBN-REPORT-8579.

69 The MathWorks Inc. (2024). Design a guidance system in MATLAB and Simulink. Documentation page. https://www.mathworks.com/help/simulink/slref/designing-a-guidance-system-in-matlab-and-simulink.html (accessed 07 March 2024).

8

Metasurfaces for Electronic Protection and Multifunctionality

The many applications we have discussed so far in this book allow an easy understanding of how we can introduce metasurfaces in the EW scenario. We also considered such applications because they provide the reader with the big picture of how the cat-and-mouse EW game, commented first in Section 1.1.1, has been unfolding over the past few decades. However, we illustrated all the concepts we studied up to this point with applications *against* radars (requiring little effort to extend them to other systems such as comms or lidars as we previously explained in Section 1.6). Naturally, a response incorporating metasurface technologies *in favor of* radars (and those other systems) is likely to show up in the theater too – and in the blink of an eye.

In essence, EP substantially differs from ES and EA. Traditionally, the activity involves protecting radars and communication systems against adversary jamming [1]. It presents a crucial role for every EW action rather like any other defensive means in the military theory:

> With growing experience, all skillful commanders sought to profit by the power of the defensive, even when on the offensive.
>
> —B. H. Liddell Hart, military theorist

We can implement EP measures at the antenna level, or with RF-front-end and even processor functions. An example at the RF-front-end level is the hopping of operating frequencies so as to avoid interferences from a fixed-frequency hostile signal [2]. At the processor level, we can mention the use of pulse compression techniques that allow long-period, low-amplitude pulses in LPI radars [3], as detailed in Section 5.1.

As it is probably obvious to the reader that followed all the previous chapters, we can use metasurfaces for EP purposes at the antenna level. By using metasurfaces in the antennas of one's own systems, one can avoid jamming signals from reaching the receiver or the system as a whole.

Metasurface-Driven Electronic Warfare, First Edition. Rafael Gonçalves Licursi de Mello.

Published 2025 by John Wiley & Sons, Inc.
Companion website: www.wiley.com/go/licursidemello/metasurfacedrivenelectronicwarfare1

In this chapter, we will see that we can employ not only traditional FSSs [4–8] but also rasorbers [9–13], studied in Chapter 4, to counter directed-energy weapons, a modern element employed against radars, communication systems, and drone swarms [14–22]. By properly planning the transmission, reflection, and absorption of electromagnetic waves throughout the spectrum, we can improve the shielding of coated systems at specific frequency ranges without bothering the functioning of their antennas.

This chapter will also discuss the use of metasurfaces against cross-polarization jamming. We can use reconfigurable polarization-conversion metasurfaces [23–26], briefly introduced in Section 6.4.2, to detect when a system is under such kind of jamming. Then, we can select the most advantageous antenna polarization accordingly. At the fundamental level, this approach is similar to the traditional one, which discriminates the polarization of received signals using auxiliary, orthogonal antennas [27]. Using metasurface here contributes to system simplicity, compactness, power efficiency, and low cost, among others, allowing for the application in missile seekers or lightweight drones. Another application opportunity for reconfigurable polarization-conversion metasurfaces is in polarimetric radars, which exploit the polarization dependence in wave scattering [28–31]. We will also provide brief comments in this matter.

One further way of performing EP with metasurfaces is through the steering and shaping of one or multiple antenna beams. By illuminating specific regions of interest and avoiding others, we can significantly improve the radio link [32]. This functionality finds application in radars, communication systems, and DRFMs, as previously stated in Section 5.2.2. The generation of multiple beams is useful for spatial multiplexing [33], multifunctional radars [34–36], and multi-target jammers [37]. The shaping of beams enables the creation of a null in the radiation pattern aiming at avoiding potentially hostile signals. Modern radars use phased-array antennas to carry out these advanced functionalities, but the costs of these antennas may be of the order of 50% of the total costs of the radar itself [38]. This chapter will also study how we can use metasurfaces as an alternative to phased arrays, which provides benefits not only in the matter of costs but also in size, weight, and power consumption once again.

The unparalleled capability of metasurfaces in the tailoring of electromagnetic waves also allows us to obtain different properties at different frequencies from a single device, as discussed in Section 3.4.1. Hence, to conclude the chapter, we will see the design of metasurface-based multifunctional antennas. Multifunctional antennas are a key element in the modern scenario because they combine the roles of multiple individual antennas into a single, integrated device. Thus, they enable the design of simple, compact, light systems operating multiple, concurrent standards [39, 40], or even different applications, like radars and communications [41, 42]. This capability is paramount in a unprecedently dynamic world where the electromagnetic scenario is increasingly dense.

8.1 Electronic Protection Against Directed-Energy Weapons

Directed-energy weapons were envisioned to physically damage the circuits of adversary radars and communication systems [14–17]. The intention is to radiate high-power electromagnetic pulses of a few nanoseconds or tens of nanoseconds toward a target system.

Much effort has been recently put into these weapons, a technology still under development, because today they are also considered a strong candidate to counter drone swarms [18–22].

8.1.1 A Brief Introduction to Directed-Energy Weapons

Introducing a high-power signal at a target system can be seen as a brute-force EA method. Damages can take place through two main mechanisms [43]:

- The exceeding of temperature limits;
- The voltage breakdown of the dielectrics used in its circuits.

Typical paths through which these signals can reach the circuitry are slots, cables, and antennas. However, dedicated weapons will attempt to match the antenna's operating frequencies because the fundamental function of this component is to collect energy from the environment [44].

We can derive the frequency-domain representation of a single pulse that a directed-energy weapon tunes at a carrier frequency f_0 by using the Fourier transform in (5.1) as:

$$\begin{aligned} V_{pulse}(f) &= \int_{-\infty}^{+\infty} V_0 \mathrm{rect}\left(\frac{t-TOA}{PW}\right) e^{j2\pi f_0 t} e^{-j2\pi f t}\, dt \\ &= V_0 PW \mathrm{sinc}[\pi(f-f_0)PW] e^{-j2\pi(f-f_0)TOA} \end{aligned} \tag{8.1}$$

The presence of the sinc function in (8.1) makes it evident that the energy is spread throughout many spectral components.

The conventional way of countering directed-energy weapons is by ensuring proper shielding of one's systems' housing boxes as well as proper grounding of the interconnecting cables therebetween [44]. We expect a good design in this matter to provide 40 dB of attenuation [45]. Obviously, the directed-energy weapon can compensate such attenuation by employing higher levels of power, to the limit of what it has available.

8.1.2 Countering Directed-Energy Weapons with Metasurfaces

We can use FSSs and rasorbers as an extra layer of EP, filtering the spectral components and allowing whereby the transmission of only those in the operating frequency ranges of the antennas. Although the maximum of (8.1) occurs at the carrier frequency f_0 of the directed-energy weapon (which can match that of the antenna), such filtering restricts the accepted power. Hence, we reduce the chances that damage through thermal effects or dielectric voltage breakdowns occurs.

We saw in Section 4.3 that we can use traditional FSSs, seen by many as precursors of metasurfaces, for the spatial filtering of electromagnetic waves [4–8]. Among the different types, bandpass FSSs are the most appropriate to counter directed-energy weapons. This type of FSS, whose transmission-line equivalent Fig. 4.2b previously represented, is typically composed of metal plates with etched slots that reflect the impinging waves except at frequencies where the slots resonate. Thus, we can tune the slots so as to allow the proper operation of the antennas of one's own system, whereas the FSS reflects out most of the energy coming from the weapon. B.A. Munk [46] provides the analysis of many FSSs, including their performance as a function of the unit cell type.

Although the FSS approach can be useful to protect a single drone, reflecting out energy is a suboptimal solution for a drone swarm. The reason is that the reflected energy can impinge over other drones in the swarm in a modern scenario, where both the swarm and the directed-energy weapon utilization may be dense. To circumvent this problem, an alternative is to exploit absorption mechanisms rather than reflections.

As opposed to FSSs, rasorbers allow the transmission of waves in a specific band while absorbing them out of band [9–13]. As Section 4.5.2 previously stated, a lossy band-stop FSS stacked over a bandpass FSS typically makes up a rasorber. That section also presented a detailed study about these metasurfaces, including the analysis of a transmission-line equivalent. It also provided the performance of prominent metasurfaces, giving an idea of what the attenuation would be when directed-energy weapons illuminate them. Finally, that section also listed their physical characteristics, such as thickness and unit cell types.

To the best of the author's knowledge, no open literature is available on the use of FSSs or rasorbers in the specific problem of directed-energy weapons. Even prominent FSS/rasorber works do not study the temperature and voltage breakdown limits of the surface itself. Such features are of primary importance for the discussed EP application because the associated energy will interact with the FSS/absorber even when the surface does not let waves pass through.

As the reader may imagine, the temperature and voltage breakdown limits of the structure will depend on the features of the used materials, that is, the dielectric and electronic components, if any. We can tackle this issue with a simple choice of the material/component considering this matter, besides its conventional properties. Another factor, of much higher complexity, is the shape of the unit cell. Depending on how distributed or compact the geometry of the unit cell is and where resonances take place therein, fields and currents will concentrate more or less at specific parts of the dielectric/component. This problem may require extensive field/current analyses using full-wave simulators.

A last point, the system's antennas can still accept the directed energy in both FSS and rasorber approaches. We should use antennas with well-behaved patterns privileging radiation over specific regions of interest to reduce damages even further. To this end, we should concurrently implement the steering and shaping of beams, which Section 8.5 will discuss.

8.2 Electronic Protection Against Cross-Polarization Jammers

Cross-polarization jammers attempt to disturb the good functioning of radars by introducing cross-polarized signals in their systems, as Section 7.3 explained. This EA measure may induce wrong calculations and deceive the target radar because the latter considers in the processing the co-polarized radiation patterns of its antennas.

As we also saw in Section 7.3, the gain in the cross-polarized radiation pattern will be lower than in the co-polarized one in most of the directions of a reasonably well-designed radar antenna (see Fig. 7.4). In Section 7.3.1, we referred to this feature from an EA perspective as a cross-polarization disadvantage $G_{x-pol}(\theta, \Phi)$, defined there as the ratio $G_{A-D,h}(\theta, \Phi)/G_{A-D,v}(\theta, \Phi)$ for exemplification purposes.

From an EP standpoint, we can exploit the ratio $G_{x-pol}(\theta, \Phi)$, which in general is the ratio between the cross- and co-polarized antenna gains at each direction, in polarization discrimination methods. Traditionally, this maneuver implies the addition of an auxiliary antenna [27], identical to the main one except for an orthogonal polarization. Further, it also requires an extra receiver (or at least a switching mechanism leading to a common receiver).

8.2.1 Countering Cross-Polarization Jammers with Metasurfaces

Rather than adding an auxiliary antenna, we can use reconfigurable metasurfaces in the design of polarization-tunable antennas [23–26]. This approach decreases complexity, size, weight, power consumption, and costs to counter cross-polarization jammers. In this way, we can finally perform the application in lightweight drones and mainly in missile seekers. The latter are particularly vulnerable to this kind of jamming [47]. Briefly introduced in Section 6.4.2, these metasurfaces are reconfigurable AMCs. They can use PIN diodes, varactors, and so forth, strategically deployed over their unit cells to obtain different reflection coefficient phases φ_r for different polarizations. As such, they can convert the polarization of reflected waves through the mechanism previously discussed in Section 4.4.3, but in a reconfigurable manner.

B. Sanz-Izquierdo et al. [23] use one such a metasurface with a coplanar waveguide-fed radiating element for a frequency-reconfigurable operation from 1.15 to 1.60 GHz. The unit cell, composed of a dual-layer square patch over a ground plane, could provide different reflection coefficient phases φ_r for different polarizations depending on the biasing of four varactors orthogonally set over the square corners. In this way, the final antenna can dynamically provide circularly polarized, unidirectional radiation patterns for LHCP or RHCP with a gain of up to 9 dBiC. The antenna provides a $G_{x-pol}(\theta, \Phi)$ ratio below −15 dB at all studied frequencies within angular ranges $-30° < \theta < +30°$ and $-30° < \Phi < +30°$. Hence, by switching the polarization of the antenna between LHCP and RHCP, the polarization of the intercepted signal will become clear due to differences in the signal intensity. The dimensions of the whole device are only $132 \times 132 \times 5.8$ mm^3 which represents $0.5 \times 0.5 \times 0.02\ \lambda^3$ at 1.15 GHz.

D. Chen et al. [24] associate another reconfigurable polarization-conversion metasurface, now using four varactors that connect neighbor patches over a ground plane, with a U-shaped dipole radiating element. This time, the final device can dynamically provide LHCP, RHCP, vertical, or horizontal polarizations in unidirectional radiation patterns that it can frequency-reconfigure from 1.75 to 2.73 GHz. The antenna provides a gain of 7.5 dBi/dBiC for all frequencies and polarizations. The measured $G_{x-pol}(\theta, \Phi)$ ratio is below −30 dB within angular ranges $-30° < \theta < +30°$ and $-30° < \Phi < +30°$. The device is also compact and low-profile, with dimensions of only $108 \times 108 \times 6.5$ mm^3, that is, $0.6 \times 0.6 \times 0.04\ \lambda^3$ at 1.75 GHz.

At this point, it is worth highlighting that these antennas [23, 24] present two reconfigurable functionalities: besides the polarization conversion, they can also dynamically change their operating frequencies. This is interesting from an EP point of view because these antennas allow a system with frequency hopping in parallel to the polarization discrimination. N. Zhang et al. [25] and H. Yu et al. [26] propose reconfigurable metasurfaces for the simultaneous conversion of polarization and steering of reflected waves. When associated with a radiating element, these metasurfaces can also provide two concurrent EP functionalities: polarization discrimination and the spatial selection of targets. Section 8.5 will study the latter in more detail.

8.3 Exercise: Simulating EP Against Cross-Polarization Jammers

Difficult level: medium

We can do an exercise to practice the concepts concerning EP against cross-polarization jamming with our EW scenario simulator. In the folder entitled *Ch08_EP_cross_pol_jam* in the companion website, the reader can find a set of files similar to those of the folder *Ch07a_EA_cross_pol_jam*, used in the exercises of Section 7.8. The difference is that, now, we can obtain the radiation patterns of the antennas A–D of the monopulse radar from two different files. We emulate a vertically polarized dipole illuminating a parabolic dish, as described in Section 7.3, by using the *parabolic_reflector_V_pol.mat* file. On the other hand, *parabolic_reflector_H_pol.mat* represents the radiation pattern of a horizontally polarized dipole illuminating a parabolic dish.

The idea of this exercise is to emulate that the radar can switch the polarization of its antennas (as if we used metasurfaces) at specific periods whereby enabling a polarization discrimination mechanism. You can do that by calling both files in the *build_mnp_rdr_parabolic* function and using the *antenna* object contained therein to create different *phased.Radiator* and *phased.Collector* objects in the radar platform. To check the radiation patterns of each antenna in both polarizations, first, set the *vis.show_radar_ant_pat* variable in the *def_visualization* function to true. Then, set the *"Polarization"* parameter as *"V"* to *"H"* when the *build_mnp_rdr_parabolic* function calls *pattern*.

In the *main.m* file, you should compile some lines of code. Within the main for-loop (the one depending on the *p* variable), create an if-statement defining which *phased.Radiator* and *phased.Collector* objects the current loop should consider. You can take the if-statement activating the cross-polarization jamming (the one depending on the *t_start_jam* variable) as a reference. Note that you should do this for all of the four antennas A–D of the monopulse radar at once.

After setting the simulator, define when the radar should switch from a main, initial polarization to the other polarization. Register the received power level before and after the switching. Then, define criteria to adopt a specific polarization according to a comparison between the registered power levels.

8.4 Metasurfaces for Radar Polarimetry

When Section 4.1 introduced the principles of RCS, we saw that the interactions between electromagnetic waves and a scatterer depend, among others, on the polarization of both the impinging and the scattered waves. Putting it differently, when a vertically polarized wave illuminates an object, the latter scatters a specific amount of energy with vertical polarization depending on the term RCS_{vv}, and with horizontal polarization depending on RCS_{hv}. Likewise, the object scatters a certain amount of energy in the horizontal and vertical polarizations depending on RCS_{hh} and RCS_{vh} when the impinging wave is horizontally polarized. Equation (4.2) expresses all these terms explicitly.

Radars can use this polarization dependence as a means of discriminating features of the target in the environment. In general, such technique, denominated radar polarimetry [28–31], requires that the radar be capable of transmitting and receiving waves in both polarizations. To this end, we require two antennas/RF front ends (or at least a mechanism switching both antennas to a common RF front end).

In this context, antennas with reconfigurable polarization-conversion metasurfaces [23–26] can be useful again. To be more precise, dynamically switching the antenna polarization allows a polarimetric radar to transmit and receive waves with all polarization combinations using a single antenna and RF front end.

These metasurface-based antennas, detailed in Section 8.2.1, contribute once again for a lower system complexity, size, weight, power consumption, and costs. Hence, they enable the application in small satellites, missile seekers, and lightweight drones.

8.5 Metasurfaces as an Alternative to Phased Arrays

In Section 5.2.2, we introduced phased arrays as a key element of modern DRFMs for their capabilities of generating multiple antenna beams that we can electronically steer and shape in a dynamic manner. These functionalities actually find applications in other EW systems too, such as communications and radars.

In communications, the dynamic beamsteering provides a better one-way radio link using (6.2). In other words, we can steer the maximum of the realized gain $G_{t,m}$ and/or $G_{r,m}$ of each antenna toward its counterpart while avoiding directions containing interferers. In the receiving mode, this maneuver means that we can substantially improve the signal-to-noise-plus-interference ratio (SNIR) in a dynamic manner. In the transmitting mode, the advantage is also to mitigate the waste of energy sent toward unconcerned directions [32]. In radars and advanced EA systems, beamsteering avoids the need of moving the antenna mechanically. In this way, it allows the convenient positioning in an optimized compartment of the platform as well as an excellent beam agility.

For communication purposes, the generation of multiple beams allows a signal multiplexing [33]. In radars, the concurrent operation of diverse modes in parallel is possible, with the illumination of different regions with different signals [34–36]. Similarly, in EA, advanced systems can address multiple targets at once [37].

For both communications and radars, the shaping of beams enables the creation of a null in the radiation pattern. We can conveniently place this null at the direction of an interferer, which consists in a signal filtering maneuver in the spatial domain.

AESAs are one type of phased array that use transmit/receive modules with PAs, LNAs, and phase shifters throughout the array at the element level [48]. In this way, they can easily provide the necessary currents to generate the desired antenna beams. They are useful not only for EP but for EW as a whole. However, as Section 5.2.2 previously stated, the complexity related to AESAs leads to up to 50% of the costs of a military radar [38]. Hence, we require alternatives for applications with low costs and also size, weight, and power consumption.

In the sequel, we will study the basic principles of phased arrays and see how we can use metasurfaces as an alternative thereto in these matters. Metasurfaces are also capable of performing the shaping of beams [49–52]. However, for the sake of clarity, we will focus hereafter on one functionality, namely the beamsteering.

8.5.1 Working Principle of Phased-Array Antennas

The traditional way of interpreting how a phased array operates is by thinking about the total fields generated by the array as the superposition of the fields due to each array element.

Consider in Fig. 8.1 a linear array with $n = 1, 2, \ldots, N$ identical, arbitrary radiating elements equally spaced in the z dimension. The distance between the phase centers of the elements is d. Next, consider an observation point $P(r, \theta, \Phi)$ located at the far-field region of the array, that is, $r \gg Nd$. The element $n = 1$ is located at

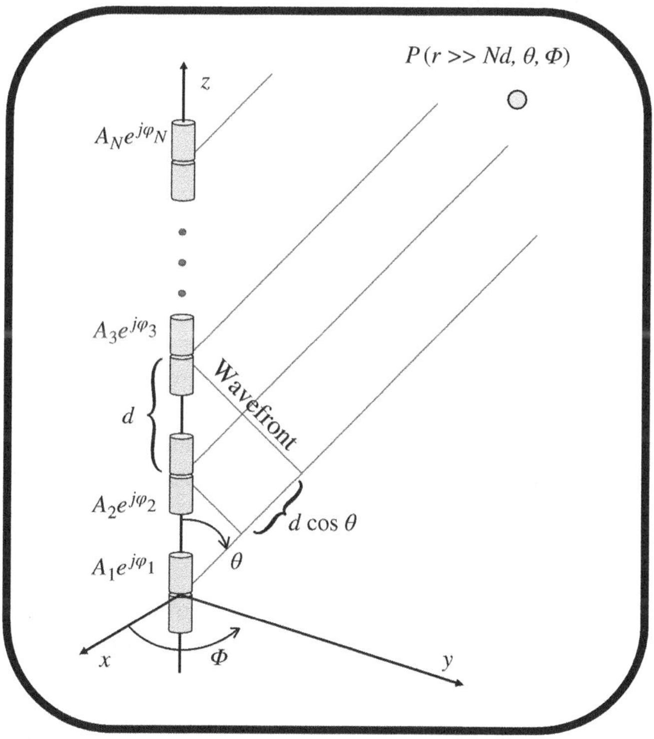

Fig. 8.1 Linear array composed of N identical, arbitrary radiating elements

the origin and distanced from this point by $r_1 = r$. The others are distanced therefrom by $r_n = r-(n-1)d\cos\theta$. Thus, the electric field due to the array is [53]:

$$\begin{aligned} \boldsymbol{E}_{array}(r,\theta,\Phi) &= \sum_{n=1}^{N} \boldsymbol{E}_n(r,\theta,\Phi) \\ &= KF(\theta,\Phi)\sum_{n=1}^{N} A_n e^{j\varphi_n}\frac{e^{-jkr_n}}{r_n} \end{aligned} \tag{8.2}$$

where, for simplicity, we define that:

- $\boldsymbol{E}_n(r, \theta, \Phi)$ is the electric field due to the element n [V/m];
- K is a proportionality constant [unitless];
- $\boldsymbol{F}(\theta, \Phi)$ is a function whose spatial dependence accounts for the radiation pattern of the type of element in the array [unitless];
- A_n and φ_n are, respectively, the amplitude and phase of an equivalent excitation voltage for the element n, [V and rad].

The term e^{-jkr_n}/r_n in (8.2) accounts for the propagation until the observation point $P(r, \theta, \Phi)$. Because $r \gg Nd$, we can consider r_n equal to r for amplitude purposes (but not for phase). Consequently, (8.2) becomes:

$$\begin{aligned} \boldsymbol{E}_{array}(r,\theta,\Phi) &= KF(\theta,\Phi)\frac{e^{-jkr}}{r}\sum_{n=1}^{N} A_n e^{j\varphi_n} e^{j(n-1)kd\cos\theta} \\ &= KF(\theta,\Phi)\frac{e^{-jkr}}{r} AF \end{aligned} \tag{8.3}$$

where $AF = \sum_{n=1}^{N} A_n e^{j\varphi_n} e^{j(n-1)kd\cos\theta}$ is known as the array factor.

In the above development, AF depends only on the elevation dimension θ because the array is strictly distributed over the z dimension, being thus one-dimensional. If the array were also spread over the y or x dimensions, AF would also depend on the azimuthal dimension Φ.

We can see in (8.3) that the array factor AF is a quantity that multiplies the function $\boldsymbol{F}(\theta, \Phi)$, which in turn expresses the spatial dependence due to the type of radiating element used in the array. Hence, it is fair to say that the array factor AF spatially modulates the radiation pattern of the array elements. This is the main takeaway of what we have developed so far.

As (8.3) also made evident, the array allows the obtaining of arbitrary radiation patterns (limited by $\boldsymbol{F}(\theta, \Phi)$, of course) under the condition that we provide proper excitation amplitudes and phases A_n and φ_n. To be more precise, the array allows the shaping and steering of the radiation pattern. However, we will illustrate this capacity only with the beamsteering functionality. To this end, consider uniform excitation amplitudes, that is, $A_1 = A_2 = \cdots = A_N$. Moreover, we vary the excitation phases linearly, that is, $\varphi_n = (n-1)\varphi_0$, where φ_0 is the phase difference between adjacent elements. In this case, the array factor in (8.3) becomes [54]:

$$\begin{aligned} AF &= A_1 \sum_{n=1}^{N} e^{j(n-1)\varphi_0} e^{j(n-1)kd\cos\theta} \\ &= A_1 \sum_{n=1}^{N} e^{j(n-1)(kd\cos\theta + \varphi_0)} \end{aligned} \tag{8.4}$$

By manipulating (8.4), we have:

$$AFe^{j(kd\cos\theta + \varphi_0)} = A_1 \sum_{n=1}^{N} e^{jn(kd\cos\theta + \varphi_0)} \tag{8.5}$$

Following, subtracting (8.5) from (8.4) leads to:

$$\begin{aligned} AF\left[1 - e^{j(kd\cos\theta + \varphi_0)}\right] &= A_1\left[1 - e^{jN(kd\cos\theta + \varphi_0)}\right] \\ AF &= A_1 \frac{1 - e^{jN(kd\cos\theta + \varphi_0)}}{1 - e^{j(kd\cos\theta + \varphi_0)}} \end{aligned} \tag{8.6}$$

Finally, by further manipulating (8.6), we have:

$$\begin{aligned} AF &= A_1 \frac{e^{jN\frac{kd\cos\theta + \varphi_0}{2}}}{e^{j\frac{kd\cos\theta + \varphi_0}{2}}} \left(\frac{e^{-jN\frac{kd\cos\theta + \varphi_0}{2}} - e^{jN\frac{kd\cos\theta + \varphi_0}{2}}}{e^{-j\frac{kd\cos\theta + \varphi_0}{2}} - e^{j\frac{kd\cos\theta + \varphi_0}{2}}} \right) \\ AF &= A_1 e^{j(N-1)\frac{kd\cos\theta + \varphi_0}{2}} \frac{\sin\left(N\frac{kd\cos\theta + \varphi_0}{2}\right)}{\sin\left(\frac{kd\cos\theta + \varphi_0}{2}\right)} \\ AF &= A_1 e^{j(N-1)\frac{kd\cos\theta + \varphi_0}{2}} BS(\varphi_0) \end{aligned} \tag{8.7}$$

In (8.7), $BS(\varphi_0) = \sin[N(kd\cos\theta + \varphi_0)/2]/\sin[(kd\cos\theta + \varphi_0)/2]$ represents a beamsteering function that depends on the phase difference φ_0 between adjacent elements in the array. The maximum value of $BS(\varphi_0)$ occurs when $kd\sin\theta + \varphi_0 = 0$. Hence, we can control the angle θ where the maximum occurs through φ_0. Specifically:

$$\theta = \arccos\left(-\frac{\varphi_0}{kd}\right) = \arccos\left(-\frac{\lambda}{2\pi}\frac{\varphi_0}{d}\right) \tag{8.8}$$

Fig. 8.2 sketches the absolute value of the beamsteering function $BS(\varphi_0)$ versus θ we calculated with (8.7) for three different values of phase difference between adjacent elements, namely $\varphi_0 = 0°$, $\varphi_0 = 40°$, and $\varphi_0 = 80°$. Here, we use $N = 10$ elements spaced by $d = \lambda/2$. The maximum of $BS(\varphi_0)$ shifts from 90° to 102.8°, and then to 116.4°, which is coherent to (8.8).

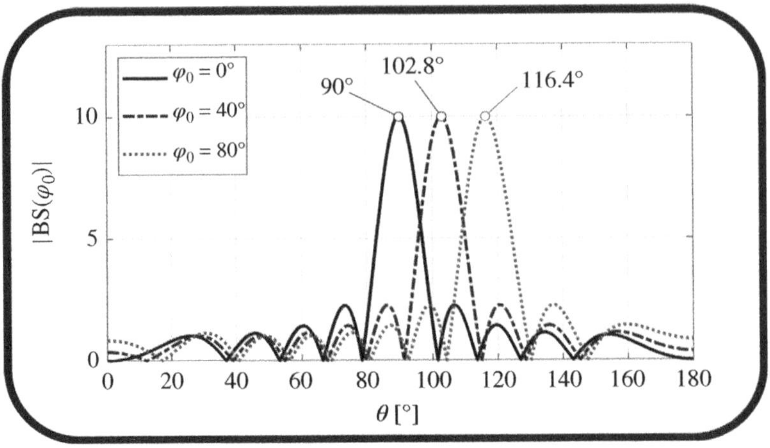

Fig. 8.2 Beamsteering function $BS(\varphi_0)$ in (8.7) for three different values of phase difference φ_0 between $N = 10$ elements

Note that even though obtaining a desired steering seems trivial with the explained method, providing the required excitations to each element of the array can be a challenge. The reason is that neighbor elements may radiate toward one another, and surface waves may travel throughout the array (see Section 2.1.7). In this way, the elements induce spurious currents over one another due to a mutual coupling [55].

Many approaches exist to reduce such effect. Among them, the use of corrugations [56], resonant mushroom-like structures [57], or even adding extra coupling paths to cancel the original mechanisms [58]. AESAs have advantages in this matter because they can generate an independent excitation for each element using the transmit/receive modules. In this way, although mutual coupling effects also exist, AESAs may tame these phenomena through optimization methods. Naturally, increased complexity, costs, power consumption, size, and weight related to the PAs, LNAs, and phase shifters present in each of the many modules is what allows such performance.

Another way of interpreting AESAs, coherent with what we saw in Section 3.2.5, is by using Love's equivalence principle [59], a theorem that quantitatively formalizes Huygens' principle. By considering the AESA as an aperture where electric and magnetic surface currents $\boldsymbol{J}_s$ and $\boldsymbol{M}_s$ exist (in this case, impressed over many radiating elements by a complex feeding system), we can create an arbitrary distribution of fields $\boldsymbol{E}_0$ and $\boldsymbol{H}_0$ (naturally, limited by the degrees of freedom of the currents in the elements). Such interpretation makes the link with the working principle of metasurfaces that can work as an alternative to phased arrays.

8.5.2 Beamsteering with Phase-Gradient Metasurfaces

The study in Section 8.5.1 concerning beamsteering with phased arrays may have given the reader the feeling of having already had contact with a similar subject previously in this book. Indeed, Section 3.2.7 introduced several phase-gradient metasurfaces capable of steering electromagnetic beams [60–62]. D.F. Sievenpiper et al. [60] perform the functionality with a reflective metasurface exploiting the generalized law of reflection (3.35). C. Pfeiffer et al. [61] and J.P.S. Wong et al. [62] do so with transmissive metasurfaces exploiting the generalized law of refraction (3.36).

These laws, formally derived by N. Yu et al. [63], predict the steering of electromagnetic beams subjected to a linear phase gradient when reflected upon or transmitted through a surface/interface. They are an update of the several-century-old Snell's laws for the reflection and refraction of waves, (2.61) and (2.62), derived in Section 2.1.5, which we can actually see as a special case when the phase gradient is zero.

What catches the eye when comparing the beamsteering with phased arrays or metasurfaces is the resemblance between the mechanisms. A quick look at (8.8), (3.35), and (3.36) reveals that a phase variation throughout a spatial dimension is what activates the steering in all of them. Naturally, the discrete radiating elements in the antenna array also discretized the phase variation. On the other hand, the tiny unit cells in metasurfaces spatially homogenizes such variation. Nonetheless, at the fundamental level, these devices work similarly.

Now, the question is why we should choose one approach rather than the other to perform beamsteering. The fact of the matter is that if, on the one hand, the steering mechanism is the same, that is, the spatial phase variation, on the other, the way of obtaining such phases may substantially differ.

As previously stated, AESAs can easily provide a desired excitation at the element level through their expensive transmit/receive modules. This is not the case, however, of passive phased-array architectures. Passive phased arrays use a well-known radiating element with high-gain, unidirectional patterns (typically a horn antenna) illuminating a reflectarray made of unit cells whose size typically does not exceed half a wavelength. We can control the reflection coefficient of the unit cells so as to translate the illumination of the radiating element into a specific radiation pattern [64]. Although much cheaper and simpler than AESAs, the positioning of the radiating element illuminating the reflectarray makes the passive phased array bulky and inconvenient to many applications – including small satellites, missile seekers, and lightweight drones. In the metasurface approach, all these matters are pushed to the limit, leading to simple, cheap, low-profile, lightweight antenna designs.

8.5.3 Beam-Steerable Antennas with Reflective Metasurfaces

In their seminal work, Sievenpiper et al. [65] demonstrated that mushroom-like AMCs work well when placed in the vicinity of a radiating element, allowing for extremely low-profile antenna designs. Sievenpiper et al. [60] demonstrate that we can use varactors in the same kind of metasurface to provide an individual, reconfigurable reflection coefficient phase φ_Γ to each of its unit cells. Also, they fabricate and measure a prototype containing 1152 individually controlled cells. Next, they implement phase gradients to steer reflected beams in a range of $\theta_r = \pm 45°$. In the experiments, they validate concepts by placing the metasurface in the far-field region of a horn antenna. Although Sievenpiper et al. [60] do not demonstrate beamsteering by associating the structure with a closely spaced radiating element, other works do so [32].

For example, J.A. Vásquez-Peralvo et al. [66] propose a mushroom-like AMC spaced by a quarter wavelength from a printed dipole. Similarly to Sievenpiper et al. [60], varactors that can be fed through the mushroom vias connect neighbor unit cells in one of the dimensions. As such, we can vary the unit cell reflection coefficient phase φ_Γ in a range of up to 330°. They simulate, fabricate, and measure a metasurface composed of 22×14 unit cells. The prototype can steer the antenna beam in a range of $\pm 40°$ at 2.0 GHz. They demonstrate the beamsteering only in the plane of the varactors, namely the H-plane, in steps of 10°. This one-dimensional beamsteering requires varactors of the same row to receive the same bias voltage, which implies only 22 control inputs. They measure a realized gain of around 7.0 dBi when the beamsteering is not performed (i.e., when the beam points in the broadside direction). This gain drops to around 6.0 dBi when they steer the beam to 40°. The total thickness of the antenna is 87 mm, or 0.58 λ at 2.0 GHz, significantly lower than a typical reflectarray antenna.

B. Ratni et al. [52] propose a patch antenna operating at 10.0 GHz placed 30 mm over a 1.52-mm thick metasurface, which leads to a total thickness of 31.5 mm or 1.05 λ at 10.0 GHz. A couple of parallel metal lines over a grounded slab makes up the unit cell. These lines can bias a varactor placed between them. In this way, we can fully vary the unit cell reflection coefficient phase φ_Γ (range of 360°). Hence, an arbitrary, continuous phase gradient along the surface is possible in one of the dimensions. A metasurface made up of 30×30 unit cells is simulated, fabricated, and measured. The feeding of the parallel lines conveniently biases the varactor from the side of the metasurface. As such, the device requires only 30 control inputs. The prototype steers the antenna beam in a range of ±55° in the plane of the varactors, the E-plane this time.

8.5.4 Beam-Steerable Antennas with Transmissive Metasurfaces

The previously mentioned antennas [52, 60, 66] carry out beamsteering through a reflection mechanism. However, we may also implement such functionality with transmissive metasurfaces working as antenna superstrates [32].

W. Pan et al. [67] directly cover a horn antenna operating at 5.3 GHz with a varactor-loaded, phase-gradient, transmissive metasurface superstrate. The unit cells comprise a metal plate with square slots. They can vary the transmission coefficient phase φ_τ in a modest range of 90°. To achieve a full 360°-range, they stack up five layers of one such a cell, which allows an arbitrary, continuous phase gradient over the surface. They simulate, fabricate, and measure a prototype composed of 6 × 6 unit cells. In terms of control, we can independently address each row or column of varactors. Thus, we require only six controls to carry out beamsteering in each of the planes. In the E-plane, the realized gain is respectively measured as 15.4, 14.3, 13.3, and 12.2 dBi for steering angles of 0°, 10°, 20°, and 30°. For the same angles in the H-plane, 15.4, 14.5, 13.4, and 12.2 dBi are measured.

Similar to W. Pan et al. [67], J.R. Reis et al. [68] also cover a horn antenna directly with a five-layer metasurface superstrate. Each layer of the unit cell comprises a metal sheet with a double-square slot with varactors so as to reach full 360°-range in the variation of the transmission coefficient phase φ_τ at 5.20 GHz. This time, however, a sixth layer makes possible the independent biasing of each unit cell. As such, we can steer the antenna beam toward angles outside the main planes. The fabricated prototype is made of 5 × 5 unit cells implying a total of 25 control inputs. In the measurements, the realized gain is 15.6 dBi without the beamsteering and 12.1 dBi for a steering of 21° in azimuth and 20° in elevation. For a beamsteering of 15° in azimuth and 12° in elevation, 14.1 dBi is measured.

At this point, it is worthwhile to note that decreasing the number of controls from 1152 [60] to 30 and 22 [52, 66], and then 6 and 25 [67, 68] represents a remarkable advancement in the maturity level of the technology in the matter of simplicity and feasibility. This contributes to a system with lower costs, size, weight, and power consumption. However, in terms of thickness, the complete devices comprising a horn plus a superstrate [67, 68] still fall far short of what most of the constrained applications require. The antennas of B. Ratni et al. [52] and J.A. Vásquez-Peralvo et al. [66] present thicknesses in the order of the wavelength – which is already reasonable for extremely demanding applications such as small satellites, missile seekers, and lightweight drones. However, these antennas are not planar. As it will become more obvious in the next sections, metasurfaces have the potential to push all these frontiers even more. However, to achieve an extremely lightweight, compact, planar, low-profile design, we require alternative solutions for the antenna topology.

8.5.5 Fabry–Perot Resonances

At first glance, turning the antennas of [67, 68] into low-profile designs seems straightforward by replacing the horn with a planar radiating element with a high-gain, unidirectional radiation pattern. The problem is that a planar radiating element providing these patterns typically employs a planar reflector, as thoroughly studied in Sections 6.3 and 6.4. Hence, in the end, the low-profile topology will present a transmissive metasurface superstrate arranged in parallel to a reflector – which may prove problematic as it will become more obvious to the reader in the sequel.

Even though we design the unit cells of transmissive metasurfaces to transmit most of the energy of impinging waves, in practice the story is different. A finite-array transmissive metasurface will present some degree of reflectivity when placed at the near-field of a radiating element. Reasons for that are analogous to those listed in items i–iii of Section 6.4.5 for the finite-array AMC: we simulate the unit cell considering specific conditions of angle of incidence and polarization that do not match those in the complete device – which leads to a different performance as discussed in Section 3.4.2; further, surface waves, explained in Section 2.1.7, may travel over the finite, planar structure, causing extra resonances when reflecting upon its edges [69]; lastly, the proximity of the metasurface superstrate with other structures (the reflector and radiating element) may give birth to extra couplings the standalone simulations do not consider.

To put it more precisely, in general, we should consider transmissive metasurfaces as partially reflective surfaces. Well, since the 1950s, it is well known that a cavity formed by a reflector and a partially reflective surface makes internal resonances arise when excited by electromagnetic waves [70]. Such resonances are known as Fabry–Perot resonances [71].

To understand the consequences of such phenomenon, consider in Fig. 8.3 a cavity formed by a superstrate with partially reflective behavior and a reflector. An isotropic radiating element lies between the reflector and the superstrate, in a region hereafter we will call Fabry–Perot cavity. When the element radiates at a frequency $f = f_1$, the partially reflective superstrate transmits some energy. The superstrate reflects back the rest of the energy, which bounces inside the cavity, and impinges again upon the superstrate. Then, the superstrate transmits another portion of energy, and the cycle repeats. Outside the cavity, interference patterns result in one or more beams with specific widths and directions depending on the superstrate properties and the phase shift between subsequent bouncing waves. Figure 8.3a depicts Fabry–Perot resonances for a case where constructive

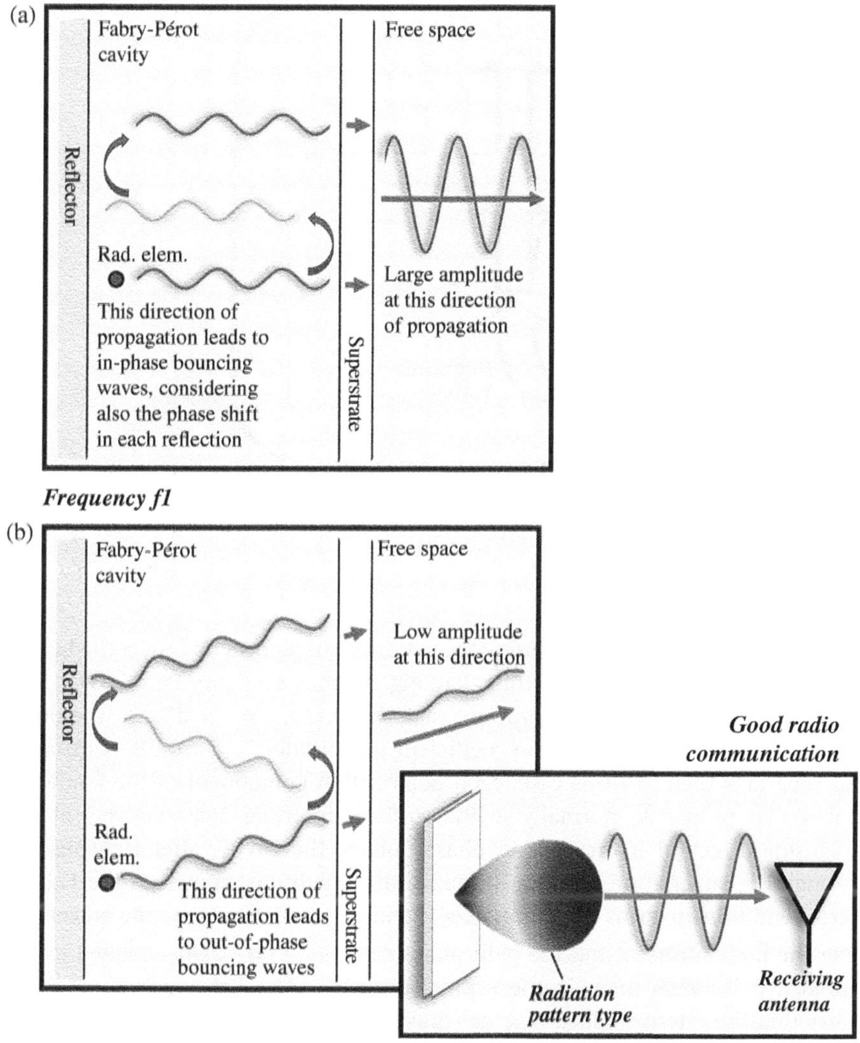

Fig. 8.3 Operation of the Fabry–Perot cavity at a frequency $f = f_1$: (a) constructive interferences occur in the broadside direction; (b) destructive interferences occur in other directions. "Gonçalves Licursi de Mello [72]/Springer Nature/CC BY 4.0."

interferences occur at the broadside direction. Concurrently, destructive interferences occur in other directions, as shown in Fig. 8.3b [72].

The configuration of Fig. 8.3 leads to a radiation pattern with maximum gain at the broadside direction. For this to happen, we should adjust the Fabry–Perot

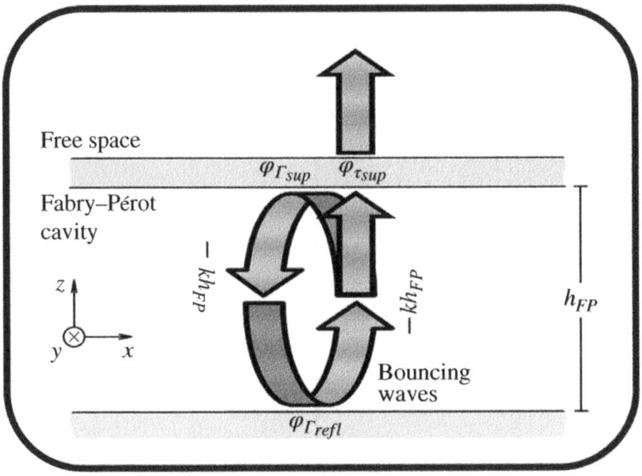

Fig. 8.4 A Fabry–Perot cavity within which bouncing waves experience a phase deviation φ_{bounce}."Gonçalves Licursi de Mello [32]."

cavity. Otherwise, spurious interference patterns can deform or deflect the beams in undesired directions. To illustrate that, Fig. 8.4 shows a Fabry-Perot cavity. The superstrate has a reflection coefficient magnitude $|\Gamma_{sup}| \neq -\infty$ dB. On the other hand, the reflector has reflection coefficient magnitude $|\Gamma_{refl}| = 0$ dB. A distance h_{FP} separates them. In this case, each polarization component of the bouncing waves that propagate normally to the surfaces interfere constructively at the broadside direction if they are in phase outside the cavity, after transmission through the superstrate. We can formalize this condition, known as the Fabry–Perot condition, in terms of the phases $\varphi_{\Gamma_{sup}}$ and $\varphi_{\Gamma_{refl}}$ associated with the reflections over the both surfaces, and the path phase delay $\varphi_p(f) = -2kh_{FP}$, related to the round trip between them in the normal direction. The Fabry–Perot condition states that these terms added together must result in a phase difference between subsequent bounces [70]:

$$\varphi_{bounce} = \varphi_{\Gamma_{sup}} + \varphi_{\Gamma_{refl}} + \varphi_p \tag{8.9}$$

that equals $2N\pi$, where $N \in \mathbb{Z}$. The transmission coefficient phase $\varphi_{\tau_{sup}}$ of the superstrate does not influence (8.9) because every amount of transmitted energy is subjected to it.

Following, we consider that neither the superstrate nor the reflector converts the polarization of waves for the sake of simplicity. Then, for a specific polarization component, the electric field E_T outside the Fabry–Perot cavity is the superposition of every bounced wave originated by a driving electric field E_0 inside the cavity and somewhen transmitted through the superstrate. For the broadside direction [70]:

$$E_T = E_0 \frac{\sqrt{1-|\Gamma_{sup}|^2}}{1-|\Gamma_{sup}|e^{j\varphi_{bounce}}} e^{j\varphi_{\tau sup}} \tag{8.10}$$

From (8.10), we can express the magnitude $|E_T|$ of the electric field outside the Fabry–Perot as:

$$|E_T| = |E_0|\sqrt{\frac{1-|\Gamma_{sup}|^2}{1+|\Gamma_{sup}|^2 - 2|\Gamma_{sup}|\cos\varphi_{bounce}}} \tag{8.11}$$

Now, it is evident from (8.11) that a phase difference $\varphi_{bounce} = 2N\pi$ maximizes the magnitude $|E_T|$. Note that $|\Gamma_{sup}|$ weights the sensitiveness of $|E_T|$ to $\cos\varphi_{bounce}$. To illustrate that, we depict in Fig. 8.5 the normalized electric field magnitude $|E_T|/|E_0|$ outside the Fabry–Perot cavity as a function of $|\Gamma_{sup}|$ for $\varphi_{bounce} = 2N\pi$. We can see that radiation at the broadside direction is enhanced by 7.7 dB for a moderately transmitting superstrate with $|\Gamma_{sup}| = -3.0$ dB. The improvement reaches more than 12 dB for a less transmissive superstrate with $|\Gamma_{sup}| = -1.0$ dB.

Fabry–Perot resonances improve radiation in directions other than the broadside when the phase difference $\varphi_{bounce} \neq 2N\pi$, which may lead to deformed radiation patterns. In particular, resonances efficiently destroy radiation in the broadside direction when $\varphi_{bounce} = (2N+1)\pi$. In this case, the magnitude $|E_T|$ is minimized [32]. Figure 8.6 depicts such a scenario considering a hypothetical case where the same Fabry–Perot cavity of Fig. 8.4 is now operating at a frequency

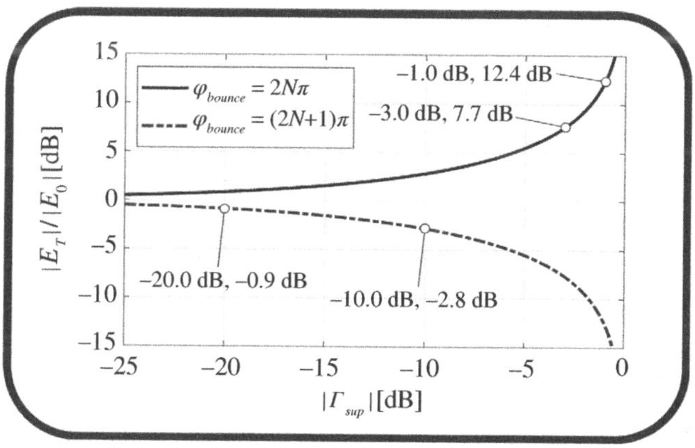

Fig. 8.5 Normalized electric field magnitude $|E_T|/|E_0|$ outside the Fabry-Perot cavity as a function of $|\Gamma_{sup}|$ "Gonçalves Licursi de Mello [32]."

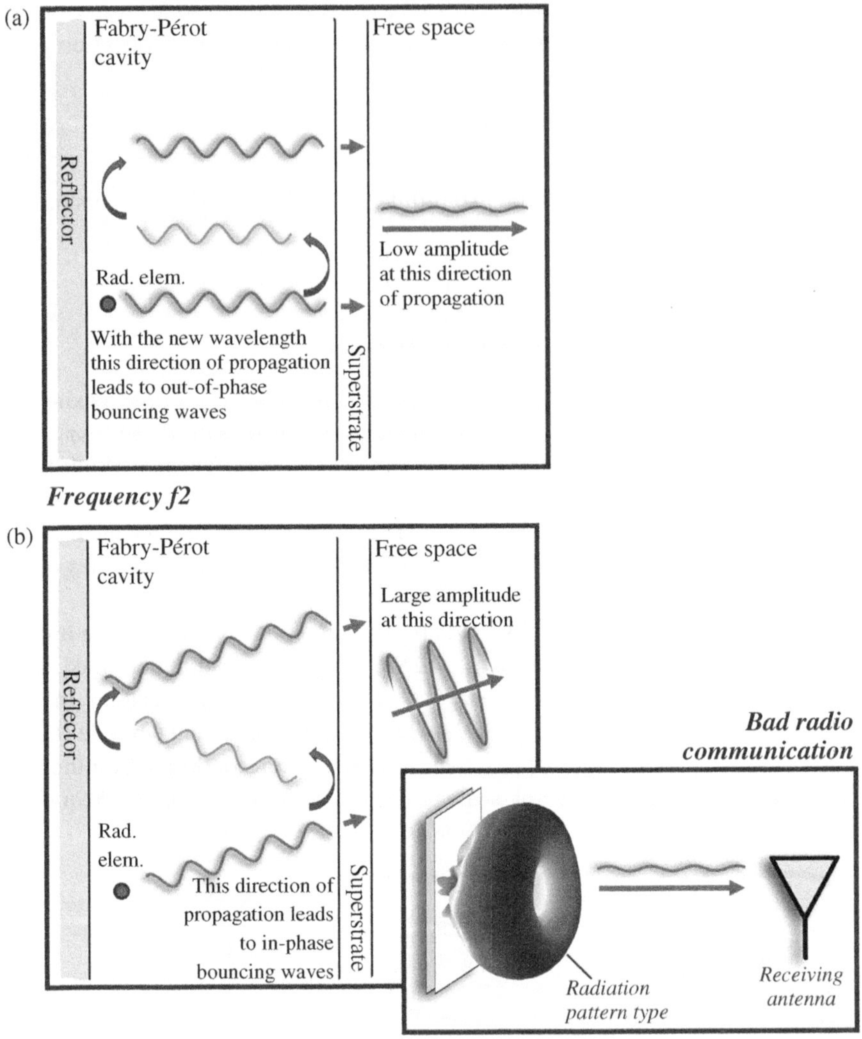

Fig. 8.6 Operation of the Fabry–Pérot cavity but now at a frequency $f = f_2$: (a) destructive interferences occur in the broadside direction; (b) constructive interferences occur in other directions. "Gonçalves Licursi de Mello [72]/Springer Nature/CC BY 4.0."

$f = f_2$. The change of frequency has an unavoidable impact in the path phase delay $\varphi_p(f) = -2kh_{FP}$ [72].

Figure 8.5 also shows $|E_T|/|E_0|$ for $\varphi_{bounce} = (2N+1)\pi$. It is evident that a slightly reflective superstrate with a reflection coefficient magnitude as low as $|\Gamma_{sup}| = -10.0$ dB is sufficient to impact radiation at the broadside direction in -2.8 dB. We can see some disturbance even when $|\Gamma_{sup}| = -20.0$ dB.

The above discussion aims to make it clear to the reader that, even though the topology in question allows us to exploit the functionality of a transmitting metasurface in a lightweight, compact, low-profile design, Fabry–Perot resonances cannot be neglected in the design procedure at all.

8.5.6 Beamsteering with Radiating Element Surrounded by Metasurfaces

A vast literature concerning beamsteering in antenna topologies consisting of a radiating element surrounded by a reflector and a metasurface superstrate is available. Sometimes, the reflector is also a metasurface. Often, these works do not cite the presence of Fabry–Perot resonances explicitly. The reason is that the Fabry–Perot cavity can be unintentionally built [32].

Let us start with an example representing exactly what we suggested at the beginning of Section 8.5.5. J.R. Reis et al. [73], change the antenna topology of J.R. Reis et al.[68] by replacing the horn with a PEC reflector-backed array of patch radiating elements. They use again the same type of metasurface superstrate, whose unit cell is a metal sheet with a double-square slot with varactors, in a five-layer configuration. However, the operation is at 3.60 GHz this time. They simulate, fabricate, and measure a prototype with 5×5 unit cells, shown in Fig. 8.7a. Each cell is again independently biased, allowing the steering of beams in a range of $\pm 20°$, including angles outside the main planes, and utilizing 25 control inputs. In the simulations, the realized gain is 8.2 dBi at the broadside direction and degrades up to 3 dB within the steering range. These values drop by about 5 dB in the measurements, whose setup Fig. 8.7b shows. The use of a lossy substrate in the fabrication to reduce implementation costs justifies such performance drop. Fig. 8.7c depicts the radiation pattern for a steering angle of 15° in elevation. The total thickness is 60 mm, or 0.72 λ at 3.60 GHz.

P. Xie et al. [74] form a Fabry–Pérot cavity with a patch radiating element surrounded by a PEC reflector and a metasurface superstrate. Each metasurface unit cell has two PIN diodes, providing four operating states. Lines coming from the top layer of the metasurface bias one of them while vias connected to the bottom layer bias the other. They simulate, fabricate, and measure a finite metasurface containing 6×6 cells. The metasurface is divided into two parts where the diodes can be differently set. Hence, 16 combinations exist for the operating state of the whole metasurface. The spacing between the metasurface and the PEC reflector is optimized to $h_{FP} = 30.5$ mm. In the end, the total thickness is 33.5 mm, or 0.56 λ at 5.0 GHz. After optimization, 11 state combinations provide high-gain, unidirectional radiation patterns. The respective steering angles are measured as 0°, $\pm 6°$, $\pm 18°$, $\pm 24°$, $\pm 46°$, and $\pm 54°$ in the H-plane. The corresponding values for the realized gain are measured as 11.1, 9.8, 8.7, 7.6, 6.6, and 6.3 dBi.

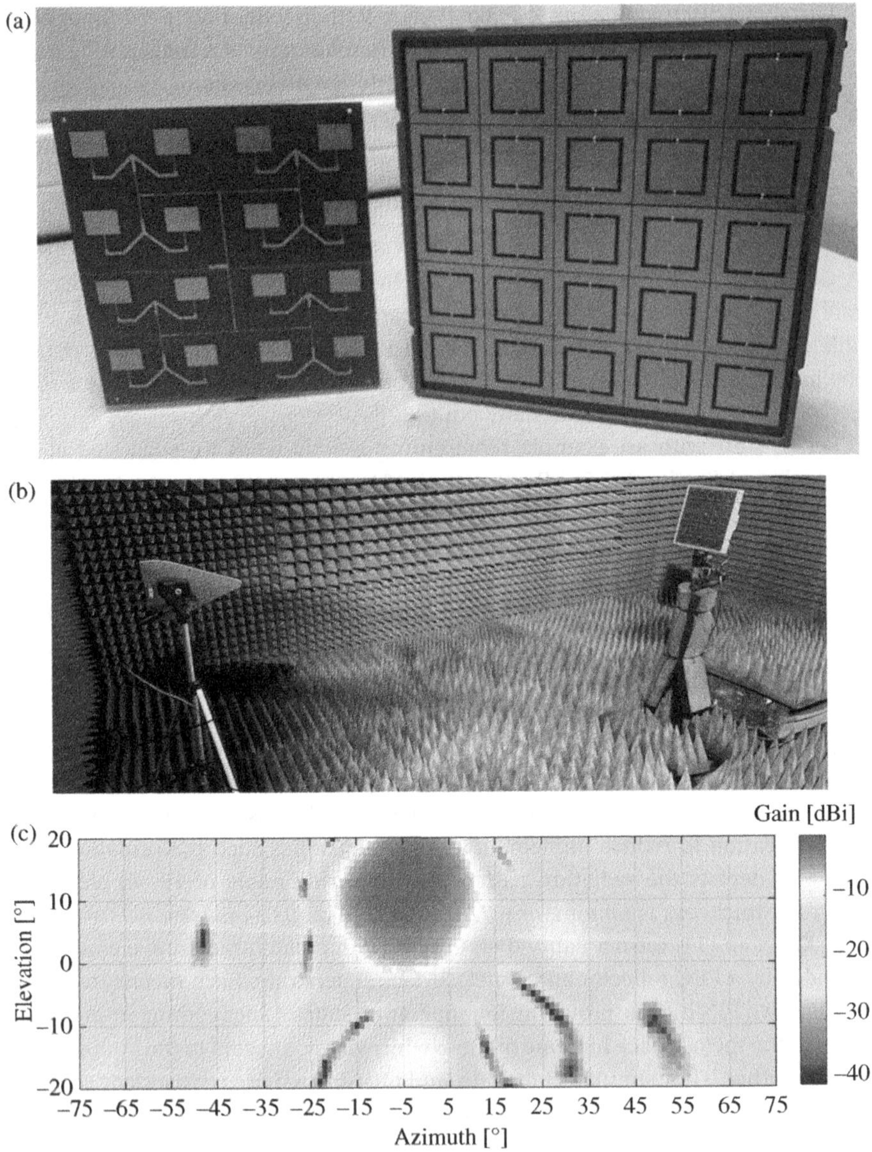

Fig. 8.7 Antenna of [73]: (a) patch array (left) and superstrate (right); (b) measurement setup; (c) radiation pattern for steering toward 15° in elevation. "Reproduced from Reis et al. [73]/MDPI/CC BY 4.0."

T. Hongnara et al. [75] propose a topology comprising a PEC surface with two slot radiating elements, and a metasurface superstrate composed of 5 × 5 square-ring unit cells. Here, we can perform a beamsteering of up to ±30° in the H-plane by adjusting the relative phase between the radiating elements, which operate at 2.45 GHz. We can control such phase through the feeding lines of the elements, located in the back of the structure. A spacing $h_{FP} =$ 8.5 mm separates the PEC surface and metasurface superstrate. The total thickness is only 10.9 mm, that is, 0.09 λ at 2.45 GHz. The realized gain is measured around 9.0, 8.0, 6.5, and 6.0 dBi for steering angles of 0°, 10°, 20°, and 30°, respectively.

M.U. Afzal et al. [76] stack up two phase-gradient metasurface superstrates above a high-gain slot array antenna operating at 19.5 GHz. The total thickness is around 50 mm, which represents 3.25 λ at that frequency. In the near-field region, the slot array antenna does not provide an ideal, uniform phase distribution. Thus, they adjust the gradient in one of the metasurface superstrates to carry out a phase compensation in this matter. The angular disposition of the two phase-gradient metasurfaces provides a sequential-phase transformation. We can mechanically rotate the metasurfaces to perform beamsteering in both azimuth and elevation accordingly. The angular limits are ±40.6°. A maximum value of 30.9 dBic is measured at 19.5 GHz for the realized gain when the beam points in the broadside direction. Such gain drops to 26.7 dBic when we steer the beam toward 15° in azimuth and 40.6° in elevation.

Last, M.S. Rabbani et al. [77] propose a beam-steerable Fabry–Pérot leaky-wave antenna for operation at 35 GHz. The topology comprises a feeding dipole surrounded by an AMC with 91 × 91 unit cells and a metasurface superstrate, as Fig. 8.8a illustrates. What is interesting about this work is that it is the AMC that provides reconfigurability, not the superstrate. A particular mechanism, namely the variation of the spacing between the AMC patches and its ground plane, provides such reconfigurability. Such mechanism is possible because the patches and the ground plane are not embedded in the same dielectrics, as we can see in Fig. 8.8a. A piezoelectric actuator performs an electro-mechanical variation of such spacing whereby controlling the reflection coefficient phase $\varphi_{\Gamma_{refl}}$ of the AMC. In this way, we can vary the phase difference φ_{bounce} between subsequent bounces inside the cavity, expressed by (8.9). Consequently, the Fabry–Pérot resonances improve radiation in a new direction. The spacing between the AMC and the superstrate is $h_{FP} = 3.5$ mm. When the spacing between the patches and the ground plane of the AMC is maximum, the total thickness of the structure is 5.5 mm (or 0.64 λ). Figure 8.8b shows the side view of the fabricated prototype installed in the mast of the anechoic chamber. Varying a single voltage that controls the actuator carries out reconfiguration. In this way, the cavity can steer the beam in the H-plane up to 27° (from 14° to 41°). Figure 8.8c shows a panoramic view of the measurement setup, which attains where a maximum gain of 23.1 dBi. Figure 8.8d shows the measured radiation patterns at 35 GHz.

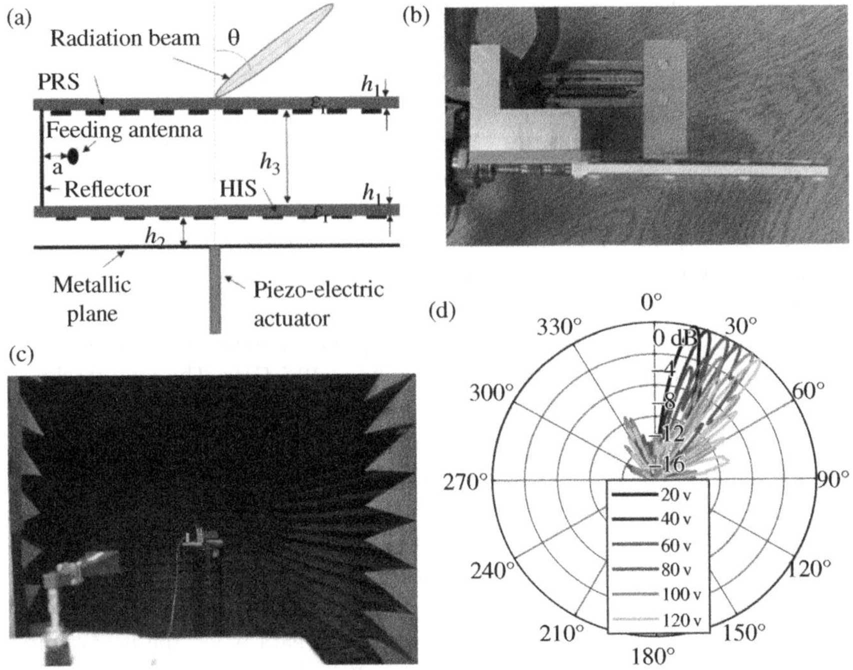

Fig. 8.8 Antenna of [77]: (a) schematic; (b) fabricated prototype; (c) measurement setup; (d) radiation patterns at 35 GHz. "Reproduced from Rabbaini et al. [77]/IEEE/CC BY 4.0."

The last work [77] is particularly interesting because it concurrently exploits the functionalities of two metasurfaces. This work intentionally uses the metasurface superstrate to provide some reflectivity and create constructive interferences outside the cavity through Fabry–Pérot resonances. It improves whereby radiation as a function of $|\Gamma_{sup}|$ in (8.11). On the other hand, the AMC is used to control the direction in which the constructive interferences take place through φ_{bounce} in (8.11). It steers thereby the high-gain antenna beam activated by the superstrate.

Actually, if we consider that we can combine all the functionalities studied so far in this book in dual-metasurface topologies, we can envision many innovative designs. Further, if the metasurfaces provide, each one, different functionalities at different frequency ranges, the possibilities are countless, as we will see next.

8.6 Multifunctional Antennas with Metasurfaces

We saw in Chapter 3 that the working principle of metasurfaces, based on the resonance of the unit cells, results in an inherently frequency-dependent performance. From many perspectives, that of Section 3.4.1 in particular, we can see such behavior as a limitation. However, we can also exploit this feature for multifunctionality purposes.

The ability to carry out multiple functions with a single device allows us to simultaneously operate multiple standards [39, 40], multiple applications [41, 42], or multiple tasks in cognitive radios [78] and radars [79].

In fact, some of the metasurface-based antennas we already studied are multifunctional. For example, we originally conceived the antennas that illustrated the use of the multiband-AMC reflection scheme for ES applications in Sections 6.4.3 and 6.4.4 to operate multiple communication standards in parallel [80–82]. On the other hand, the frequency/polarization-reconfigurable antennas discussed in Sections 6.4.2 and 8.2.1 can serve multiple standards in an interchangeable fashion [23, 24]. Likewise, some metasurfaces briefly commented in Section 8.2.1 [25, 26] can dynamically steer/shape beams and change the polarization when illuminated by a radiating element, finding applications in joint radar and communications.

Today, multifunctionality at the antenna level is a hot issue, with metasurfaces adding to the euphoria. The use of two metasurfaces surrounding a radiating element for a single-band operation [77] is already thought-provoking, with new degrees of freedom opening possibilities for distinctive, combined functionalities. Nonetheless, to fully exploit metasurface's capabilities, we can use their frequency-dependent properties in such topology to obtain multifunctionality in diverse bands. H. Lei et al. [83] employ two metasurfaces surrounding a radiating element to generate independent beams with different polarizations and directions at 12.0 and 15.0 GHz. R.G.L. Mello et al. [72] exploit such topology (two metasurfaces surrounding a radiating element) to cover four different frequency bands with stable, high-gain, unidirectional radiation patterns with an independent beamsteering in one of the bands.

These works lead to the envision of a generic, multiband, topological scheme incorporating arbitrary functionalities, as Fig. 8.9 illustrates. By placing a metasurface substrate at a distance h_{sup} over a radiating element spaced by a distance h_{refl} from a metasurface reflector, we can obtain a different functionality for each band. The performance in each band depends on the corresponding behavior of the metasurfaces, considered arbitrary in Fig. 8.9. Further, the radiating element should provide stable radiation patterns with gain maxima at each broadside direction in all the addressed bands, that is, it should be wideband – or at least multiband.

8.6.1 Challenges in Multiband/Multifunctional Antennas with Metasurfaces

So promising is the idea illustrated in Fig. 8.9 but the implementation challenges must not be ignored. After studying Section 8.5.5, the reader may realize that obtaining a good performance at multiple bands with a radiating element

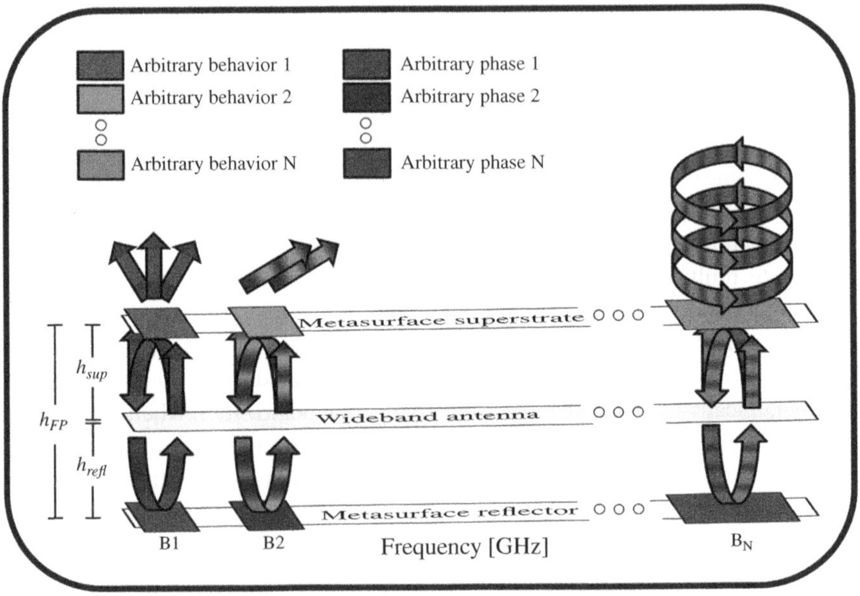

Fig. 8.9 Metasurface-based, multiband, multifunctional, topological scheme: in each band of interest, the metasurfaces present an arbitrary behavior, providing a specific functionality to the complete device. "Gonçalves Licursi de Mello [32]."

surrounded by two metasurfaces actually means handling Fabry–Pérot resonances at multiple bands. Fig. 8.9 represents such challenge with the bouncing waves in each band. Recalling the Fabry–Pérot condition (8.9), we can see that an inevitably frequency-dependent term is present, namely $\varphi_p(f) = -2kh_{FP}$. In other words, if frequency changes, the interferences outside the formed Fabry–Pérot cavity can also change due to a different electrical path within the cavity, as Fig. 8.6 previously illustrated.

As made evident throughout Sections 8.5.5 and 8.5.6, this difficulty arises independently of whether we intend the device to be a Fabry–Pérot antenna or not. M.S. Rabbani et al. [77] explicitly exploit Fabry–Pérot resonances to have a higher gain and steer the antenna beam. However, even though H. Lei et al. [83] do not desire Fabry–Pérot resonances, their metasurface superstrate was not perfectly transparent – which means that the resonances were there. To obtain a good operation at both 12.0 and 15.0 GHz, H. Lei et al. [83] separate the superstrate and reflector with a distance that sub-optimally satisfy the Fabry–Pérot condition at both frequencies simultaneously. This strategy is possible because the ratio between the highest and lowest operating frequencies is close to one ($f_{ratio} = 1.25$).

If f_{ratio} is not close to one, then we should exploit other possibilities to meet the Fabry–Pérot condition; otherwise, spurious interferences can deform, or deflect the radiation pattern to undesired directions. Fortunately, (8.9) presents terms other than the path phase delay φ_p, specifically the reflection coefficient phases $\varphi_{\Gamma_{sup}}$ and $\varphi_{\Gamma_{refl}}$ of the metasurfaces.

Now, metasurfaces' frequency dependence becomes an advantage because appropriate values for $\varphi_{\Gamma_{sup}}$ and $\varphi_{\Gamma_{refl}}$ can compensate for the change in φ_p thereby satisfying the Fabry–Pérot condition in each band. From an analytical point of view, it does not matter whether we explicitly use the reflection coefficient phase of the superstrate $\varphi_{\Gamma_{sup}}$ or the reflector $\varphi_{\Gamma_{refl}}$ in this process. However, it is unavoidable to recall at this point that AMCs can provide arbitrary phases, as discussed in Section 3.2.4. Further, the fact that we can build multiband AMCs allows us to adjust the Fabry–Pérot resonances in each band.

Before advancing, it is worth taking some time to compare Fig. 8.9, which also illustrates the arbitrary phases of the metasurface reflector, with Fig. 6.12, which represented earlier a reflection scheme exploiting both near-PMC and near-PMC behaviors of an AMC.

8.6.2 Obtaining Stable Radiation Patterns in Multiple Bands

In the generic topology of Fig. 8.9, a reflector backs a radiating element and forms a Fabry–Pérot cavity with the superstrate. Thus, we identify three conditions to obtain stable, high-gain, unidirectional radiation patterns at the addressed frequency ranges.

The first is the condition for the enhancement of the broadside gain of a reflector-backed antenna (6.22) [84]. The second, naturally, is the Fabry–Pérot condition (8.9) [70]. Because the radiating element is located between the superstrate and the reflector, we identify the following as the third condition:

$$h_{FP} > h_{refl} \tag{8.12}$$

In a multiband operation, the path phase φ_p delay of both (6.22) and (8.9) may considerably vary from one band to another because the wavenumber $k = \omega\sqrt{\mu\varepsilon}$ is frequency-dependent, whereas the spacings h_{refl} and h_{FP} are physically fixed. Hence, to obtain stable, high-gain, unidirectional radiation patterns in multiple bands, the metasurfaces should provide phases $\varphi_{\Gamma_{sup}}$ and $\varphi_{\Gamma_{refl}}$ that meet conditions (6.22), (8.9), and (8.12) at the center frequency of each band given h_{refl} and h_{FP}.

R.G.L. Mello et al. [72] propose a step-by-step methodology to effectively accomplish this task, which is as easier as $\varphi_{\Gamma_{sup}}$ and $\varphi_{\Gamma_{refl}}$ are adjustable in each band of operation. However, the functionalities that one may wish to extract from the metasurfaces can constrain $\varphi_{\Gamma_{sup}}$ and $\varphi_{\Gamma_{refl}}$. To be more precise, a polarization conversion in the metasurface reflector will still leave some degree of freedom to $\varphi_{\Gamma_{refl}}$. In this case, the superstrate will be free to carry out no matter which functionality because such a versatile $\varphi_{\Gamma_{refl}}$ can compensate the Fabry–Pérot resonances. However, a beam shaping in the reflector perhaps will constrain $\varphi_{\Gamma_{refl}}$. In this case, it will be up to the superstrate to compensate the Fabry–Pérot resonances with $\varphi_{\Gamma_{sup}}$.

8.6.3 Example of Multiband/Multifunctional Antenna with Metasurfaces

In what follows, we exemplify the implementation of the idea Fig. 8.9 illustrates with a particular design but without any loss of generality with respect to arbitrary functionalities and operating frequency ranges.

In Section 3.2.7, we saw that Huygens metasurfaces can provide a good magnitude $|\tau_{TEM}|$ and control of phase $\varphi_{\tau_{sup}}$ in the transmission of waves. R.G.L. Mello et al. [72] use a reconfigurable Huygens metasurface akin to the one Fig. 3.10 represented earlier as antenna superstrate to dynamically steer the antenna beam in the 5G (range from 3.50 to 3.65 GHz, hereafter called B2 band). At the same time, the metasurface do not disturb a stable, high-gain, unidirectional radiation patterns for the other addressed standards (4G and Wi-Fi 2.4/5/6E, frequency ranges 2.40–2.70, 5.17–5.83, and 5.93–6.45 GHz, hereafter called B1, B3, and B4 bands). Thus, the ratio between the highest and lowest operating frequencies was $f_{ratio} = 2.69$. We use a dual-band AMC [80] as a key element to tame the Fabry–Pérot resonances. Such AMC, similar to the one used in Section 6.4.3, also provides low-profile features. The two metasurfaces form an unintentional Fabry–Pérot cavity surrounding a grooved bow-tie antenna [85] rather like the one used in Section 6.4.3. The radiating element covers all the addressed bands stable radiation patterns with gain maxima at each broadside direction.

R.G.L. Mello et al. [72] originally envision the operation of multiple communication standards in parallel. However, nothing prevents us from adjusting the device to carry out a concurrent communication and radar operation, using the beamsteering for the latter while maintaining the former unbothered. But more importantly, what matters in this section is the illustration of concepts that we can generalize to antennas with arbitrary functionalities at multiple bands, as depicted in Fig. 8.9.

Fig. 8.10 depicts the unit cell of the Huygens metasurface, which is inspired in a previous work [49]. Fig. 8.10a illustrates one of its layers, which comprises two symmetrical split rings, each one with a varactor of tunable capacitance C. Fig. 8.10b shows the other layer, which comprises a cut wire with a capacitor-type structure and three varactors, each one of value C. In the simulations, we set a TEM wave whose the orientation of propagation the arrow in Fig. 8.10c indicates. In this configuration, the unit cell assumes a grating disposition.

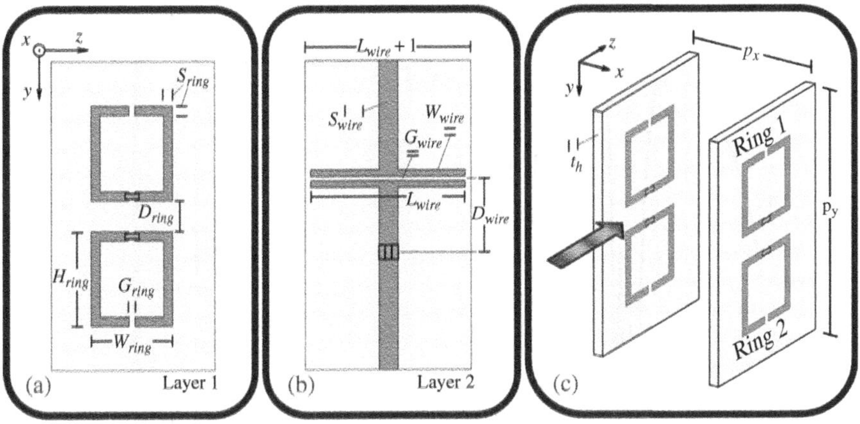

Fig. 8.10 Reconfigurable Huygens unit cell (varactors are highlighted): (a) layer 1: split rings; (b) layer 2: cut wire with capacitor-type structure; (c) orientation of the unit cell when illuminated by a TEM wave. "Gonçalves Licursi de Mello [72]/Springer Nature/CC BY 4.0."

R.G.L. Mello et al. [86] demonstrate that a unit cell with asymmetric rings in layer 1 leads to more degrees of freedom for the independent adjustment of the transmission coefficient τ_{sup} in multiple bands. R.G.L. Mello et al. [72] adjust such unit cell to provide the desired response in the B1–B4 bands. Fig. 8.11 shows the magnitude $|\tau_{sup}|$ and phase $\varphi_{\tau_{sup}}$ for two values of capacitance ($C = 0.03$ pF and $C = 0.20$ pF) we simulate with commercial software (refer to [72] for detailed parameters). For this TEM-idealized case, we see a reasonable level of transmittance in $|\tau_{sup}|$ for all the bands. In the B2 band, the transmission coefficient phase $\varphi_{\tau_{sup}}$ varies in 148° allowing the steer of electromagnetic beams. We can vary such phase through the biasing of the varactors, which may provide different values of capacitance C. In this way, we may impose a gradual phase shift between adjacent unit cells according to the generalized law of refraction (3.36). Note that the phase variation is only 8° at 2.40 GHz when the capacitance C varies. This means that the beamsteering in the B1 will be negligible. In the B3 and B4 bands, only some weak resonance-like phenomena vary when we change C.

R.G.L. Mello et al. [72] also obtain the reflection coefficient magnitude $|\Gamma_{sup}|$ and phase $\varphi_{\Gamma_{sup}}$ of the Huygens unit cell through simulations. Then, we calculate the spacings h_{refl} and h_{FP} and the phases $\varphi_{\Gamma_{refl}}$ of the AMC in each band by using the proposed methodology to simultaneously meet conditions (6.22), (8.9), and (8.12) at the center frequency of each band.

R.G.L. Mello et al. [72] optimize together the complete device, which comprises the grooved bow-tie radiating element surrounded by a 7 × 8-cell dual-band AMC

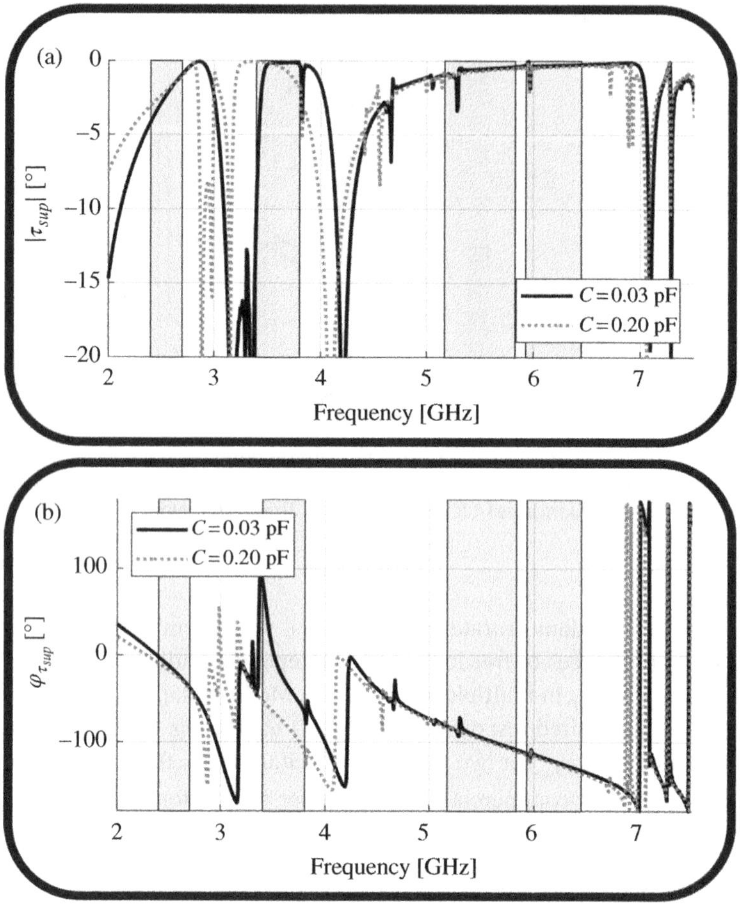

Fig. 8.11 Transmission coefficient of the asymmetric unit cell of [72] when the capacitance of the varactors in changed: (a) magnitude; (b) phase. "Gonçalves Licursi de Mello [72]/ Springer Nature/CC BY 4.0."

and a 4 × 4-cell Huygens metasurface. Such optimization accounts for the introduction of mechanical supports and feeding lines for the Huygens metasurface. The thicknesses of the AMC and Huygens metasurfaces are respectively 4.9 and 12.2 mm, and the spacing therebetween is h_{FP} = 19.5 mm. Thus, the total thickness is 36.6 mm, or only 0.29 λ at 2.4 GHz. Figure 8.12a shows the complete prototype.

To perform beamsteering in the B2 band, we bias the varactor columns with different voltages. That means that the functionality utilizes only four control inputs. For illustration purposes, we consider five beamsteering states, respectively, denominated −2, −1, 0, +1, and +2 from left to right. For the sake of clarity, Fig. 8.12b shows the broadside realized gain for three of them (−2, 0, and +2).

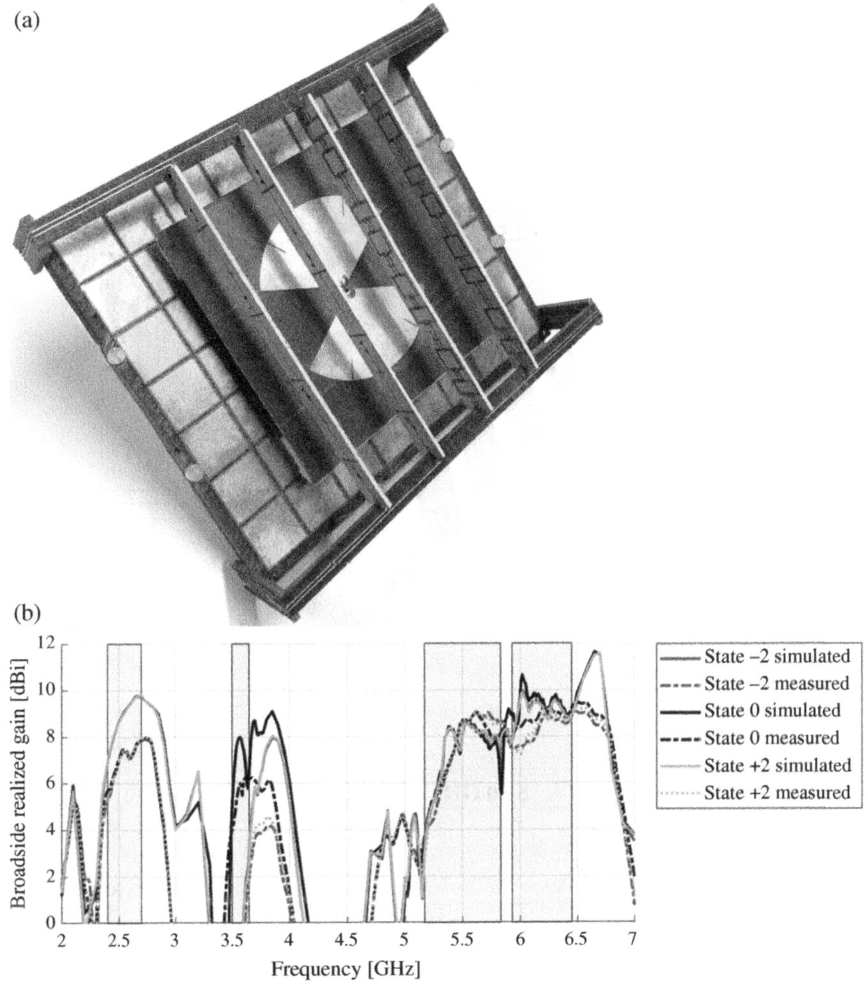

Fig. 8.12 Low-profile, multiband, multifunctional antenna performing an independent beamsteering in one of the bands: (a) fabricated device; (b) broadside realized gain. "Gonçalves Licursi de Mello [72]/Springer Nature/CC BY 4.0."

The B1–B4 bands are highlighted. For the B1 band, the peak gain is simulated as 9.8 dBi and measured as 7.9 dBi for the states −2, 0, and +2. In the B2 band, the peak gain of the state 0 is simulated as 8.0 dBi and measured as 6.0 dBi (note that the other states mostly radiate at other directions). For the B3 and B4 bands, peak gains are respectively simulated as 8.7 and 10.6 dBi and measured as 9.0 and 9.2 dBi.

Figure 8.13 shows the radiation patterns at the center frequencies of the B1–B4 bands in the H- and E-planes for the five states. In the B2 band, the H-plane uses a

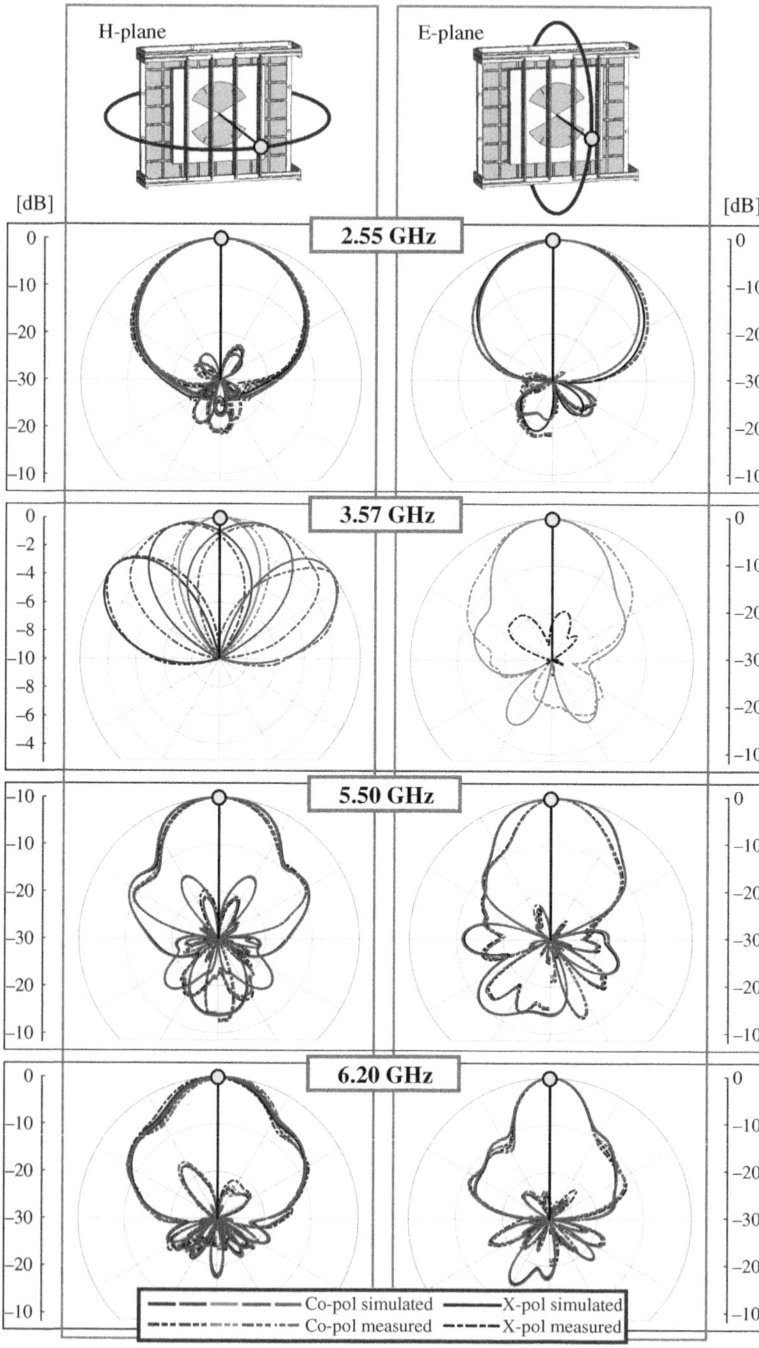

Fig. 8.13 Radiation patterns of the multiband, multifunctional antenna in the B1–B4 bands; dots indicate the broadside direction. "Gonçalves Licursi de Mello [72]/Springer Nature/CC BY 4.0."

scale of 10 dB to highlight the beamsteering, and the E-plane presents only the state 0. We see again a very good agreement between simulations and measurements. For all the cases, the cross-polarization levels are low and the front-to-back ratio large in the simulations and measurements.

In the above results, the radiation patterns of the B1, B3, and B4 bands practically do not change when we perform the beamsteering in the B2 band. Therefore, we indeed achieve multifunctionality in multiple bands. Besides, all the patterns are unidirectional and do not present any deformation associated to Fabry–Pérot resonances which indicates that we effectively tame such mechanism [72]. The thickness of the complete device is only 0.29 λ at the lowest operating frequency. Moreover, control simplicity is another advantage: beamsteering is performed by utilizing only four control inputs. More generally, these results indicate that a couple of metasurfaces surrounding a radiating element can indeed lead to the design of low-profile, multiband, multifunctional antennas concurrently serving to multiple standards or even applications.

8.6.4 Roadmap to New Metasurface-Based Multifunctional Antennas

The possibility of designing highly versatile, low-profile antennas for a multiband, multifunctional operation is of first importance to attain low complexity, costs, size, weight, and power consumption in the system. All these features corroborate with the fact that extremely constrained applications such as lightweight drones, missile seekers, and small satellites can benefit from the topology consisting of two metasurfaces, one as antenna substrate and the other as reflector, surrounding a radiating element.

We designed the particular antenna that Section 8.6.3 illustrated for concurrent multi-standard communications, namely for the 5G/4G and Wi-Fi 2.4/5/6E, operating in the ranges of 2.40–2.70, 3.50–3.65, 5.17–5.83, and 5.93–6.45 GHz. However, by changing the shape dimensions, we can adjust the same design for other frequency ranges. The scaling principle [87], explained earlier in Section 6.2.1 in the context of frequency-independent antennas, allows maneuver.

If we desire a relative shift between the operating frequency bands, then we should adjust the metasurfaces accordingly. R.G.L. Mello [32] proposes a detailed methodology to adjust the reconfigurable Huygens metasurface to this end. Also, R.G.L. Mello et al. [80] provide guidelines to adjust the dual-band AMC.

The beamsteering functionality of the design that Section 8.6.3 discusses is also useful for radar applications. As such, we can carry out joint radar and communications, with the static radiation patterns performing the latter. However, it turns out that many radar applications require a directivity substantially higher than the one seen in the 3.57-GHz, steerable patterns of Fig. 8.13. To obtain an antenna with higher directivity, one possibility is to use an array instead of a single radiating element as Section 8.5.1 previously commented. Of course, in this case, the metasurfaces will have to be larger in order to fit the array. Because the array elements are wideband, their size may be relatively large for the upper frequencies in terms of wavelength. This can avoid the compact spacing therebetween, ultimately leading to the appearance of grating lobes [88].

Now, if we envision completely new antennas for other functionalities and frequency bands, we can use the roadmap proposed in Section 6.4.5 with some adaptations.

Once again, the project starts by identifying the specs in terms of frequency range, polarization, gain, and so forth. However, a significant difference already at this very stage is the fact that now we can envision different functionalities for each band – and choose the metasurfaces accordingly, as well as a compatible radiating element.

In terms of radiating element, issues similar to those listed in the items (a)–(d) of Section 6.4.5 are relevant (available literature, geometry simplicity, acknowledged compatibility with the adopted metasurfaces, and polarization matters). In this context, the element size adds to the difficulty in the case of arrays.

After the radiating element, we should conceive the unit cells of the metasurface superstrate and reflector to provide the desired functionality in each band. Adjusting each band/functionality may be hard. However, it is not worth overly optimizing the isolated unit cells at this point. The main reason is that we will need to readjust them later on to provide specific reflection coefficient phases $\varphi_{\Gamma_{sup}}$ and $\varphi_{\Gamma_{refl}}$ when we carry out the methodology to tame the Fabry–Pérot resonances. Another point is that the performances of the finite-array metasurfaces and the radiating element may change when we put them together. The reasons are those listed in items i–iii of Section 6.4.5 (different angle and polarization conditions in the simulations of the unit cell and the complete device, presence of surface waves, and extra coupling phenomena due to the proximity of the structures).

The final stage of the design is to optimize the complete device after putting all the pieces together. This time, however, a good understanding not only of each structure is crucial but also of the Fabry–Pérot resonances therebetween.

8.7 Chapter Final Remarks

We can apply different EP measures at the antenna, RF-front-end, and processor levels to protect radars and communication systems against hostile signals. Metasurfaces allow innovative implementation at the antenna level, leading to simple, compact, power-efficient, and low-cost devices that can easily fit the fuselage of air/spacecraft or embed drones and missile seekers.

In this context, we can regard metasurfaces as the EP response in the cat-and-mouse EW game to directed-energy weapons, envisioned recently as a strong candidate to neutralize drone swarms [18–22]. Against cross-polarization jamming, the technology is particularly interesting because this EA measure commonly targets missile seekers [47].

Reconfigurable metasurfaces also allow low-profile, cheap antennas to perform dynamic functionalities once exclusive to phased arrays, namely the shaping and steering of the antenna beam. Even more impactfully, the gathering of multiple functionalities in a single, compact, planar device is also possible in multiple frequency ranges. In this way, lightweight platforms that could not carry out these functionalities will soon be able to do so. Hence, we expect new behaviors in the EW scenario, which will bring dynamism to the next level and require operators and AI models to adapt accordingly.

References

1 Adamy, D. (2001). *EW 101: a first course in electronic warfare.* Norwood, MA: Artech House, Chapter 1, pp. 4.

2 Adamy, D. (2001). *EW 101: a first course in electronic warfare.* Norwood, MA: Artech House, Chapter 7.2, pp. 125–129.

3 Adamy, D. (2001). *EW 101: a first course in electronic warfare.* Norwood, MA: Artech House, Chapter 7.3, pp. 129–134.

4 Lee, S.-W., Zarrillo, G., Law, C.-L. (1991). Simple formulas for transmission through periodic metal grids or plates. *IEEE Transactions on Antennas and Propagation* 30 (5): 904–909. https://doi.org/10.1109/TAP.1982.1142923

5 Whitbourn, L.B., Compton, R.C. (1985). Equivalent-circuit formulas for metal grid reflectors at a dielectric boundary. *Applied Optics* 24 (2): 217–220. https://doi.org/10.1364/AO.24.000217

6 Whites, K.W., Mittra, R. (1996). An equivalent boundary-condition model for lossy planar periodic structures at low frequencies. *IEEE Transactions on Antennas and Propagation* 44 (12): 1617–1629. https://doi.org/10.1109/8.546248

7 Munk, B.A. (2000). *Frequency selective surfaces: theory and design.* Hoboken, NJ: Wiley.

8 Munk, B.A. (2003). *Finite antenna arrays and FSS.* Hoboken, NJ: Wiley, Chapter 2, pp. 15–55.

9 Han, Y., Chang, Y., Che, W. (2022). Frequency-selective rasorbers: a view of frequency-selective rasorbers and their application in reducing the radar cross sections of antennas. *IEEE Microwave Magazine* 23 (2): 86–98. https://doi.org/10.1109/MMM.2021.3125463

10 Costa, F., Monorchio, A. (2012). A frequency selective radome with wideband absorbing properties. *IEEE Transactions on Antennas and Propagation* 60 (6): 2740–2747. https://doi.org/10.1109/TAP.2012.2194640

11 Wu, Y., Xiong, J., Deng, F. et al. (2021). A miniaturized frequency selective rasorber with high selectivity passband and wideband absorption properties. *EPJ Applied Metamaterials* 8 (5): 1–7. https://doi.org/10.1051/epjam/2020017

12 Wang, Z., Zeng, Q., Fu, J. et al. (2018). A high-transmittance frequency-selective rasorber based on dipole arrays. *IEEE Access* 6: 31367–31374. https://doi.org/10.1109/ACCESS.2018.2843795

13 Shi, R., Yu, S., Kou, N. (2022). A miniaturized frequency-selective rasorber with absorption bands on two sides of passband for antenna dome. *IEEE Antennas and Wireless Propagation Letters* 21 (11): 2161–2165. https://doi.org/10.1109/LAWP.2022.3187958

14 Weinberg, G.V. (2022). Performance prediction of directed energy weapons. *Progress in Electromagnetics Research Letters* 108: 79–88. http://doi.org/10.2528/PIERM21111201

15 Neri, F. (2018). *Introduction to electronic defense systems*, 3rd ed. Boston, MA: Artech House, Section 7.5, pp. 532–535.

16 De Martino, A. (2018). *Introduction to modern EW systems*, 2nd ed. Norwood, MA: Artech House, Section 1.4, pp. 20–22.

17 Adamy, D. (2001). *EW 101: a first course in electronic warfare*. Norwood, MA: Artech House, Section 9.2.3, pp. 185–186.

18 Weinberg, G.V., Kracman, M. (2023). Swarm UAV defeat modelling through lifetime distribution analysis. *Progress in Electromagnetics Research Letters* 111: 71–78. http://doi.org/10.2528/PIERL23022808

19 Weinberg, G.V. (2022). Prediction of UAV swarm defeat with high-power radio frequency fields. *IEEE Transactions on Electromagnetic Compatibility* 64 (6): 2157–2162. https://doi.org/10.1109/TEMC.2022.3193881

20 Zhao, M., Chen, Y., Zhou, X. et al. (2022). Investigation on falling and damage mechanisms of UAV illuminated by HPM pulses. *IEEE Transactions on Electromagnetic Compatibility* 64 (5): 1412–1422. https://doi.org/10.1109/TEMC.2022.3187017

21 Zhang, J., Han, K., Zhang, P. et al. (2023). A survey on joint-operation application for unmanned swarm formations under a complex confrontation environment. *Journal of Systems Engineering and Electronics* 34 (6): 1432–1446. https://doi.org/10.23919/JSEE.2023.000162

22 Esteves, J.L., Cottais, E., Kasmi, C. (2018). Unlocking the access to the effects induced by IEMI on a civilian UAV. *International Symposium on Electromagnetic Compatibility (EMC EUROPE)*, Amsterdam, Netherlands (27–30 August 2018), New York, US: IEEE. https://doi.org/10.1109/EMCEurope.2018.8484990

23 Sanz-Izquierdo, B., Liang, B., Parker, E.A. et al. (2016). An application of active frequency selective surface to reconfigurable antenna technology. *Proceedings of the Active and Passive RF Devices Seminar*, London, UK (17 February 2016), New York, US: IEEE. https://doi.org/10.1049/ic.2016.0006

24 Chen, D., Yang, W., Che, W. et al. (2019). Polarization-reconfigurable and frequency-tunable dipole antenna using active AMC structures. *IEEE Access* 7: 77792–77803. https://doi.org/10.1109/ACCESS.2019.2919518

25 Zhang, N., Chen, K., Zhao, J. et al. (2022). A dual-polarized reconfigurable reflectarray antenna based on dual-channel programmable metasurface. *IEEE Transactions on Antennas and Propagation* 70 (9): 7403–7412. https://doi.org/10.1109/TAP.2022.3165872

26 Yu, H., Zhang, Z., Su, J. et al. (2023). Quad-polarization reconfigurable reflectarray with independent beam-scanning and polarization switching capabilities. *IEEE Transactions on Antennas and Propagation* 71 (9): 7285–7298. https://doi.org/10.1109/TAP.2023.3291460

27 Neri, F. (2018). *Introduction to electronic defense systems*, 3rd ed. Boston, MA: Artech House, Chapter 6, pp. 477–503.

28 Hajnsek, I., Desnos, Y.-L. (2021). *Polarimetric synthetic aperture radar: principles and applications.* Cham, Switzerland: Springer.

29 Ryzhkov, A.V., Zrnic, D.S. (2019). *Radar polarimetry for weather observations.* Cham, Switzerland: Springer.

30 Zyl, J.V., Kim, Y. (2011). *Synthetic aperture radar polarimetry.* Hoboken, NJ: Wiley.

31 Elachi, C., Ulaby, F.T. (1990). *Radar polarimetry for geoscience applications.* Boston, MA: Artech House.

32 Gonçalves Licursi de Mello, R. (2022). Active and passive metasurfaces: methodology for the design of a low profile, beam-steerable, multiband, and wideband antenna. PhD Thesis. Institut Polytechnique de Paris. https://www.theses.fr/2022IPPAT025 (accessed 12 October 2024).

33 Björnson, E., Wymeersch, H., Matthiesen, B. et al. (2022). Reconfigurable intelligent surfaces: a signal processing perspective with wireless applications. *IEEE Signal Processing Magazine* 39 (2): 135–158. https://doi.org/10.1109/MSP.2021.3130549

34 Jeong, N.-H., Kim, M., Choi, J.-H. et al. (2023). Beam scheduling of maritime multifunctional radar based on binary integration. *IEEE Access* 11: 35796–35807. https://doi.org/10.1109/ACCESS.2023.3265801

35 Kellett, J., Dawber, W., Wallace, W. et al. (2022). Multifunction maritime radar and RF systems—technology challenges and areas of development. *IEEE Aerospace and Electronic Systems Magazine* 37 (4): 6–20. https://doi.org/10.1109/MAES.2021.3081168

36 Klemm, R., Nickel, U., Gierull, C. (2017). *Novel radar techniques and applications volume 1: real aperture array radar, imaging radar, and passive and multistatic* *radar*. Stevenage, UK: Scitech Publishing, Chapter 3, pp. 135–171. https://doi.org/10.1049/SBRA512F

37 Xing, H., Xing, Q., Wang, K. (2023). A joint allocation method of multi-jammer cooperative jamming resources based on suppression effectiveness. *MDPI Mathematics* 11 (826): 1–25. https://doi.org/10.3390/math11040826

38 Herd, J.S., Conway, M.D. (2016). The evolution to modern phased array architectures. *Proceedings of IEEE* 104 (3): 519–529. https://doi.org/10.1109/JPROC.2015.2494879

39 Sharma, S.K., Chieh, J.-C.S. (2021). *Multifunctional antennas and arrays for wireless communication systems*. New York, NY: Wiley-IEEE Press.

40 Saha, C., Siddiqui, J., Antar, Y. (2019). *Multifunctional ultrawideband antennas*, 1st ed. Boca Raton, FL: CRC Press.

41 Liu, F., Masouros, C., Petropulu, A.P. et al. (2020). Joint radar and communication design: applications, state-of-the-art, and the road ahead. *IEEE Transactions on Communications* 68 (6): 3834–3862. https://doi.org/10.1109/TCOMM.2020.2973976

42 Ouedraogo, S., Sáenz, I.D.H., Guinvarc'h, R. et al. (2019). Design and experimental validation of multifunction antenna with direct modulation for radar and communication. *Progress in Electromagnetics Research* 164: 17–25. http://doi.org/10.2528/PIER18061804

43 Neri, F. (2018). *Introduction to electronic defense systems*, 3rd ed. Boston, MA: Artech House, Section 7.5.3.1, pp. 534.

44 Neri, F. (2018). *Introduction to electronic defense systems*, 3rd ed. Boston, MA: Artech House, Section 7.5.2, pp. 532.

45 Neri, F. (2018). *Introduction to electronic defense systems*, 3rd ed. Boston, MA: Artech House, Section 7.5.3.2, pp. 535.

46 Munk, B.A. (2000). *Frequency selective surfaces: theory and design*. Hoboken, NJ: Wiley, Chapter 2, pp. 26–82.

47 Global Business Media. (2011). Electronic countermeasures. *Special Report*. https://issuu.com/magazineproduction/docs/gbm_electronic_countermeasures_ezine/1 (accessed 17 November 2023).

48 Brown, A.D. (2022). *Active electronically scanned arrays: fundamentals and applications*. Norwood, MA: Artech House, Chapter 4, pp. 103–134. https://doi.org/10.1002/9781119749097.ch4

49 Chen, K., Feng, Y., Monticone, F. et al. (2017). A reconfigurable active Huygens' metalens. *Advanced Materials* 29 (17): 1–7. https://doi.org/10.1002/adma.201606422

50 Raeker, B.O., Grbic, A. (2019). Compound metaoptics for amplitude and phase control of wave fronts. *Physics Review Letters* 122 (11): 1–6. https://doi.org/10.1103/PhysRevLett.122.113901

51 Emara, M.K., Kundu, D., Macdonell, K. et al. (2023). Dynamic metasurface reflectors based on coupled resonators for simultaneous magnitude and phase control. *IEEE Access* 11: 129552–129565. https://doi.org/10.1109/ACCESS.2023.3333869

52 Ratni, B., de Lustrac, A., Piau, G.-P. et al. (2018). Reconfigurable meta-mirror for wavefronts control: applications to microwave antennas. *Optics Express* 26 (3): 2613–2624. https://doi.org/10.1364/OE.26.002613

53 Balanis, C.A. (2016). *Antenna theory: analysis and design*, 4th ed. New York, NY: Wiley, Section 6.3, pp. 293–297.

54 Balanis, C.A. (2016). *Antenna theory: analysis and design*, 4th ed. New York, NY: Wiley, Section 6.3.3, pp. 302–304.

55 Balanis, C.A. (2016). *Antenna theory: analysis and design*, 4th ed. New York, NY: Wiley, Section 8.7, pp. 474–480.

56 Kildal, P.-S. (1990). Artificially soft and hard surfaces in electromagnetics. *IEEE Transactions on Antennas and Propagation* 38 (10): 1537–1544. https://doi.org/10.1109/8.59765

57 Zhang, L., Castaneda, J.A., Alexopoulos, N.G. (2004). Scan blindness free phased array design using PBG materials. *IEEE Transactions on Antennas and Propagation* 52 (8): 2000–2007. https://doi.org/10.1109/TAP.2004.832516

58 Chen, L., Zhang, T., Moghaddam, S.M. et al. (2019). A method of reducing mutual coupling using an extra coupling path. *13th European Conference on Antennas and Propagation (EuCAP)*, Krakow, Poland (31 March–05 April 2019), New York, US: IEEE. https://doi.org/10.1109/APUSNCURSINRSM.2018.8608579

59 Love, A.E.H. (1901). The integration of the equations of propagation of electric waves. *Philosophical Transactions of the Royal Society, Series A* 197: 1–45, 1901.

60 Sievenpiper, D.F., Schaffner, J.H., Song, H.J. (2003). Two-dimensional beam steering using an electrically tunable impedance surface. *IEEE Transactions on Antennas and Propagation* 51 (10): 2713–2722. https://doi.org/10.1109/TAP.2003.817558

61 Pfeiffer, C., Grbic, A. (2013). Metamaterial Huygens' surfaces: tailoring wave fronts with reflectionless sheets. *Physics Review Letters* 110 (197401): 1–5. https://doi.org/10.1103/PhysRevLett.110.197401

62 Wong, J.P.S., Epstein, A., Eleftheriades, G.V. (2015). Reflectionless wide-angle refracting metasurfaces. *IEEE Antennas and Wireless Propagation Letters* 15: 1293–1296. https://doi.org/10.1109/LAWP.2015.2505629

63 Yu, N., Genevet, P., Kats, M.A. et al. (2011). Light propagation with phase discontinuities: generalized laws of reflection and refraction. *Science* 334 (6054): 333–337. https://doi.org/10.1126/science.1210713

64 Nayeri, P., Yang, F., Elsherbeni, A.Z. (2018). *Reflectarray antennas: theory, designs, and applications.* New York, NY: Wiley-IEEE Press.

65 Sievenpiper, D.F., Zhang, L., Broas, R.F.J. et al (1999). High-impedance electromagnetic surfaces with a forbidden frequency band. *IEEE Transactions on*

Microwave Theory and Techniques 47 (11): 2059–2074. https://doi.org/10.1109/22.798001

66 Vásquez-Peralvo, J.A., Fernández-González, J.M., Rigelsford, J.M. (2020). Beam steering using active artificial magnetic conductors: a 10-degree step controlled steering. *IEEE Access* 8: 177964–177975. https://doi.org/10.1109/ACCESS.2020.3027141

67 Pan, W., Huang, C., Chen, P. et al. (2013). A beam steering horn antenna using active frequency selective surface. *IEEE Transactions on Antennas and Propagation* 61 (12): 6218–6223. https://doi.org/10.1109/TAP.2013.2280592

68 Reis, J.R., Caldeirinha, R.F.S., Hammoudeh, A. et al. (2017). Electronically reconfigurable FSS-inspired transmitarray for 2-D beamsteering. *IEEE Transactions on Antennas and Propagation* 65 (9): 4880–4885. https://doi.org/10.1109/TAP.2017.2723087

69 Costa, F., Luukkonen, O., Simovski, C.R et al. (2011). TE surface wave resonances on high-impedance surface based antennas: analysis and modeling. *IEEE Transactions on Antennas and Propagation* 59 (10): 3588–3596. https://doi.org/10.1109/TAP.2011.2163750

70 Trentini, G.V. (1956). Partially reflecting sheet arrays. *IRE Transactions on Antennas and Propagation* 4 (4): 666–671. https://doi.org/10.1109/TAP.1956.1144455

71 Pérot, A., Fabry, C. (1899). On the application of interference phenomena to the solution of various problems of spectroscopy and metrology. *Astrophysical Journal* 9: 87. https://adsabs.harvard.edu/full/1899ApJ.....9...87P (accessed 15 February 2024).

72 Gonçalves Licursi de Mello, R., Lepage, A.C., Begaud, X. (2023). Taming Fabry–Pérot resonances in a dual-metasurface multiband antenna with beam steering in one of the bands. *Scientific Reports* 13 (9871): 1–10. https://doi.org/10.1038/s41598-023-36828-4

73 Reis, J.R., Vala, M., Oliveira, T.E. et al. (2021). Metamaterial-inspired flat beamsteering antenna for 5G base stations at 3.6 GHz. *MDPI Sensors* 21 (23): 1–14. https://doi.org/10.3390/s21238116

74 Xie, P., Wang, G., Cai, T. et al. (2017). Novel Fabry-Pérot cavity antenna with enhanced beam steering property using reconfigurable meta-surface. *Applied Physics A* 123 (7): 462–467. https://doi.org/10.1007/s00339-017-1060-9

75 Hongnara, T., Chaimool, S., Akkaraekthalin, P. et al. (2018). Design of compact beam-steering antennas using a metasurface formed by uniform square rings. *IEEE Access* 6: 9420–9429. https://doi.org/10.1109/ACCESS.2018.2799551

76 Afzal, M.U., Esselle, K.P., Koli, M.N.Y. (2022). A beam-steering solution with highly transmitting hybrid metasurfaces and circularly polarized high-gain radial-line slot array antennas. *IEEE Transactions on Antennas and Propagation* 70 (1): 365–377. https://doi.org/10.1109/TAP.2021.3111522

77 Rabbani, M.S., Churm, J., Feresidis, A.P. (2022). Continuous beam-steering low-loss millimeter-wave antenna based on a piezo-electrically actuated metasurface. *IEEE Transactions on Antennas and Propagation* 70 (4): 2439–2449. https://doi.org/10.1109/TAP.2021.3137248

78 Saha, C., Siddiqui, J., Antar, Y. (2019). *Multifunctional ultrawideband antennas*, 1st ed. Boca Raton, FL: CRC Press, Section 4.3, pp. 132–138.

79 Gurbuz, A.C., Cetiner, B. (2018). Multifunctional reconfigurable antennas for cognitive radars. *IEEE Radar Conference (RadarConf18)*, Oklahoma City, OK, USA (23–27 April 2018), New York, US: IEEE. https://doi.org/10.1109/RADAR.2018.8378790

80 Gonçalves Licursi de Mello, R., Lepage, A.C., Begaud X. (2023). A low-profile, triple-band, and wideband antenna using dual-band AMC. *MDPI Sensors* 23 (4): 1920. https://doi.org/10.3390/s23041920

81 Gonçalves Licursi de Mello, R., Lepage, A.C., Begaud X. (2022). Réflecteur multibande à métamatériaux. *Speech presented at the Journées Scientifiques URSI France 2022* in Palaiseau, France (22–23 March 2022).

82 Gonçalves Licursi de Mello, R., Lepage, A.C., Begaud X. (2023). A Compact, quad-band, and wideband antenna using triple-band AMC. *Proceedings of 13th International Conference on Metamaterials, Photonic Crystals and Plasmonics (Meta2023)*, Paris, France (18–21 July 2023), Paris, France: Metaconferences. https://metaconferences.org/META23/files/meta23_proceedings.pdf

83 Lei, H., Liu, Y., Jia, Y. et al. (2022). A low-profile dual-band dual-circularly polarized folded transmitarray antenna with independent beam control. *IEEE Transactions on Antennas and Propagation* 70 (5): 3852–3857. https://doi.org/10.1109/TAP.2021.3125419

84 Djoma, C., Jousset, M., Lepage, A.C. et al. (2014). Maximal bandwidth of an Archimedean spiral antenna above a reflector. *IEEE Antennas and Wireless Propagation Letters* 13: 333–336. https://doi.org/10.1109/LAWP.2014.2303858

85 Gonçalves Licursi de Mello, R., Lepage, A.C., Begaud X. (2022). The bow-tie antenna: performance limitations and improvements. *IET Microwaves, Antennas & Propagation* 16 (5): 283–294. https://doi.org/10.1049/mia2.12242

86 Gonçalves Licursi de Mello, R., Lepage, A.C., Begaud, X. (2021). Metasurface de Huygens multibande avec dépointage du faisceau dans la bande 5G 3.4–3.8 GHz. *Proceedings of the 9ème Conf. Plenière du GDR ONDES*, Lille, France (01 December 2021).

87 Rumsey, V.H. (1966). *Frequency independent antennas*. New York, NY: Academic Press, Section 1.1, pp. 1–3.

88 Munk, B.A. (2000). *Frequency selective surfaces: theory and design*. Hoboken, NJ: Wiley, Section 1.9, pp. 23–25.

9

Emerging Use Cases of Metasurfaces in EW

The applications discussed hitherto can reach the theater of operation in a quite reasonable time. From the coating of platforms to the use in antennas, most of the metasurfaces we proposed in the previous chapters for stealth, ES, EA, and EP purposes require just a few adjustments to become operational. In other words, the job to be done is much more about engineering than research.

Understanding the core of each application we discussed so far was not trivial and, certainly, following this path may have been challenging for many. However, as foreshadowed in Section 3.5, such effort was essential – not only because it provided knowledge about the techniques we detailed but, rather, because it is going hereafter to allow a glimpse of the scenario of the future.

This chapter is the most conceptual one in this book. We reserved it for those metasurface applications that we cannot consider mature for the real world yet but are going to be key in the battlefield of tomorrow. We still need to address many open questions before they can reach the theater with a reasonable performance. However, we expect unprecedented levels of dynamism when it is time for this to come. In the end, all the applications we comment here aim at enabling the execution of EW tasks from lightweight, cheap drones – and perhaps now the reader figures out why we have been highlighting throughout the whole book the conveniences of metasurfaces as opposed to conventional EW systems in this matter.

The motivation for this chapter arose from many reports of the latest conflicts that indicate that the character of war is changing again. A focus on the extensive use of drones and the dispersal of EW assets rather than their concentration in major platforms has recently become evident [1–6]. Further, prominent technical studies also confirm that massive EW networks provide remarkable advantages for perceiving and interacting with the EW environment [7–9].

The first application this chapter will discuss concerns the DOA estimation using a single metasurface instead of multiple antennas as traditionally done

Metasurface-Driven Electronic Warfare, First Edition. Rafael Gonçalves Licursi de Mello.

Published 2025 by John Wiley & Sons, Inc.
Companion website: www.wiley.com/go/licursidemello/metasurfacedrivenelectronicwarfare1

[10–16]. The principles behind these direction-finding metasurfaces [17–20] actually exploit the working mechanism of space-time modulated metasurfaces, first studied in Section 4.4.5. Such capability is key to enable ES tasks in lightweight, cheap drones which in turn may support two hard EW problems. The first one is the deinterleaving of radar pulses in extremely dense environments. By using a drone swarm to estimate the DOA of a single pulse from different perspectives, a distributed system can instantaneously estimate the position of emitters (instead of their direction) through triangulation [21]. This maneuver spatially filters intercepted pulses facilitating deinterleaving even more. The other problem concerns the complete characterization of multifunctional AESA radars. These radars can concurrently operate diverse modes in parallel, illuminating different regions with different antenna beams and signals [22–24], as discussed in Section 8.5. To put it more specifically, ES systems located at different positions would perceive them differently. Moreover, a single ES system would not be able to fully characterize them, giving birth to the need for an ES-capable drone swarm.

The second application this chapter envisions is the execution of cognitive jamming based on a multi-agent reinforcement learning approach [25–37] using EW networks distributed in autonomous drone swarms. Cognitive jamming requires the perceiving of the environment followed by a corresponding action. For perceiving, the technique requires in the swarm the ES capabilities of the previous paragraph. For the actions, it also demands EA capabilities. Performing cognitive jamming with an autonomous swarm is particularly interesting against multifunctional radars. The reason is not only that the swarm can fully characterize these radars but also that the distributed EA assets would be able to plentifully interact with them.

The third application regards the use of RISs to control the access to the electromagnetic spectrum. As Section 5.4.4 briefly commented, we can set up the reflection coefficient phase φ_Γ of each unit cell of a RIS to control interferences at a specific direction. Conventionally, we can associate RISs with AI algorithms in the context of multipath propagation to *maximize* signals at a specific area through the constructive interference of indirect paths [38–45]. However, nothing avoids us to *minimize* signals by setting the RISs to cause destructive interferences at specific areas, as a couple of the very latest works [46, 47] propose. For open areas, where multipath phenomena may be negligible, we envision a RIS-equipped drone swarm to cancel signals propagating therethrough.

Doubtless, after reading this chapter, many will think that metasurface-equipped autonomous drone swarms are a quite avant-gardist application (not to say unfeasible). However, the question they should ask themselves is not *if* these applications are reasonable, but rather *when* they are going to be so. As we previously stated, the drone warfare is imposing itself as a crucial element in the battlefield – whether we want it or not. In parallel, the technological maturity levels of metasurfaces, including the RIS, are developing at full steam – and right before our very eyes. That is why the civilian community already selected it as one essential element for the next generation of mobile communications, the 6G [48]. Therefore, achieving these capabilities before opponents is paramount – and this chapter is a starting point.

9.1 Metasurface-Driven ES Capabilities in Drone Swarms

The significance of the DOA parameter in an ES context catches the eye when we assess all the complexity to estimate it with traditional amplitude-based methods [10–16]. We saw at the beginning of Chapter 6 that typical deinterleaving approaches use the DOA parameter to start. The reason is that, even though some radars can vary the parameters of their signals, their platforms cannot vary their angular position with respect to the ES system substantially in the time frame of the radar pulses. Hence, we can consider the ability to estimate DOA essential to enable ES capabilities in a platform.

Fig. 6.1 illustrated earlier the need for multiple antennas with further processing in multiple channels in an ES system. Such need can be prohibitive for many modern applications, in particular for lightweight, cheap drones. However, the latest research proposes single-channel direction-finding methods with space-time modulated metasurfaces, which can fill this gap.

9.1.1 Direction Finding with Metasurfaces

Space–time modulated metasurfaces, first studied in Section 4.4.5, are quite avant-gardist structures, with the first interesting experimental demonstrations showing up mainly from the late 2010s [17]. Among their many applications, they can simultaneously steer electromagnetic beams and shift their frequencies.

At this moment, the reader may be wondering how they could use one such a metasurface for direction-finding purposes. We saw in Section 4.4.5 that the beam-steering this kind of metasurface carries out at each frequency $f_0 \pm nf_{\Gamma m}$ exploits the generalized law of reflection [49] previously studied in Section 3.2.7. Referring to (3.35), we can see that the final position of the reflected beams depends not only on the phase gradient over the surface but also on the angle of incidence of the impinging waves. Hence, at a particular observation point, the beams at the different frequencies $f_0 \pm nf_{\Gamma m}$, illustrated earlier in Fig. 4.10b, will present an amplitude imbalance that depends on these two variables. Since the system has information about the phase gradient at each frequency, it can resolve for the angle of incidence if a probe somewhere over the metasurface intercepts the reflected signals – and this, only with the spectral information contained in a single channel.

X. Fang et al. [18] exploit this idea to estimate the DOA of a single impinging wave. They divide a 5.2-mm thick metasurface with two reconfigurable states for φ_Γ differing 180° at f_0 = 3.3 GHz in supercells. They modulate the states of two types of supercells using quasi-square waveforms with frequency $f_{\Gamma m}$ = 1 MHz. The waveforms present a delay in relation to one another. In the end, two reflected beams result at 3.3 GHz ± 1 MHz at θ directions. We can place a horn antenna over the metasurface to receive the reflected signals and measure the amplitude of the harmonics. Full-wave simulations show an accuracy better than 0.4° in the θ dimension. X. Fang et al. [18] also present the formulation to extend the method to both θ and Φ dimensions simultaneously.

D. Xia et al. [19] also exploit the concept of space–time modulated metasurface but in a transmissive way, with the horn antenna located behind the metasurface. A 3.3-mm thick unit cell comprises two metal patches on its both sides, one to receive impinging waves and the other to retransmit them to the horn at f_0 = 5.0 GHz. A metal via connects the patches. The unit cell uses an MMIC amplifier with an area of 2.6×4.4 mm^2 based on InGaP HBT technology. Such amplifier enhances the accuracy of DOA estimations, providing 9.8 dB of gain. Its 1-dB compression point is 12.2 dBm, and its noise figure is 4.5 dB. Rather than the phase φ_τ of the transmission coefficient, its magnitude $|\tau_{TEM}|$ modulates the unit cells. To this end, an FPGA turns the amplifiers on and off through a bias circuit with different duty cycles. Again, supercells divide the metasurface. From the spectral analysis within the single receiver, a signal model allows the estimation of the signals arriving at each unit cell of the metasurface. Next, a direct application of direction-finding algorithms conventionally used for antenna arrays is possible. In this manner, we can simultaneously estimate the DOA of multiple impinging waves in both θ and Φ dimensions using a single channel. Measurements of a prototype show an accuracy around 0.25°.

M. Huang et al. [20] also use a transmissive space–time modulated metasurface for direction-finding purposes. Figure 9.1a depicts a 4.6-mm thick, multilayer unit cell. At f_0 = 9.5 GHz, two varactors reconfigure the transmission coefficient phase φ_τ shown in Fig. 9.1b through a bias voltage. An MCU where we can set different phase profiles biases the metasurface. This time, however, an AI model composed of an ensemble of random forests associate the fields measured by a simple probe behind the metasurface to the angle of incidence. Figure 9.1c illustrates this concept. We can train the AI model for angles within a range of ±60° with a resolution of 0.1°, that is, for 1201 different angles. Then, when new signals arrive, each random forest predicts an incident angle. The final estimate is the average of the estimates from all random forests. Fig. 9.1d shows the 20×40-unit cell metasurface whose fabrication and measurements validate the idea. The spacing between

(a)

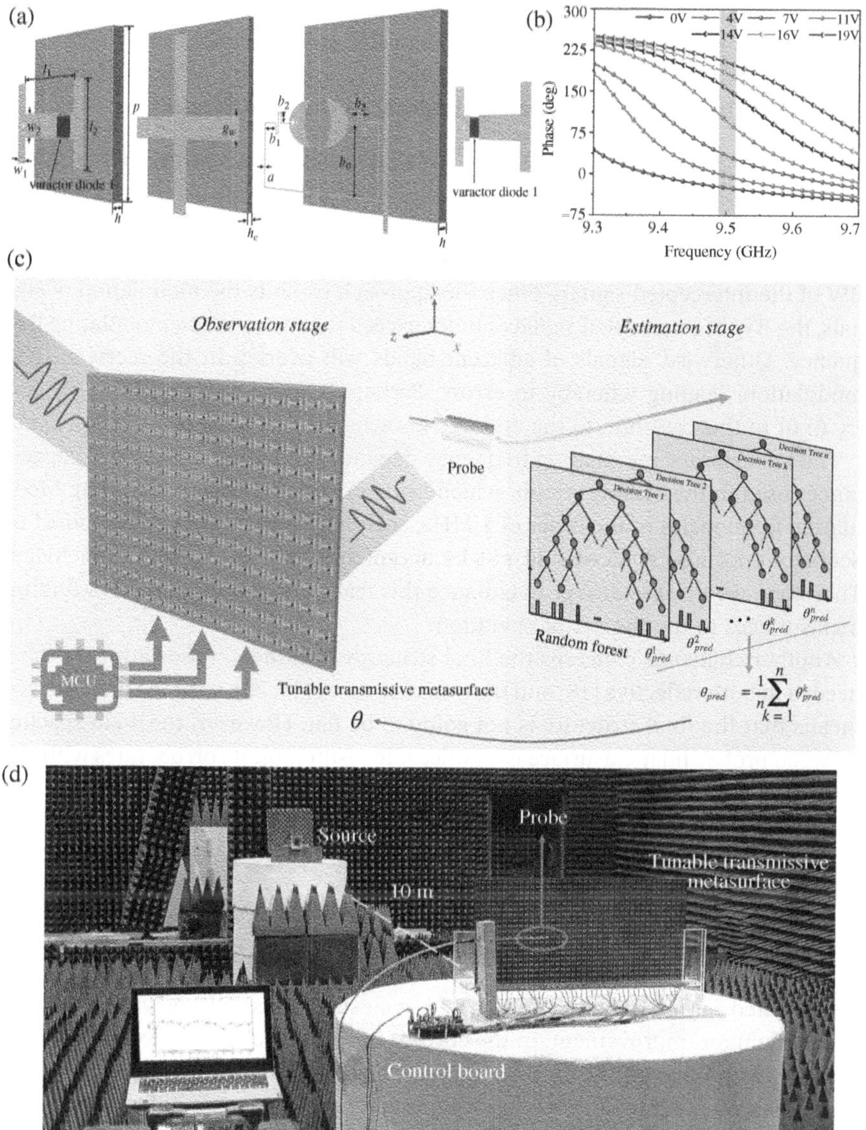

Fig. 9.1 Metasurface of [20]: (a) unit cell; (b) transmission phases; (c) direction finding conceptual illustration; (d) measurement setup. "Reproduced from Huang et al. [20]/Walter de Gruyter GmbH/CC BY 4.0."

the probe and the metasurface is only 4 cm, which shows that the device can be thin. Results reveal estimate errors less than 0.5° at 95% of cases.

9.1.2 Challenges to Bring the Technology to the Real World

In the works Section 9.1.1 studied, the BW of the metasurface depends mainly on the range of frequencies where the unit cells can provide the proper response when reconfigured. Within this range, we can use a single metasurface to estimate the DOA of different signals. However, a critical limitation exists in the matter of the BW of the intercepted signals. Since the approach exploits the modulation of signals, the BW of intercepted signals cannot exceed the metasurface modulating frequency. Otherwise, signals of adjacent bands will overlap in the receiver after modulation, leading whereby to errors. Perhaps some AI approaches [20] can try to fit to this case but, to the author's knowledge, this has not been done yet.

The metasurface modulating frequency depends on the resistances and capacitances used in the biasing circuit, which leads to an *RC* time constant [50]. Modulating frequencies of the order of 1 MHz, seen here and in the works studied in Sections 4.4.4 and 4.4.5, would not be acceptable for a real application today. Therefore, we must endeavor to enhance this feature and bring direction-finding metasurfaces to the theater of operation.

Another challenge concerns the final structure bulkiness. Associating a probe, needed in both reflective [18] and transmissive cases [19, 20], with the metasurface means that the final structure is not going to be flat. However, the 4-cm spacing between probe and metasurface is a noticeable result [20]. It corroborates with the possibility of using the technology in lightweight drones. For other types of structure [18, 19], we can install the probe over a convenient part of the platform (like a wing tip) but in this case the technology is limited to drones where this is indeed possible.

9.1.3 Functionalities Unlocked by ES-Capable Drone Swarms

As we stated earlier, direction finding with metasurfaces is still a new area but with much room for improvement. In the current state, the technology would be interesting only in very specific missions and scenarios with lightweight, cheap drones. Nevertheless, enabling the DOA estimation in these drones can unlock the implementation of extensive, distributed ES networks with autonomous swarms in the future.

In extremely dense environments, many emitters can lie at the same angular position in relation to a single ES system. In this case, a deinterleaving whose first step is based on DOA can be impossible, even for the latest AI approaches. As we already stated, estimating DOA from different perspectives with a distributed ES network makes it possible the instantaneous triangulation of the position of radars

[21]. From a deinterleaving standpoint, if the DOA parameter resolves signals impinging from different directions (see comments around Fig. 6.1), the estimation of the position the signals come from is unable to deinterleave only signals from radars concentrated in the same geographical spot. In this way, we can instantly separate all the pulses into smaller sets – dramatically reducing the complexity of the next deinterleaving steps. This capability is key in the modern scenario where the number of emitters is increasing each day.

Another functionality unlocked by ES-capable, autonomous drone swarms is the ability to fully characterize multifunctional AESA radars. The multiple antenna beams that these radars use can carry out diverse, parallel tasks through the transmission of independent signals, each one characterizing an operating mode [22–24]. The dispersal of ES assets throughout the scenario allows the resolving for each of the concurrent modes of multifunctional AESA radars. Furthermore, the AESA sidelobes sometimes transmit pulses from a specific mode to unintended directions, substantially hindering the resolving of other modes [51]. The distributed ES network allows the perceiving of these pulses from other perspectives, allowing the identification and association thereof with their original modes.

To conclude, ES-capable, autonomous drone swarms are also useful for cognitive jamming using multi-agent reinforcement learning, which we will study right next.

9.2 Metasurface-Driven Cognitive Jamming in Drone Swarms

Cognitive jamming is the EW discipline that processes vestiges of the EA impact over a target radar to optimize the applied measures accordingly in a observe-orient-decide-act (OODA) loop.

Among many examples of OODA loop, we can cite the natural, cognitive process a crew follows when trying to use EA measures to evade a missile. The crew observes the sensor indicators and interprets what is happening according to its expertise and instincts. Next, it takes the decision of insisting in the current measure or switching to another one finally acting accordingly. An experienced pilot would take, let us say, three seconds to iterate over such OODA loop. Cognitive jamming automates this process, considerably reducing the loop time and allowing either the test of more EA measures or the parametric optimization of a specific one.

Cognitive jamming is becoming increasingly necessary in a scenario where the operation of highly dynamic radars is each day oftener, in particular cognitive radars. These radars use advanced AI models to interact with the environment and optimize their performance by varying not only their receiver features, but also their waveforms and even the tracking rules [52–54]. In this context, the effectiveness of EA measures is increasingly difficult to guarantee since the internal functioning of the radar a priori is hidden from the EA system [55].

9.2.1 Cognitive Jamming with Reinforcement Learning

Recently, reinforcement learning has been shown to be a strong candidate for cognitive jamming [25–37]. Its definition fits the application like a glove: reinforcement learning is a type of machine learning where an AI agent learns from interacting with the environment. We design the algorithm to apply penalties to the agent for bad actions and rewards for good ones. The goal is to maximize the cumulated reward. In EW systems, we can use the AI agent to apply a specific EA measure over a target radar (that is, to perform an *action*) and to check right after how the radar signals change (that is, the *state* of the target radar).

Figure 9.2 illustrates this concept with a surveillance radar illuminating an aircraft. Through its ES system, the aircraft intercepts the radar pulses, each of which we can characterize with a PDW composed of a carrier frequency f_0, amplitude, PW, DOA, TOA, and modulation type. Through a sequence of PDWs from the radar in question, the ES system can recognize patterns of occurrence and antenna scanning, among others [56]. Then, the system can classify the radar operating mode, finally assigning a threat level thereto [57].

Such operating mode is the radar state $\mathbf{s} = (f_0, PW, \ldots, scan, mod)$ in the reinforcement learning terminology, a vector containing all the information we need to

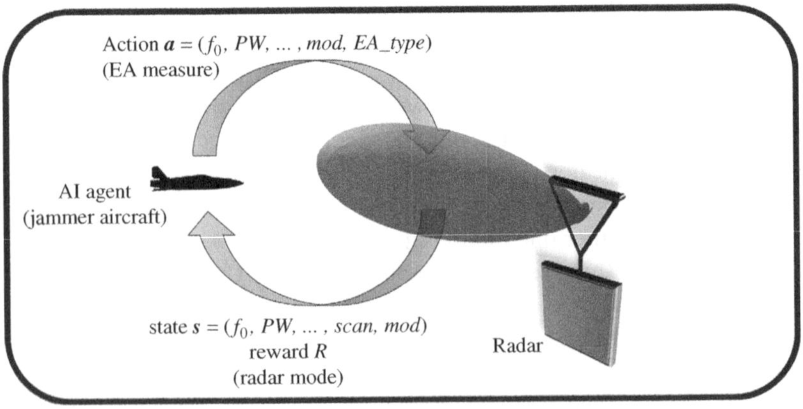

Fig. 9.2 A single-agent reinforcement learning model correctly perceives one mode of a conventional radar

describe the radar functioning. Of course, for an accurate estimation of $\boldsymbol{s}$, the ES system should deinterleave the sequence of PDWs from other radar signals.

The AI agent in the aircraft carries out specific EA techniques over the radar, each of which we may describe in the reinforcement learning terminology as an action $\boldsymbol{a} = (f_0, PW, \ldots, mod, EA\_type)$. Prior to learning, the AI agent defines the actions $\boldsymbol{a}$ through trial and error, that is, it addresses EA measures to the radar at random. However, the ES system assesses again the state $\boldsymbol{s}$ of the radar after each action $\boldsymbol{a}$.

The AI agent perceives a reward R depending on how the new state relates to the old one according to predefined criteria. For example, the AI agent of L.-W. Feng et al. [27] receives a positive reward when the radar switches to a lower threat-level mode after the EA measure. Analogously, the agent receives a penalty when the radar switches to a higher threat-level one. Then, an AI model associates the actions $\boldsymbol{a}$ to the perceived rewards R, allowing the agent to learn what are the best actions to take to achieve specific goals.

So promising cognitive jamming is, we cannot ignore the implementation challenges. One of them is the fact that the environment (the target radar in our case) is only partially observable [58]. For example, consider the scenario where, rather than a strictly surveillance, we have a multifunctional AESA radar, a high-performance radar widely employed today. Such radar can concurrently run multiple tasks, such as surveillance, target tracking, weather monitoring, among others. To this end, its phased-array antenna employs multiple, independent beams, each one transmitting a different signal [22–24].

Figure 9.3 represents signals of only one of these modes, denominated mode 1, illuminating an aircraft with an AI agent. In this situation, the ES system in the

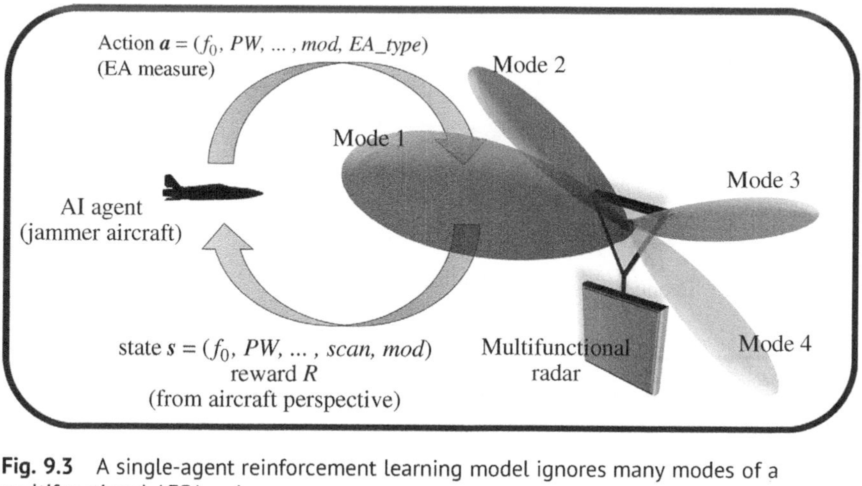

Fig. 9.3 A single-agent reinforcement learning model ignores many modes of a multifunctional AESA radar

aircraft simply ignores the many other modes that the multifunctional AESA radar may operate. In this case, the estimated radar state $\boldsymbol{s}$ contains only partial information of what is happening. A complete characterization of the radar state needs to consider all of its modes. That is where the drone swarm comes into play.

9.2.2 Cognitive Jamming with Drone Swarms

Figure 9.4 depicts a scenario where, instead of a single aircraft, an autonomous swarm composed of N drones surrounds the multifunctional AESA radar of Fig. 9.3. Consider that the swarm is somehow ES-capable. As such, we can estimate a higher-dimensional radar state $\boldsymbol{s}$ as:

$$\boldsymbol{s} = \begin{pmatrix} \boldsymbol{s}_1 \\ \boldsymbol{s}_2 \\ \vdots \\ \boldsymbol{s}_{N-1} \\ \boldsymbol{s}_N \end{pmatrix} \tag{9.1}$$

where $\boldsymbol{s}_1, \boldsymbol{s}_2, \ldots, \boldsymbol{s}_{N-1}, \boldsymbol{s}_N$ are the radar states as perceived from the perspective of each of the $\boldsymbol{n}$ drones. Note that the swarm is able to fully characterize the radar if each mode illuminates at least one drone – for which a massive swarm is essential.

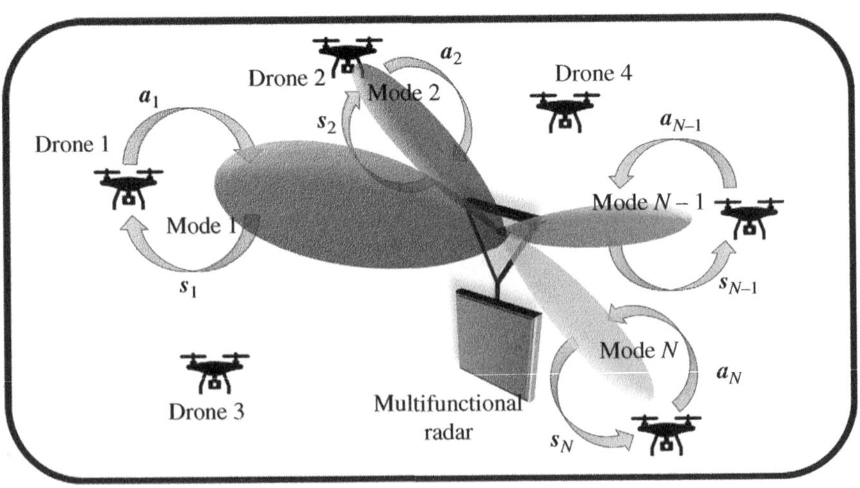

Fig. 9.4 A multi-agent reinforcement learning model perceives all the modes of the multifunctional radar as a higher-dimensional state and performs higher-dimensional actions

Consider that the drone swarm is also EA-capable, that is, each of the N drones is able to create co-polarized and cross-polarized copies of the radar signals, in a dynamic manner. Hence, each drone can play with the respective co- and cross-polarized AESA lobes that are illuminating it, imposing a specific EA measure thereover. In this way, the swarm carries out a higher-dimensional action $\boldsymbol{a}$ over the multifunctional radar:

$$\boldsymbol{a} = \begin{pmatrix} \boldsymbol{a}_1 \\ \boldsymbol{a}_2 \\ \vdots \\ \boldsymbol{a}_{N-1} \\ \boldsymbol{a}_N \end{pmatrix} \tag{9.2}$$

where $\boldsymbol{a}_1$, $\boldsymbol{a}_2$, ..., $\boldsymbol{a}_{N-1}$, $\boldsymbol{a}_N$ are the EA measures imposed by each of the N drones.

Equations (9.1) and (9.2) allow the implementation of a multi-agent reinforcement learning technique. The multi-agent reward methodology can exploit either the isolate rewards R_1, R_2, ..., R_{N-1}, R_N perceived by each of the N drones or a general, more complex reward R perceived by the whole swarm. The former provides a high degree of independence to the drones, a case where each drone is responsible for achieving specific goals over the respective radar mode illuminating it. On the other hand, the latter enables a shared learning and combined knowledge where we prioritize a global objective. Various architectures are possible, each one with specific advantages and levels of implementation complexity. Fortunately, multi-agent reinforcement learning is a hot topic today, and the open literature provides works where the agent shares information [59], actions [60], and rewards [61]. To finish, the reader can refer to L. Wang et al. [28] and M.A. Aref et al. [29] to obtain specific insight about the use of multi-agent reinforcement learning in EW.

9.2.3 The Role of Metasurfaces in Cognitive Jamming

From the discussed on cognitive jamming so far, we can conclude that all that we need to perform the application in the modern scenario is a massive EW network with ES and EA capabilities. However, the following question becomes unavoidable in this matter: how can one build a massive EW network in a scenario where resources are limited due to practical, logistic, economic, and even political reasons?

Similar to the case Section 7.7 commented, here, a *massive* swarm comprises hundreds or even thousands of drones deployed in the EW environment. This is necessary to increase the chances that each of the modes of the multifunctional AESA radar illuminates at least one drone, which will ensure that the higher-dimensional state $\boldsymbol{s}$ precisely represents the radar functioning in its plentifulness.

Analogously, in this situation, we increase the chances that the higher-dimensional action $\boldsymbol{a}$ is able to fully interact with all of the radar modes from every input possibility thereof.

We are once again coming across the fact that DRFMs are not suitable to embed lightweight, cheap drones – and so are not multi-channel ES systems, such as the one Fig. 6.1 illustrated earlier.

All that said, we can summarize the role of metasurfaces in cognitive jamming with multi-agent reinforcement learning in two points:

- First, metasurfaces should provide ES capabilities to each of the drones in the swarm, as Section 9.1 envisioned; the fact that the ES network would be particularly effective against multifunctional AESA radar, as Section 9.1.3 explained, is noteworthy here;
- Second, metasurfaces should make each drone in the swarm EA-capable; to this end, we can exploit the concepts Chapter 7 discussed, mainly those about the coating of drones with co- and cross-polarized power-amplifying metasurfaces; in this way, the drones will be able to amplify reflected waves no matter their polarization, in a dynamic manner, allowing thereby the swarm to play with the co- or the cross-polarized lobes of each of the multifunctional AESA radar.

In terms of challenges to bring this application from the domain of ideas to the theater of operation, with regard to metasurfaces, they are the same as those we commented in Chapter 7 and Section 9.1.2. Most of the other challenges relate to the reinforcement learning itself and to the autonomous drone swarm, such as how to establish the proper communication and synchronization of the drones, the resilience of such communication to jamming, the definition of the best spatial arrangement for the swarm, and so forth.

9.3 Metasurface-Controlled Spectrum Access

Since the very first lines of this book, we have been highlighting that metasurfaces can create unprecedently dynamic scenarios including smart electromagnetic environments.

Indeed, because metasurfaces can provide all the different electromagnetic responses we have been studying hitherto, researchers recently had the glimpse of a new concept in the physical layer of wireless systems called RIS [62–66]. Originally, they envisioned the RIS approach mainly to shape complex field distributions within disordered, enclosed environments with a certain degree of randomness [67–69]. Lately, they have exploited this concept mainly in the context of multipath propagation channels, or channels where a line of sight between Tx and Rx antennas does not exist. However, as it will become more obvious in this section, we can also exploit the potential of RISs in open environments.

In this concept, reconfigurable surfaces deployed in the environment, such as coating building walls and ceilings or even aerial platforms like balloons or drones [70], create extra, controllable propagation paths. Because a software module can govern it (that can even be an AI model as it will also be clearer soon), such surfaces have appeared under different names like software-controlled metasurface, intelligent reflecting surface, reconfigurable intelligent surfaces, and many others, from which we take the acronym RIS.

For the sake of simplicity, we will neglect all polarization issues in the analyses of this section.

9.3.1 Multipath Propagation Channel

Consider in Fig. 9.5 a radio link established in a scenario where obstacles are present, like buildings, mountains, platforms, and people, among others. In the illustrated case, there is a line of sight between the Tx and Rx antennas through which electromagnetic waves can propagate. If the Tx antenna transmits a signal $x(t)$, the signal $y_{dir}(t)$ the Rx antenna receives through this direct path is:

$$y_{dir}(t) = x(t) * h_{dir}(t) \tag{9.3}$$

where $h_{dir}(t)$ is the channel response accounting for the propagation through the direct path. However, the waves also propagate from the Tx antenna toward other directions in the scenario of Fig. 9.5. In this way, many obstacles present there reflect, diffract, or scatter them, making them propagate through indirect paths

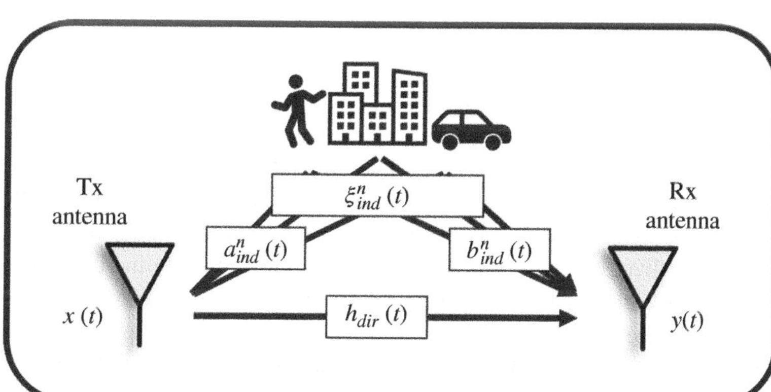

Fig. 9.5 Uncontrolled scenario and simplistic representation of direct and indirect paths traveled by an electromagnetic wave between a Tx and Rx antenna therein

before arriving at the Rx antenna. In a multipath situation where N obstacles lead to relevant indirect paths, the final signal $y(t)$ the Rx antenna receives is:

$$y(t) = y_{dir}(t) + y_{ind}(t) \tag{9.4}$$

where

$$y_{ind}(t) = x(t) * h_{ind}(t) \tag{9.5}$$

$$h_{ind}(t) = \sum_{n=1}^{N} a_{ind}^{n}(t) * \xi_{ind}^{n}(t) * b_{ind}^{n}(t) \tag{9.6}$$

Here, $\xi_{ind}^{n}(t)$ is the electromagnetic response of each of the N obstacles, which we can model with a reflection, transmission, or absorption coefficient. Moreover, $a_{ind}^{n}(t)$ and $b_{ind}^{n}(t)$ are the channel responses accounting for the propagation until and from each of these obstacles.

What (9.3) through (9.6) are telling us is that the final signal $y(t)$ the Rx antenna receives is the sum of attenuated and delayed versions of the original signal $x(t)$. Specifically, they are attenuated and delayed throughout the direct path and all the indirect paths where ξ_{ind}^{n} is not neglectable [63]. In this way, the waveform of such sum can significantly differ from that of $x(t)$.

Naturally, we can also represent (9.4) in the frequency domain by using the Fourier transform of each time-domain function and replacing the convolution operations by simple multiplications as:

$$y(f) = y_{dir}(f) + y_{ind}(f) \tag{9.7}$$

where

$$y_{dir}(f) = x(f)h_{dir}(f) \tag{9.8}$$

$$y_{ind}(f) = x(f)h_{ind}(f) \tag{9.9}$$

$$h_{ind}(f) = \sum_{n=1}^{N} a_{ind}^{n}(f)\xi_{ind}^{n}(f)b_{ind}^{n}(f) \tag{9.10}$$

We explicitly passed (9.4) to the frequency domain here to remind the reader that each spectral component of the related quantities can suffer different effects in the channel. From a frequency-domain analysis perspective, the multipath response can yield diverse changes in the spectral content of $y(f)$ in relation to $x(f)$, corroborating with the fact that the waveform of the received signals can significantly differ from that of the original ones.

Note in the above development that a significant obstacle between the Tx and Rx antennas can obstruct the direct propagation path in some cases. In these cases, we can consider the direct channel response $h_{dir} = 0$ in (9.3) and (9.8). Further, many of the obstacles in the environment, like vehicles, people, and doors, among others, may move. Besides, atmospheric conditions affecting propagation may change. Hence, the final channel response is dynamic. From the standpoint of the wireless system, the channel dynamism can be so aggressive and complex that we can actually consider it random in many cases [63]. It is true that a good planning for the antenna location throughout the scenario – when possible – can help in this matter. However, this cannot ensure that significant fluctuations in the channels are not going to disturb the radio link. What is worse, the channel response is uncontrollable and cannot be deliberately optimized in the conventional scenario of Fig. 9.5. This is because there is no term neither in the direct nor indirect channel responses h_{dir} and h_{ind} where a network controller can act upon.

Nowadays, we can optimize the radio performance by acting over the Tx and Rx systems as a function of feedback between the latter and a network controller. Such feedback allows for a channel estimation, that is, the obtaining of an estimated channel state information (CSI) representing in some way the responses h_{dir} and h_{ind}. As such, we can adapt the behavior of the Tx and Rx systems accordingly. For example, by electronically steering or shaping the antenna beams, either as Section 8.5 explained or with multiple-input–multiple-output (MIMO) approaches [71, 72], the system can find some combinations of Tx and Rx states providing a better performance than others. In this case, a specific combination of multipath propagation may present low signal degradation. Nonetheless, in the current, operational radio technology the channel is still outside the performance optimization problem.

9.3.2 Introduction to Reconfigurable Intelligent Surfaces

After the validation of the many features that metasurfaces have recently enabled for the tailoring of electromagnetic waves, envisioning a concept like the RIS was just a matter of time.

Specifically, AMCs can provide an arbitrary reflection coefficient phase φ_Γ depending on the properties of the unit cell, such as the equivalent capacitance C_{FSS}, as we saw in Section 3.2.4. Likewise, metasurfaces can also provide an arbitrary transmission coefficient phase φ_τ, as seen in Section 3.2.6. By deploying reconfigurable components over the unit cells, like varactor or PIN diodes, among others [73], a dynamic phase control is available. If a biasing line independently addresses each unit cell, we can obtain a local, independent reflection or transmission coefficient phase.

In terms of amplitude, the miniaturization of amplifiers allowed their use in the tiny unit cells of transmissive [74–77] and reflective metasurfaces [78–82], as Section 5.4 discussed. As such, we could dynamically control the magnitudes $|\tau_{TEM}|$ and $|\Gamma_{TEM}|$ of the transmission and reflection coefficients, with the possibility of values greater than 1. Such possibility allowed for the subsequent design of power-amplifying RISs [42–45]. The unit cells of reconfigurable metasurface absorbers [83–86] can also provide some control to the amplitude of reflected waves but with a lower complexity (see Section 7.5.3).

These highly versatile characteristics yielded the development of new types of metasurfaces with independent control of transmitted and reflected fields, simultaneously [87–89]. The possibility of plentifully controlling electromagnetic waves recently provoked a discussion in the scientific community where we assume that an arbitrary surface, the RIS, exists, providing a desired, reconfigurable electromagnetic response. In the analysis of RIS-assisted environments, the exact features of the employed metasurfaces are abstracted away from the problem [66].

One way of changing the radio performance of the scenario of Fig. 9.5 would be to use a metasurface imposing a linear phase gradient over impinging waves. According to the generalized laws of reflection or refraction [49], seen in Section 3.2.7, the metasurface could appropriately orient the energy coming from the Tx antenna toward the Rx antenna. In the meantime, it could orient elsewhere energy from interfering signals. However, in a complex, multipath propagation channel, it is often more convenient to simply think about the metasurface as a set of scatterers, each one creating a new, controllable, indirect propagation path. Consider that a metasurface made of M reconfigurable unit cells is deployed somewhere in the scenario of Fig. 9.5 to perform the role of RIS. In this case, the new set of indirect paths, illustrated in Fig. 9.6 and previously represented by (9.6) and (9.10), should now account for the RIS as:

$$h_{ind}(t) = \sum_{n=1}^{N} a_{ind}^{n}(t) * \xi_{ind}^{n}(t) * b_{ind}^{n}(t) + \sum_{m=1}^{M} a_{ind}^{m}(t) * \xi_{ind}^{m}(t) * b_{ind}^{m}(t) \tag{9.11}$$

$$h_{ind}(f) = \sum_{n=1}^{N} a_{ind}^{n}(f)\xi_{ind}^{n}(f)b_{ind}^{n}(f) + \sum_{m=1}^{M} a_{ind}^{m}(f)\xi_{ind}^{m}(f)b_{ind}^{m}(f) \tag{9.12}$$

where ξ_{ind}^{m} is the electromagnetic response of each of the M unit cells of the RIS. Also, a_{ind}^{m} and b_{ind}^{m} are the channel responses accounting for the propagation until and from each of these unit cells, respectively.

Now, remember that each RIS unit cell is reconfigurable and may present different values for their reflection, transmission, or absorption coefficients. Hence, the response ξ_{ind}^{m} of each of the M unit cells in (9.11) and (9.12) is controllable – and so is the final indirect channel response h_{ind}.

More detailed signal models for multipath propagation in RIS-assisted environments are available in the literature [63, 66]. The key point here is that the presence of the RIS allows us to include the channel in the optimization problem, as

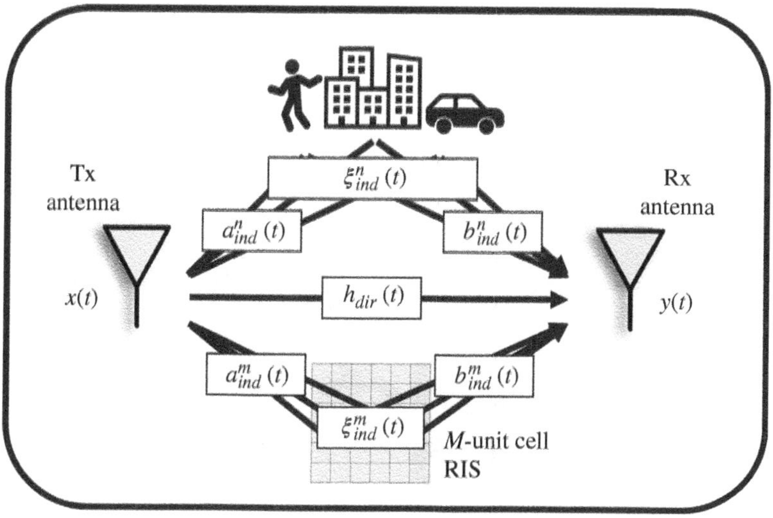

Fig. 9.6 RIS-based, controlled scenario and simplistic representation of direct and indirect paths traveled by an electromagnetic wave between a Tx and Rx antenna therein

evident in (9.11) and (9.12). To be more specific, we can now integrate into the radio performance optimization a channel we considered uncontrolled and of random dynamism in Section 9.3.1.

9.3.3 Improving the Radio Link Between Assets of Friendly Forces

Optimizing the radio performance through feedback between Rx systems and a network controller can provide an improvement in the spectrum access to friendly forces. However, this time the optimization may include not only adapting the behavior of the Rx and Tx systems but also that of a RIS. Specifically, we can reconfigure the electromagnetic responses ξ^{m}_{ind} of each of the M unit cells of a RIS deployed somewhere in the environment, such as coating a building or drone.

We typically interpret the benefits yielded by the use of RISs in terms of a reflection probability (probability that the RIS can reflect signals from the Tx to the Rx antenna), an outage probability (probability that the SNIR is enhanced by the RIS above a specific threshold), a bit error probability (probability that the decoded information differs from the original one), an ergodic capacity (concerning the channel capacity), or a transport capacity (concerning the aggregated data rate of the entire wireless system) [63]. To facilitate our discussions on the application of RISs in EW, we will simplify the pursued goal in terms of received power, that is,

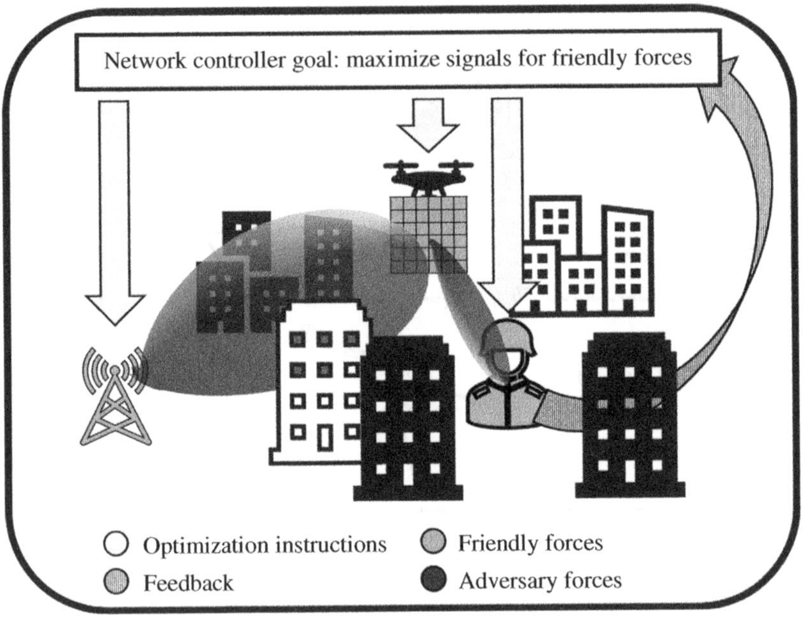

Fig. 9.7 RIS-assisted network when we integrate the RIS and both Tx and Rx systems with the controller

the pursuit of a maximum or minimum power at the Rx antenna given a specific power at the Tx antenna.

Consider in Fig. 9.7 a scenario where a Tx system under control of friendly forces transmits signals toward an also friendly force-controlled Rx system. The systems are in an urban scenario with many static and moving obstacles. In this way, dynamic multipath phenomena are relevant and/or the line of sight between Tx and Rx systems is blocked, as Section 9.3.1 previously discussed. To maximize signals in the Rx system, we embed a RIS over a drone [70]. In this scenario, a network controller optimizes the Tx and Rx systems together with the RIS. Like the case without the RIS, studied in Section 9.3.1, such optimization is possible because the Rx system provides feedback about the received signals to the network controller. Such feedback can again enable an estimation about the CSI which in turn we can use in the optimization. This time, however, the controller sends instructions not only to the Tx and Rx systems, so that they perform an electronic steering/shaping of their antenna beams. It also sends commands to the RIS, allowing a reconfiguration of the electromagnetic response of each of its unit cells.

The literature studies many cases for the optimization of radio environments assisted by a single [90–95] or multiple RISs [96, 97]. In the sequel, our discussions will focus on examples with a single RIS for the sake of simplification, without any loss of generality. It is fair to say that the community is already tackling the optimization problem of the scenario of Fig. 9.7. Such problem is analogous to the one of the 6G, where the physical layer of a network integrates Tx and Rx systems together with RISs[48]. A typical approach is to use alternating optimization methods to iteratively solve a joint beamforming problem for the best parameters for the Tx and Rx antennas and the RIS. S. Gong et al. [63] present a panoramic overview of how this problem has been conventionally tackled. Recently, many AI approaches have also drawn attention [98–100].

Wen, E. et al. [101] represent all the RIS concepts we discussed so far. They present a 1.6-mm thick, reflective, real-data-driven metasurface, which may assume a flat or conformal shape to work as a RIS at 4.6 GHz. As such, the structure is appropriate to coat diverse types of shaped objects and platforms, such as building corners or lightweight drones. Figure 9.8a,b illustrates the conformal version including the fabricated device. The inset highlights the AMC-type unit cell over which varactors are deployed. Figure 9.8c,d, respectively, depicts the reflection coefficient magnitude and phase measured when a normally incident wave illuminates a flat version of such metasurface under various bias voltages. Figure 9.8e shows the experimental setup, where an absorber blocks the direct path between a Tx and an Rx antenna. The received signals are input to a neural network which is first trained with data obtained in real measurements and then can pursue specific goals through the biasing of the unit cells in real scenarios. This approach ensures that the final operation takes into account all the imprecisions that may arise in the design stage or during manufacturing.

As the reader may have realized, the current shape of the RIS problem has considered feedback essential for proper operation. In a real application, the feedback link can take place in many forms, such as using the own radio link under optimization (as long as it has achieved minimal data rate) or through other communication standards running in parallel [102].

M. Usman et al. [103] propose a 5.0-mm thick, reflective metasurface for the role of RIS at 3.9 GHz in indoor applications using Wi-Fi as the feedback link. Figure 9.9a depicts the wireless scheme where the control link from the Rx system to the RIS is explicit. Figure 9.9b shows the fabricated metasurface. The unit cell is again of the AMC type and employs three PIN diodes to allow the reconfiguration of the reflection coefficient phase in discrete states. Figure 9.9c shows a multi-floor scenario where experiments take place and validate the proposed concepts.

One of the main takeaways from the discussed so far in this section is the importance of feedback in the typical RIS-aided environments. As it will become clear in the next sections, such feedback will not always be available.

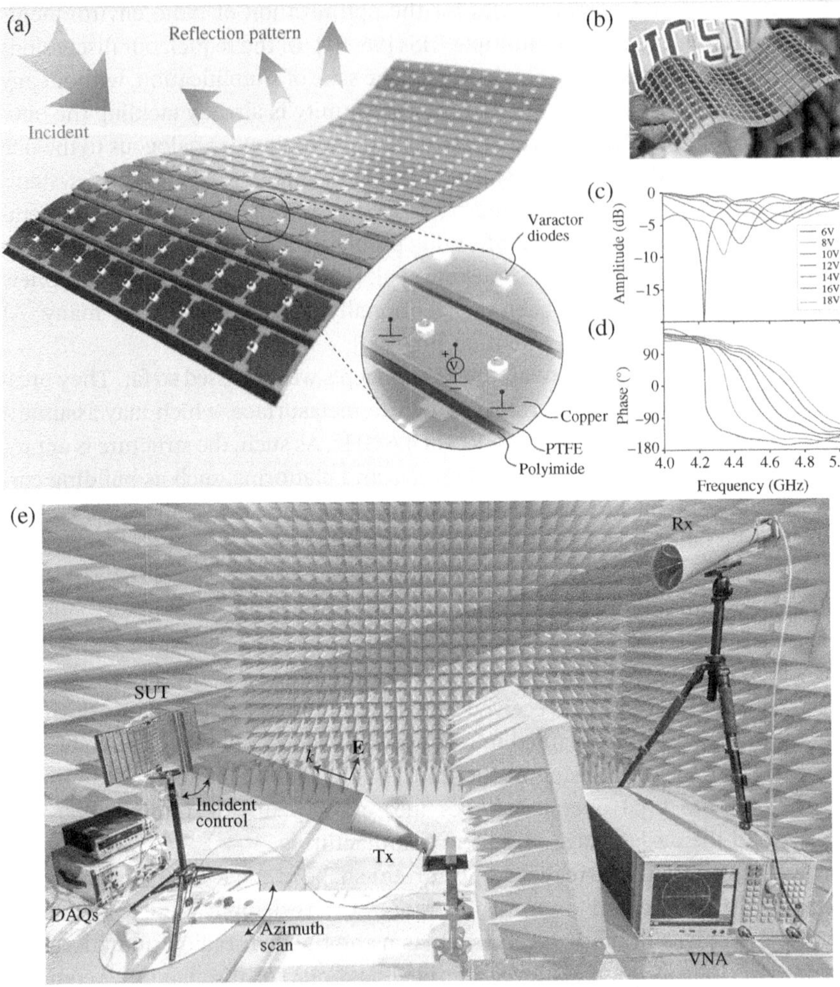

Fig. 9.8 Metasurface of [101]: (a) unit cells arranged in a conformal shape; (b) fabricated device; (c) reflection coefficient magnitude and (d) phase under different varactor bias voltages; (e) measurement setup. "Wen et al. [101]/Springer Nature/CC BY 4.0"

9.3.4 Improving Signal Interception by Friendly Forces

Now, we are interested in the case where we intend to intercept signals from adversary forces. We can do that for SIGINT or ES purposes, among others. Figure 9.10 depicts such scenario.

If the problem looks similar to that of Fig. 9.7 at first glance, they are not the same at the fundamental level. On the one hand, the Rx system and the network

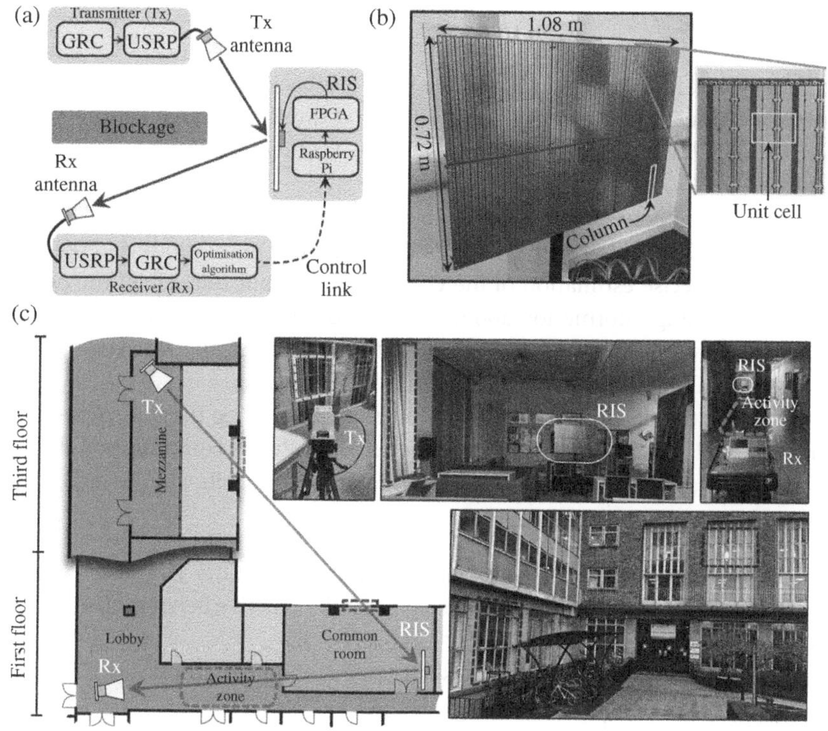

Fig. 9.9 Metasurface of [103]: (a) wireless scheme where the feedback link is explicit; (b) fabricated device; (c) multi-floor validation scenario. "Reproduced from Usman et al. [103]/Springer Nature/CC BY 4.0."

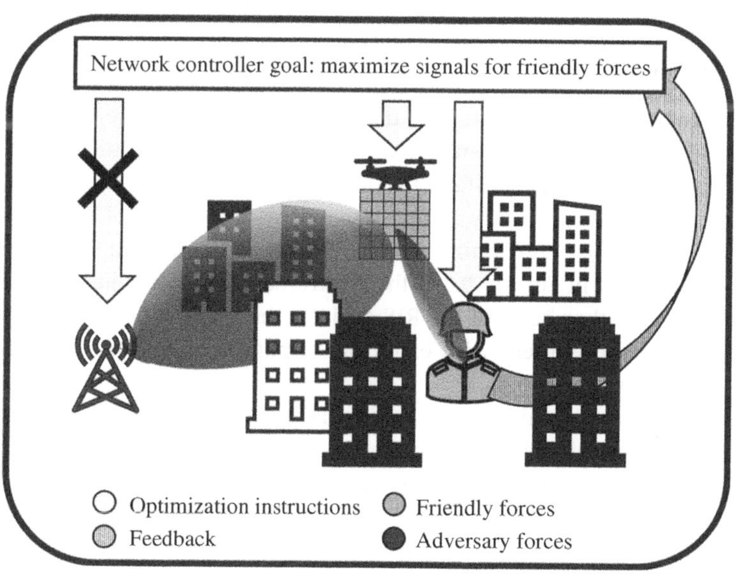

Fig. 9.10 RIS-assisted network when the Tx system belongs to the adversary and we only integrate the RIS and Rx system with the controller

controller are again integrated. Hence, we have feedback, and we can carry out a joint beamforming optimization with the RIS and the Rx antenna. On the other hand, the Tx system belongs to the adversary, leading thereby to some extra difficulties.

The first one, of course, is related to the fact that the beamforming of the Tx antenna is out of control of friendly forces. The second is more subtle and regards the fact that a precise estimation of the CSI may be complicated, albeit the possibility of feedbacking information about intercepted signals. This is due to the fact that we usually obtain the CSI through operations involving the received signals and the original ones [104] – a priori unknown in this case.

At this point, the peculiarities discussed here for improving the interception of adversary signals with RISs do not suggest that feasibility is compromised. However, they leave some open questions deserving future research, among which we can identify:

- What is the best way to perform the optimization without the Tx system, which means less degrees of freedom and even a potential reactive behavior from such system?
- What is the most effective way to do such optimization regardless of a CSI, that is, directly based on the received signals?
- What can we expect in terms of theoretical and practical performances of one such RIS-assisted environment?

The last point is particularly important because it will determine the amount of resources that is reasonable to employ in one such endeavor in the next years.

9.3.5 Avoiding Signal Interception by Adversary Forces

Now, let us make an analysis of what form the problem takes when we intend to use RISs to avoid adversary forces from receiving our signals. This is a quite advantageous, operational ability that substantially differs from EA measures in nature. As we studied in Chapter 7, in EA, we impose signals within the opponents' systems in order to disturb the functioning thereof. On the other hand, in the situations described here and in Section 9.3.6, the electromagnetic waves just do not arrive at the systems, either because the RIS deviates the original paths arriving there or because it cancels them with extra, out-of-phase paths. This is much harder to detect than EA measures.

From a RIS-analysis standpoint, the first particularity in this kind of problem is that optimization is toward a mitigation rather than a maximization of signals at locations where the opponents' Rx systems are deployed [46, 47].

We can envision two main cases to avoid the receiving of signals by adversary forces. Here in this section, we start with the simplest one which is when friendly

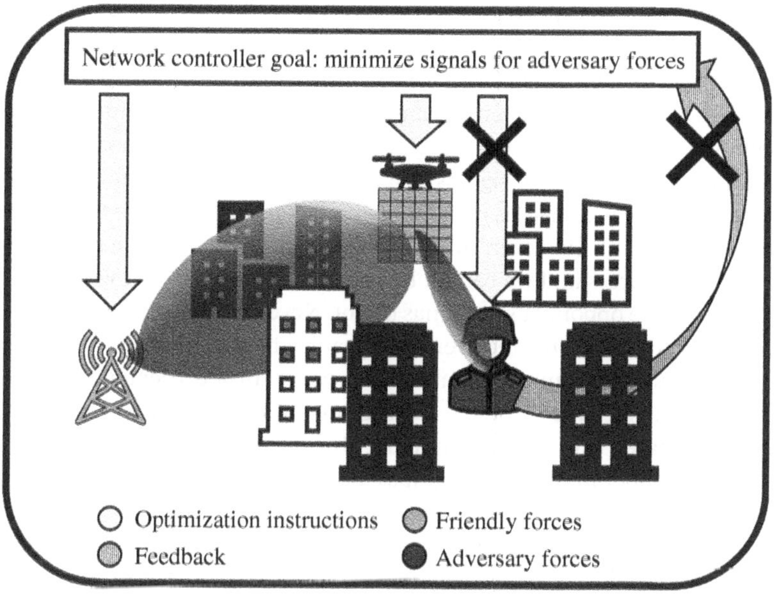

Fig. 9.11 RIS-assisted network when the Rx system belongs to the adversary and we only integrate the RIS and Tx system with the controller

forces transmit signals and the goal is to avoid the interception thereof, disturbing the adversary SIGINT and ES capabilities. In this situation, friendly forces still have control over the RIS and the Tx system, as Fig. 9.11 illustrates. However, the fact that the Rx system is under control of the opponent has the drastic consequence of disrupting the feedback with the network controller – at least the specific feedback from the target Rx system. Thus, the absence of the Rx system not only reduces the degrees of freedom in the optimization problem. It also obliges us to do the optimization of the Tx system and the RIS only with inferences about the signals at target locations. Z. Lin et al. [46] conduct a study about the signal degradation with the use of a malicious RIS but considering perfect CSI. On the other hand, S. Rivetti et al. [47] make an analysis on how resilient the signal degradation is toward uncertainties in the CSI, indicating that the malicious RIS can be reasonably successful even without precise channel information.

Although obtaining any feedback in the described situation may look challenging at first glance, the solution can be straightforward under specific circumstances. In the described scenario, adversary forces are trying to intercept signals whose intended destinations are other than their Rx system. In this context, a consequence of the energy conservation principle is that, by maximizing signals at specific locations, a mitigation in the total power in the rest of the scenario takes place. Hence, we can actually obtain indirect feedback from receivers other than the adversary one, e.g., the Rx systems under control of friendly forces.

A canonical problem worthy of attention in this matter is the one where the optimization of the RIS and the Tx system orients the totality of the transmitted power (which is known) to Rx systems of friendly forces. This condition means that the adversary is not intercepting the signals at all – at the additional condition that anybody (the opponent included) is not amplifying signals in the middle (like with a power-amplifying RIS). However, all it takes is a part of the power not to be received by friendly forces and we cannot be sure anymore whether it was somewhere collected by the opponent's antennas or not. Nevertheless, we can set maximizing signals at friendly-force receivers as a goal in the optimization problem.

The case where the RIS cannot orient all the energy toward friendly-force Rx systems is more realistic. This problem requires methods to minimize signals at specific locations based on information about them elsewhere. If conventional feedback with friendly-force Rx systems may be useful, the association thereof with data-driven approaches looks promising. However, even though many AI approaches have recently gained attention in the RIS field [98–101], to the best of the author's knowledge, none of them tackled this specific scenario yet.

If we consider information outside the RF domain, the story is different. Thinking about another canonical problem, consider that the obstacles in the scenario of Fig. 9.11 and the conditions that could somehow affect propagation are fixed, that is, the channel is static. In this case, one can completely map the scenario with different physical quantities in the absence of adversary forces. Following, one can obtain in advance the RIS configuration leading to a mitigation of signals in each spot of the scenario, given the state of one's own Tx systems. Naturally, these are quite limiting assumptions for offensive maneuvers in new terrain, but not so for defensive purposes in one's own territory. Nevertheless, we also anticipate the use of AI in such applications to make this gap smaller.

Theoretically, we can properly address new scenarios by training an AI model with many scenarios where the RIS can indeed mitigate electromagnetic waves. The hard assumption here is that the available data is sufficiently rich to represent the environment conditions in many cases so that the AI model can generalize to new ones. To exemplify, suppose that one completely maps many scenarios with cameras, imaging radars, thermal, and ultrasound sensors, among others, each one with a sufficiently high resolution and update rate. After training the AI model with all the pertinent data, the model would obtain proper learning about how RF environments behave according to the represented conditions. Then, if one deploys sensors in an operational scenario, the model can associate the measured data with its learning.

Certainly, the use of RIS to avoid the interception of one's own signals by the adversaries looks harder than the case discussed in Section 9.3.4. Nonetheless, the obtained advantages can make all the difference in a world where urban warfare is increasingly common. Further, implementing the capability based only on feedback of RF signals received by friendly forces does not seem so difficult and looks quite feasible today. However, it is still unclear how one such technique would perform. In the above context, we can see the application of AI as a means of enhancing such performance, either by minimizing signals at specific locations based on Rx systems elsewhere or by exploiting other physical domains. All that said, many open questions exist for future research endeavors, such as:

- What strategies would be appropriate to optimize the Tx system and the RIS to mitigate signals at the adversaries' locations based only on feedback of RF signals received by friendly forces?
- What can we expect in terms of (theoretical and practical) performance in this case?
- How can AI help in the abovementioned approach?
- If one intends to add other physical quantities to the problem, which are the most reliable ones?
- Once we have already chosen the physical domains, what would be the best AI approaches to exploit them?
- What are the resolution levels required in the data representing these physical quantities? Are they reasonable for sensors currently available?
- What is the update rate required in the data representing these physical quantities to operate properly given a specific level of scenario dynamism?
- What is the expected computational workload to solve such a problem using the chosen physical domains? Is this workload reasonable considering the available computer technologies?

Without a doubt, we are going to raise many other questions when we start to investigate the above ones.

9.3.6 Denying Spectrum Access to Adversary Forces

The second case to avoid the receiving of signals by adversary forces that deserves attention is when the Tx system also belongs to them. Now, friendly forces control only the RIS, as Fig. 9.12 sketches.

Naturally, feedback coming from the target Rx system (the opponent's one) is once again unavailable. In essence, the problem here looks like a special case of the one of Section 9.3.5, with two extra questions adding to the difficulty. First, the Tx system is also out of the optimization problem, along with the target Rx system. Hence, the degrees of freedom are even more limited – they only concern the RIS. The second point is that the transmitted power, a priori, is unknown.

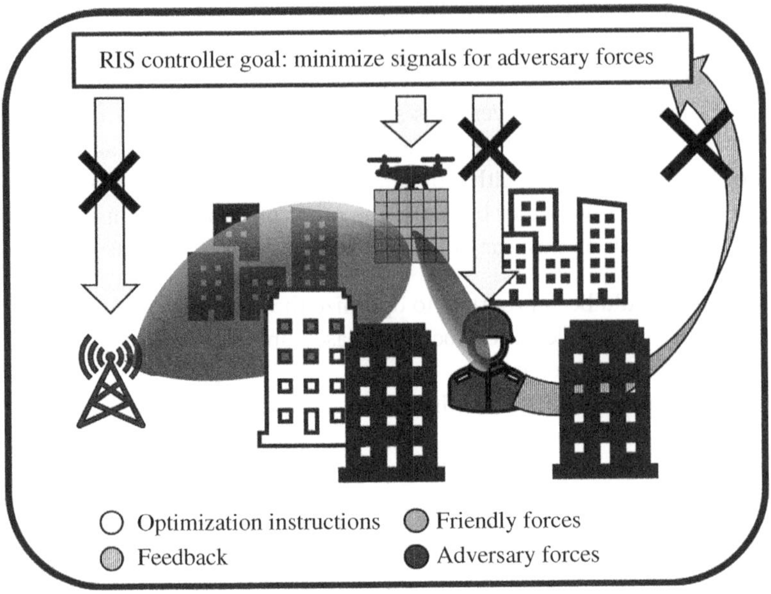

Fig. 9.12 Denying adversary forces' access to spectrum when both Tx and Rx systems belong to them and we cannot integrate them with the controller

In this context, most of the considerations in Section 9.3.5 seem reasonable here. Again, we must do the optimization only with inferences about the signals arriving at the target Rx system. Maximizing signals in the Rx systems of friendly forces seems again one good goal to pursuit. Besides, using AI to learn which RIS configurations mitigate signals at specific regions given the adversary Tx system state is also reasonable – except that, now, such state needs to be somehow estimated. The same challenges concerning availability of proper data, update rate, among others, apply once again.

Although harder than the case of Section 9.3.5, the illustrated capability is of primary importance. It ultimately denies the spectrum usage by opponents – again, by avoiding electromagnetic waves from reaching their receivers, not by introducing new signals therein like in EA. Disrupting the good functioning of the channel for the opponent can avoid its usage for communications, radars, and many other purposes. In the end, it can neutralize significant advantages provided by the wireless technology.

All the open questions of Section 9.3.5 also apply here with the addition that we should also investigate appropriate techniques to track the Tx system state with sufficiently high update rate.

9.3.7 Further Ideas for the Application of RISs in an EW Context

The ideas studied so far in this section present different levels of applicability. Perhaps, the scientific community is going soon to find some of them unfeasible for practical, logistic, or economic matters. However, the community is indicating that using RISs in an EW context is about to become a major trend. For example, the literature has recently proposed the use of RISs for EA purposes, specifically by eavesdropping and falsifying data packets exchanged in the wireless channel [105–107]. Of course, EP would not be let down in the cat-and-mouse EW game, and research about waveform strategies to avoid that has also been recently conducted [108]. Another recent work in the area of EP studies how a drone supporting wireless communications can move and escape the area of a jammer while still connected to ground devices with the aid of a RIS [109].

Although these works propose concrete EW applications using RISs, from the author's understanding, the importance of studying them today lies mainly in the fact that, certainly, the glimpse of many new ways of doing EW will become clear soon.

As a last proposal in this sense, we suggest the creation of a totally controllable, open environment with the use of a RIS-equipped drone swarm. Consider that all the drones in the swarm are ES-capable (see Section 9.1). However, instead of presenting the EA capabilities Sections 7.7 and 9.2 discussed, now each of the drones carries a RIS. Figure 9.13 illustrates such a scenario with a swarm of friendly forces surrounding an adversary radar.

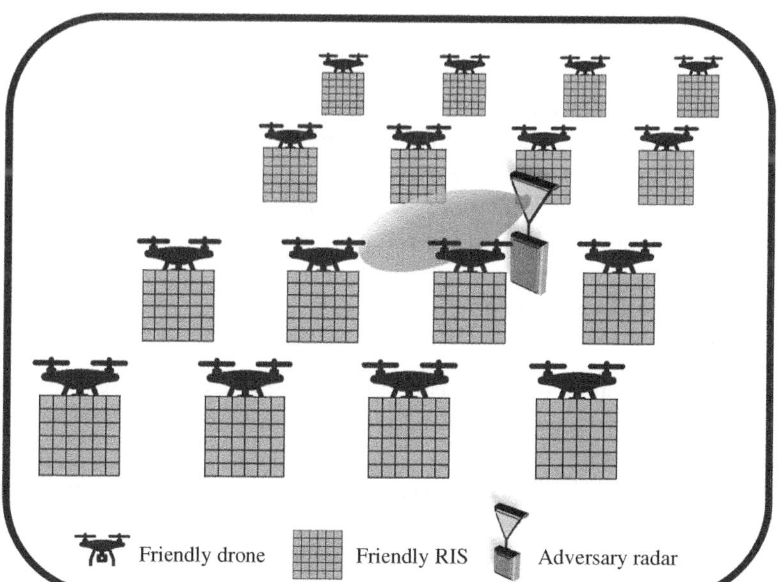

Fig. 9.13 Electromagnetic wave propagation control with an EW-capable drone swarm

There are many ways of interpreting the RISs in the analysis of such scenario. One of them, naturally, is through the multipath analysis we traditionally use in RIS problems. A second way would be to treat each RIS as a reconfigurable scatterer whose RCS is arbitrary. Also, we can see the RISs as surface impedances distributed throughout the environment.

No matter the chosen approach, theoretically, the swarm can optimize the RIS network to maximize/minimize signals at specific regions while doing the inverse elsewhere when it detects the radar using its ES capabilities.

Once again, many open questions will arise after the clear formulation of the RIS-enabled, controllable, open environment problem. Besides those already mentioned in the previous sections, we can already add:

- What is the best geometrical arrangement for the drones in the swarm to guarantee an effective, controllable multipath propagation?
- What is the minimum number of drones per volume, that is, drone density required to ensure good wave control throughout the scenario?
- How would performance degrade when we decrease the drone density?

Researchers working on the 6G will certainly answer many of these (and the previous) questions in the upcoming years as the maturity level of the related technologies becomes closer to what a real application requires.

9.4 Exercises

If you did the exercises of the previous chapters, at this point you already have many functionalities in the simulator – and a good familiarity therewith. In the sequel, we will provide ideas about advanced simulations that you can build not only to check the feasibility of the emerging use cases this chapter proposes but also to identify the related limitations. You can start by using the files relating to the previous chapters available on the companion website.

9.4.1 ES-Solving a Dense Scenario

Difficult level: high

The first interesting simulation you can build concerns a scenario with many emitters lying at the same angular position in relation to a conventional ES system, as Section 9.1.3 described.

As you may have realized, in the folders related to Chapter 7, we did not suppress the files concerning the ES functions we used in Chapter 6. In Section 6.6.1, we suggested you to write in the *process_ES_PDW* function a method to extract the features of pulses from multiple radars that may appear superimposed to one another in the same stream of data [110]. Here, you will need one such function to measure the pulse amplitude in all the channels of the ES system. In this way, you can estimate the pulse DOA using (6.23) or another method [111] in the *process_ES_DF* function. Following, implement a traditional method to cluster pulses [112]. For now, use only the DOA parameter to this end. Finally, implement a traditional approach to deinterleave pulses within the same cluster according to their TOA patterns [113] and identify the emitter accordingly.

The idea here is to run a first simulation with, say, three radars in the same region (but not exactly at the same point) and a single ES system processing all their pulses as the previous paragraph described. Use different amplitudes, PWs, and PRFs in the radar signals. Adjust your ES system to correctly identify these few radars. Right after, increase the number of radars to dozens or even hundreds until you degrade the ES performance significantly.

At this point, spread other ES systems around the radars. Consider that all ES systems feed the estimated DOAs to a single database. Implement a triangulation method [21] to estimate the locations where the pulses are emitted from. Following, adjust your clustering method to handle such locations instead of DOA. Start your simulation again with three radars. Increase again the number of radars to dozens or even hundreds observing when the ES performance will be considerably degraded.

In this exercise, you will see that deinterleaving is much easier when you triangulate because the clustering step [112] will handle fewer radars, making the final step [113] much simpler. To conclude, make an analysis on the performance of both cases. Identify the limitations, bottlenecks, and solutions for each one.

More audacious simulations would include different intrapulse modulations for each radar and the implementation of AI methods to classify intercepted pulses [114–118].

9.4.2 ES-Solving a Multifunctional Radar

Difficult level: high

As a second exercise, consider a multifunctional radar using multiple antenna beams to transmit independent signals, each one characterizing an operating mode [22–24], as Section 9.1.3 also described. The antenna beams present different scan patterns around the radar according to their functionalities.

To make your job easier, you can model the multifunctional radar as multiple different radars lying at the very same point. In a first simulation, try to characterize such radar using a single ES system. You will see that this task is still possible if the radar uses an exceptionally high power because your ES system can also receive pulses from the sidelobes of the radar antenna. Identify the order of magnitude of such power and make an analysis whether it is realistic or not.

Now, repeat your simulation using many ES systems spread around the multifunctional radar. Increase the number of systems until you achieve the full characterization of the radar, that is, until your ES network has information about all radar modes. Identify the relationship between the number and dispersal of antenna beams around the radar and the number of ES systems the full characterization of all modes requires. Again, make an analysis on the performance of both cases, identifying limitations, bottlenecks, and solutions.

9.4.3 Simulating Cognitive Jamming with a Drone Swarm

Difficult level: high

At this point, it is fair to say that you already have simulation capabilities for an ES- and EA-capable drone swarm. The reason is that you can simply adjust the simulations you did in the exercises of Sections 7.8.3 and 9.4.1 to work together. Now, it is time to take advantage of that and implement a multi-agent reinforcement learning method against a multifunctional, cognitive radar.

Hence, it is also fair to say that the challenges in this exercise are, first, the implementation of the cognitive radar and, second, the implementation of the multi-agent reinforcement learning method itself. Here, we consider that the implementation details lie outside the scope of this book. Therefore, we limit ourselves to suggesting that you build different scenarios with different cognitive radars [119–121]. Also, implement different multi-agent reinforcement learning approaches [28, 29, 59–61]. Following, make a performance analysis, identifying limitations, bottlenecks, and solutions. To conclude, identify critical points that we should solve before the application can reach the theater of operation.

9.5 Chapter Final Remarks

In the previous chapters of this book, we saw many EW applications requiring only a few adjustments in existent metasurfaces to go to the theater of operation. In this chapter, we approached other applications whose maturity levels we cannot consider sufficient for immediate development toward a real application yet.

Nonetheless, the envision of these applications is reasonable since the innovative metasurface properties have enabled many new functionalities in the physical layer. Specifically, the fact that we can use compact space-time-modulated metasurfaces for direction finding with a single channel is essential to unlock ES capabilities in drone swarms. Together with the EA capabilities Chapter 7 studied, the ES-capable drone swarm allows a glimpse of multi-agent reinforcement learning-based cognitive jamming. In parallel, we anticipate the use of RISs for the tailoring of electromagnetic wave propagation in multipath or open environments.

Of course, due to the highly visionary nature of the proposed applications, many challenges and open questions require attention. However, the key point of this chapter is not to highlight the commented applications in particular, but rather to demonstrate that new ways of doing EW with metasurfaces are just around the corner.

The proposed applications suggest that the new EW is going to provide levels of dynamism never seen before. In this way, current systems, operators, and AI models are going to require constant updates accordingly. All this makes us believe that the metasurface-driven EW will take the perpetual, unfolding cat-and-mouse EW game to an unprecedented new level.

References

1 Dixon, R. (2020). Azerbaijan's drones owned the battlefield in Nagorno-Karabakh — and showed future of warfare. *The Washington Post* (11 November 2020). https://www.washingtonpost.com/world/europe/nagorno-karabkah-drones-azerbaijan-aremenia/2020/11/11/441bcbd2-193d-11eb-8bda-814ca56e138b_story.html (accessed 08 August 2023).

2 Hecht, E. (2022). Drones in the Nagorno-Karabakh war: analyzing the data. *Military Strategy Magazine* 7 (4): 31–37. https://www.militarystrategymagazine.com/article/drones-in-the-nagorno-karabakh-war-analyzing-the-data/ (accessed 08 Aug. 2023).

3 Pietralunga, C. (2023). Russia and Ukraine take drone warfare to unprecedented scale. *Le Monde* (18 June 2023). https://www.lemonde.fr/en/international/article/2023/06/18/russia-and-ukraine-take-drone-warfare-to-unprecedented-scale_6033281_4.html (accessed 08 August 2023).

4 Cook, E. (2024). Ukraine drones learning to dodge Russian EW signal jammers. *Newsweek* (04 March 2024). https://www.newsweek.com/ukraine-fpv-drones-russia-electronic-warfare-1875577 (accessed 11 March 2024).

5 Tegler, E. (2023). A new micro kind of electronic warfare may be unfolding in Gaza. *Forbes* (03 November 2023). https://www.forbes.com/sites/erictegler/2023/11/03/a-new-micro-kind-of-electronic-warfare-may-be-unfolding-in-gaza/ (accessed 03 November 2023).

6 Watling, J., Reynolds, N. (2023). Stormbreak: fighting through Russian defences in Ukraine's 2023 offensive. *Special Report*. Royal United Services Institute. https://www.rusi.org/explore-our-research/publications/special-resources/stormbreak-fighting-through-russian-defences-ukraines-2023-offensive (accessed 10 October 2023).

7 Zhang, P., Huang, Y., and Jin, Z. (2023). An electronic jamming method based on a distributed information sharing mechanism. *MDPI Electronics* 12 (2130): 1–18. https://doi.org/10.3390/electronics12092130

8 Zhang, Y., Chen, Z., Yan, Y. (2023). Research on single target cognitive electronic reconnaissance strategy for unmanned aerial vehicle. *IET Radar, Sonar & Navigation* 17 (11): 1711–1727. https://doi.org/10.1049/rsn2.12461

9 Ren, Y., Li, B., Wang, H. et al. (2019). A novel cognitive jamming architecture for heterogeneous cognitive electronic warfare networks. *Information Science and Applications* 621: 97–104. https://doi.org/10.1007/978-981-15-1465-4_11

10 Adamy, D. (2001). *EW 101: a first course in electronic warfare*. Norwood, MA: Artech House, Section 8.4.3, pp. 156–158.

11 Al-Tarifi, M.A., Filipovic, D.S. (2018). On the assessment of antenna patterns for wideband amplitude-only direction finding. *IEEE Antennas and Wireless Propagation Letters* 17 (3): 385–388. https://doi.org/10.1109/LAWP.2018.2791199

12 Sakomura, E.S., Ferreira, D.B., Bianchi, I. et al. (2018). Compact planar two-arm compound spiral antenna for L-/X-band direction finding applications. *Proceedings of IEEE International Symposium Antenna Propagation & USNC/URSI National Radio Science Meeting*, Boston, MA, USA (09–14 July 2017), New York, US: IEEE. https://doi.org/10.1109/APUSNCURSINRSM.2018.8608579

13 Mitrović, N.M., Ponjavić, M.M. (2015). Multichannel 2-D direction finding based on differential amplitude detection. *IEEE Sensors Journal* 15 (9): 5064–5070. https://doi.org/10.1109/JSEN.2015.2434876

14 Cazden, J., Filipovic, D. (2017). Wideband amplitude-only direction finding subsystem with conical spirals. *Proceedings of IEEE International Symposium on Antennas and Propagation and USNC-URSI Radio Science Meeting*, San Diego, CA, USA (09–14 July 2017), New York, US: IEEE. https://doi.org/10.1109/APUSNCURSINRSM.2017.8072594

15 Lasser, G., Filipovic, D. (2016). A wide-band spiral based amplitude-only azimuth direction finding system. *Proceedings of IEEE International Symposium on Antennas and Propagation (APSURSI)*, Fajardo, PR, USA (26 June–01 July 2016), New York, US: IEEE. https://doi.org/10.1109/APS.2016.7696618

16 Lopez, D.G., Filipovic, D.S. (2017). On the design of milimeter-wave antennas for amplitude-only direction finding. *Proceedings of International Symposium on Phased Array Systems and Technology (PAST)*, Waltham, MA, USA (18–21 October 2016), New York, US: IEEE. https://doi.org/10.1109/ARRAY.2016.7832575

17 Zhang, L., Dai, J.Y., Moccia, M. et al. (2020). Recent advances and perspectives on space-time coding digital metasurfaces. *EPJ Applied Metamaterials* 7 (7): 1–14. https://doi.org/10.1051/epjam/2020007

18 Fang, X., Li, M., Han, J. et al. (2022). Accurate direction–of–arrival estimation method based on space–time modulated metasurface. *IEEE Transactions on Antennas and Propagation* 70 (11): 10951–10964. https://doi.org/10.1109/TAP.2022.3184556

19 Xia, D., Wang, X., Han, J. et al. (2023). Accurate 2-D DoA estimation based on active metasurface with nonuniformly periodic time modulation. *IEEE*

Transactions on Microwave Theory and Techniques 71 (8): 3424–3435. https://doi.org/10.1109/TMTT.2022.3222322

20 Huang, M., Zheng, B., Cai, T. et al. (2022). Machine–learning-enabled metasurface for direction of arrival estimation. *Nanophotonics* 11 (9): 2001–2010. https://doi.org/10.1515/nanoph-2021-0663

21 Adamy, D. (2001). *EW 101: a first course in electronic warfare*. Norwood, MA: Artech House, Section 5.6.2, pp. 98.

22 Jeong, N.-H., Kim, M., Choi, J.-H. et al. (2023). Beam scheduling of maritime multifunctional radar based on binary integration. *IEEE Access* 11: 35796–35807. https://doi.org/10.1109/ACCESS.2023.3265801

23 Kellett, J., Dawber, W., Wallace, W. et al. (2022). Multifunction maritime radar and RF systems—technology challenges and areas of development. *IEEE Aerospace and Electronic Systems Magazine* 37 (4): 6–20. https://doi.org/10.1109/MAES.2021.3081168

24 Klemm, R., Nickel, U., Gierull, C. (2017). *Novel radar techniques and applications volume 1: real aperture array radar, imaging radar, and passive and multistatic radar*. Stevenage, UK: Scitech Publishing, Chapter 3, pp. 135–171. https://doi.org/10.1049/SBRA512F

25 Pan, Z., Li, Y., Wang, S. et al. (2023). Joint optimization of jamming type selection and power control for countering multifunction radar based on deep reinforcement learning. *IEEE Transactions on Aerospace and Electronic Systems* 59 (4): 4651–4665. https://doi.org/10.1109/TAES.2023.3272307

26 Zhang, W., Ma, D., Zhao, Z. et al. (2023). Design of cognitive jamming decision-making system against MFR based on reinforcement learning. *IEEE Transactions on Vehicular Technology* 72 (8): 10048–10062. https://doi.org/10.1109/TVT.2023.3261318

27 Feng, L.-W., Liu, S.-T., Xu, H.-Z. (2022). Multifunctional radar cognitive jamming decision based on dueling double deep Q-network. *IEEE Access* 10: 112150–112157. https://doi.org/10.1109/ACCESS.2022.3214842

28 Wang, L., Song, F., Fang, G. et al. (2022). A multi-agent reinforcement learning-based collaborative jamming system: algorithm design and software-defined radio implementation. *China Communications* 19 (10): 38–54. https://doi.org/10.23919/JCC.2022.10.003

29 Aref, M.A., Jayaweera, S.K., Machuzak, S. (2017). Multi-agent reinforcement learning based cognitive anti-jamming. *IEEE Wireless Communications and Networking Conference (WCNC)*, San Francisco, CA, USA (19–22 March 2017), New York, US: IEEE. https://doi.org/10.1109/WCNC.2017.7925694

30 Zhang, C., Song, Y., Jiang, R. et al. (2023). A cognitive electronic jamming decision-making method based on Q-learning and ant colony fusion algorithm. *MDPI Remote Sensing* 15 (3108): 1–21. https://doi.org/10.3390/rs15123108

31 Liu, H., Zhang, H., He, Y. et al. (2022). Jamming strategy optimization through dual Q-learning model against adaptive radar. *MDPI Sensors* 22 (145): 1–17. https://doi.org/10.3390/s22010145

32 Xu, Y., Wang, C., Liang, J. et al. (2022). Deep reinforcement learning based decision making for complex jamming waveforms. *MDPI Entropy* 24 (1441): 1–22. https://doi.org/10.3390/e24101441

33 Zhang, S., Tian, H., Chen, X. et al. (2020). Design and implementation of reinforcement learning-based intelligent jamming system. *IET Communications* 14 (18): 3231–3238. https://doi.org/10.1049/iet-com.2020.0410

34 Slimeni, F., Chtourou, Z., Amor, A.B. (2021). Reinforcement learning based anti-jamming cognitive radio channel selection. *4th International Conference Advanced Systems and Emergent Technologies (ICASET)*, Hammamet, Tunisia (15–18 Decemeber 2020), New York, US: IEEE. https://doi.org/10.1109/IC_ASET49463.2020.9318287

35 Wang, L., Peng, J., Xie, Z. et al. (2020). Optimal jamming frequency selection for cognitive jammer based on reinforcement learning. *2nd International Conference on Information Communication and Signal Processing (ICICSP)*, Weihai, China (28–30 September 2019), New York, US: IEEE. https://doi.org/10.1109/ICICSP48821.2019.8958575

36 Haigh, K. Z., Andrusenko, J. (2021). *Cognitive electronic warfare: an artificial intelligence approach*, Boston, MA: Artech House, Chapter 7, pp. 139–158.

37 Haigh, K. Z., Mackay, A. M., Cook, M. R. et al. (2015). Parallel learning and decision making for a smart embedded communications platform. *BBN Technical Report*, BBN-REPORT-8579.

38 Okogbaa, F.C., Ahmed, Q.Z., Khan, F.A. et al. (2022). Design and application of intelligent reflecting surface (IRS) for beyond 5G wireless networks: a review. *MDPI Sensors* 22 (2436): 1–24: https://doi.org/10.3390/s22072436

39 Puspitasari, A.A., Lee, B.M. (2023). A survey on reinforcement learning for reconfigurable intelligent surfaces in wireless communications. *MDPI Sensors* 23 (2554): 1–23. https://doi.org/10.3390/s23052554

40 Gros, J.-B., Popov, V., Odit, M.A. et al. (2021). A reconfigurable intelligent surface at mmWave based on a binary phase tunable metasurface. *IEEE Open Journal of the Communications Society* 2: 1055–1064. https://doi.org/10.1109/OJCOMS.2021.3076271

41 Popov, V., Odit, M.A., Gros, J.-B. et al. (2021). Experimental demonstration of a mmWave passive access point extender based on a binary reconfigurable intelligent surface. *Frontiers in Communications and Networks* 2 (733891): 1–8. https://doi.org/10.3389/frcmn.2021.733891

42 Long, R., Liang, Y.-C., Pei, Y. et al. (2021). Active reconfigurable intelligent surface-aided wireless communications. *IEEE Transactions on Wireless Communications* 20 (8): 4962–4975. https://doi.org/10.1109/TWC.2021.3064024

43 Zhi, K., Pan, C., Ren, H. et al. (2022). Active RIS versus passive RIS: which is superior with the same power budget? *IEEE Communications Letters* 26 (5): 1150–1154. https://doi.org/10.1109/LCOMM.2022.3159525

44 Kang, Z., You, C., Zhang, R. (2023). Active-IRS-aided wireless communication: fundamentals, designs and open issues. *ArXiv (preprint)*. https://doi.org/10.48550/arXiv.2301.04311

45 Zhang, Z., Dai, L., Chen, X. et al. (2023). Active RIS vs. passive RIS: which will prevail in 6G? *IEEE Transactions on Communications* 3 (71): 1707–1725. https://doi.org/10.1109/TCOMM.2022.3231893

46 Lin, Z., Niu, H., An, K. et al. (2024). Pain without gain: destructive beamforming from a malicious RIS perspective in IoT networks. *IEEE Internet of Things Journal* 11 (5): 7619–7629. https://doi.org/10.1109/JIOT.2023.3316830

47 Rivetti, S., Demir, O.T., Bjornson, E. et al. (2024). Malicious reconfigurable intelligent surfaces: how impactful can destructive beamforming be? *ArXiv (preprint)*. https://doi.org/10.48550/arXiv.2402.13351

48 Wang, C.-X., You, X., Gao, X. et al. (2023). On the road to 6G: visions, requirements, key technologies, and testbeds. *IEEE Communications Surveys & Tutorials* 25 (2): 905–974. https://doi.org/10.1109/COMST.2023.3249835

49 Yu, N., Genevet, P., Kats, M.A. et al. (2011). Light propagation with phase discontinuities: generalized laws of reflection and refraction. *Science* 334 (6054): 333–337. https://doi.org/10.1126/science.1210713

50 Malvino, A., Bates, D. (2016). *Electronic principles*, 8th ed. New York, NY: McGraw-Hill Ed., Section 2-14, pp. 47.

51 Yuan, S., Xu, T., Zhang, M. (2022). Extraction and sequential recognition of MFR pulse groups in intercepted pulse trains. *IEEE Access* 10: 106988–106998. https://doi.org/10.1109/ACCESS.2022.3211938

52 Haykin, S. (2006). Cognitive radar: a way of the future. *IEEE Signal Processing Magazine* 23 (1): 30–40. https://doi.org/10.1109/MSP.2006.1593335

53 Smith, G.E., Cammenga, Z., Mitchell, A. (2016). Experiments with cognitive radar. *IEEE Aerospace and Electronic Systems Magazine* 31 (12): 34–46. https://doi.org/10.1109/MAES.2016.150215

54 Gurbuz, S.Z., Griffiths, H.D., Charlish, A. et al. (2019). An overview of cognitive radar: past, present, and future. *IEEE Aerospace and Electronic Systems Magazine* 34 (12): 6–18. https://doi.org/10.1109/MAES.2019.2953762

55 Neri, F. (2018). *Introduction to electronic defense systems*, 3rd ed. Boston, MA: Artech House, Section 5.4.20, pp. 449.

56 Gonçalves Licursi de Mello, R. (2018). Modern techniques to locate and identify radars with low volume, weight, costs and processing capabilities. Master Thesis. Univ. Federal de Santa Catarina. https://repositorio.ufsc.br/handle/123456789/198162 (accessed 03 October 2023).

57 Haigh, K. Z., Andrusenko, J. (2021). *Cognitive electronic warfare: an artificial intelligence approach*, Boston, MA: Artech House, Section 7.1.1, pp. 141.

58 Haigh, K. Z., Andrusenko, J. (2021). *Cognitive electronic warfare: an artificial intelligence approach*, Boston, MA: Artech House, Section 1.3.1, pp. 7.

59 Okawa, Y., Dan, H., Morita, N. et al. (2023). Multi-agent reinforcement learning with information-sharing constrained policy optimization for global cost environment. *IFAC-PapersOnLine* 56 (2): 1558–1565. https://doi.org/10.1016/j.ifacol.2023.10.1854

60 Mishra, R.K., Vasal, D., Vishwanath, S. (2021). Decentralized multi-agent reinforcement learning with shared actions. *55th Annual Conference on Information Sciences and Systems (CISS)*, Baltimore, MD, USA (24–26 March 2021), New York, US: IEEE. https://doi.org/10.1109/CISS50987.2021.9400275

61 Guisi, D.M., Ribeiro, R., Teixeira, M. et al. (2016). Reinforcement learning with multiple shared rewards. *Procedia Computer Science* 80: 855–864. https://doi.org/10.1016/j.procs.2016.05.376

62 Di Renzo, M., Zappone, A., Debbah, M. et al. (2020). Smart radio environments empowered by reconfigurable intelligent surfaces: how it works, state of research, and the road ahead. *IEEE Journal on Selected Areas in Communications* 38 (11): 2450–2525. https://doi.org/10.1109/JSAC.2020.3007211

63 Gong, S., Lu, X., and Hoang, D.T. et al. (2020). Toward smart wireless communications via intelligent reflecting surfaces: a contemporary survey. *IEEE Communications Surveys & Tutorials* 22 (4): 2283–2314. https://doi.org/10.1109/COMST.2020.3004197

64 Martini, E., Maci, S. (2022). Theory, analysis, and design of metasurfaces for smart radio environments. *Proceedings of the IEEE* 110 (9): 1227–1243. https://doi.org/10.1109/JPROC.2022.3171921

65 Barbuto, M., Zarghani, Z.H., Longhi, M. et al. (2022). Metasurfaces 3.0: a new paradigm for enabling smart electromagnetic environments. *IEEE Transactions on Antennas and Propagation* 70 (10): 8883–8897. https://doi.org/10.1109/TAP.2021.3130153

66 Björnson, E., Wymeersch, H., Matthiesen, B. et al. (2022). Reconfigurable intelligent surfaces: a signal processing perspective with wireless applications. *IEEE Signal Processing Magazine* 39 (2): 135–158. https://doi.org/10.1109/MSP.2021.3130549

67 Kaina, N., Dupré, M., Lerosey, G. et al. (2014). Shaping complex microwave fields in reverberating media with binary tunable metasurfaces. *Scientific Reports* 4 (6693): 1–8. https://doi.org/10.1038/srep06693

68 Dupré, M., Del Hougne, P., Fink, M. et al. (2015). Wave-field shaping in cavities: waves trapped in a box with controllable boundaries. *Physical Review Letters* 115 (1): 017701-1–017701-5. https://doi.org/10.1103/PhysRevLett.115.017701

69 Del Hougne, P., Fink, M., Lerosey, G. (2019). Optimally diverse communication channels in disordered environments with tuned randomness. *Nature Electronics* 2: 36–41. https://doi.org/10.1038/s41928-018-0190-1

70 Lu, H., Zeng, Y., Jin, S. et al. (2020). Enabling panoramic full-angle reflection via aerial intelligent reflecting surface. *IEEE International Conference Communications Workshops (ICC Workshops)*, Dublin, Ireland (07–11 June 2020), New York, US: IEEE. https://doi.org/10.1109/ICCWorkshops49005.2020.9145305

71 Hu, Z., Liu, G., Xie, Q. et al. (2024). A learnable optimization and regularization approach to massive MIMO CSI feedback. *IEEE Transactions on Wireless Communications* 23 (1): 104–116. https://doi.org/10.1109/TWC.2023.3275990

72 Yu, B., Ren, Z., Tang, S. (2022). CSI-impaired beamforming optimization for dense MIMO communication networks. *MDPI Electronics* 11 (3225): 1–15. https://doi.org/10.3390/electronics11193225

73 Shao, L., Zhu, W. (2021). Electrically reconfigurable microwave metasurfaces with active lumped elements: a mini review. *Frontiers in Materials* 8 (689665): 1–7. https://doi.org/10.3389/fmats.2021.689665

74 Kim, M., Rosenberg, J.J., Smith, R.P. et al. (1991). A grid amplifier. *IEEE Microwave and Guided Wave Letters* 1 (11): 322–324. https://doi.org/10.1109/75.93899

75 Taravati, S., Khan, B.A., Gupta, S. et al. (2017). Nonreciprocal nongyrotropic magnetless metasurface. *IEEE Transactions on Antennas and Propagation* 65 (7): 3589–3597. https://doi.org/10.1109/TAP.2017.2702712

76 Chen, L., Ma, Q., Jing, H.B. et al. (2019) Space-energy digital-coding metasurface based on an active amplifier. *Physical Review Applied* 11 (054051): 1–6. https://doi.org/10.1103/PhysRevApplied.11.054051

77 Wang, X., Han, J., Tian, S. et al. (2022). Amplification and manipulation of nonlinear electromagnetic waves and enhanced nonreciprocity using transmissive space-time-coding metasurface. *Advanced Science* 9 (2105960): 1–13. https://doi.org/10.1002/advs.202105960

78 Lončar, J., Grbic, A., and Hrabar, S. (2019). Ultrathin active polarization-selective metasurface at X-band frequencies. *Physical Review B* 100 (075131): 1–7. https://doi.org/10.1103/PhysRevB.100.075131

79 Dong, Y., Kang, W., Sima, B. et al. (2022). Retroreflector with polarization isolation based on nonreciprocal metasurface. *IEEE Antennas and Wireless Propagation Letters* 21 (10): 1940–1944. https://doi.org/10.1109/LAWP.2022.3186334

80 Yang, J., Yang, W., Qu, K. et al. (2023). Active polarization-converting metasurface with electrically controlled magnitude amplification. *Optic Express* 31 (18): 28979–28986. https://doi.org/10.1364/OE.499458

81 Wang, X., Xia, D., Li, G. et al. (2023). Design and waveguide measurement of 2-bit reconfigurable amplification metasurface cell. *IEEE Antennas and Wireless*

Propagation Letters 22 (9): 2090–2094. https://doi.org/10.1109/LAWP.2023.3275558

82 Taravati, S., Eleftheriades, G.V. (2021). Full-duplex reflective beamsteering metasurface featuring magnetless nonreciprocal amplification. *Nature Communications* 12 (4414): 1–11. https://doi.org/10.1038/s41467-021-24749-7

83 Wang, W., Xu, C., Yan, M. et al. (2019). Broadband tunable metamaterial absorber based on U-shaped ferrite structure. *IEEE Access* 7: 150969–150975. https://doi.org/10.1109/ACCESS.2019.2947140

84 Wang, W., Wang, J., Yan, M. et al. (2018). Dual band tunable metamaterial absorber based on cuboid ferrite particles. *Journal of Physics D: Applied Physics* 51 (31): pp. 1–13. https://doi.org/10.1088/1361-6463/aacf09

85 Zhao, J., Cheng, Q., Chen, J. et al. (2013). A tunable metamaterial absorber using varactor diodes. *New Journal of Physics* 15, (043049): pp. 1–11. https://doi.org/10.1088/1367-2630/15/4/043049

86 Xu, X., Jiang, J., Miao, L. et al. (2012). Design of tunable metamaterial absorbers based on PIN diodes. *IEICE Electronics Express* 9 (17): 1408–1413. https://doi.org/10.1587/elex.9.1408

87 Xu, J., Liu, Y., Mu, X. et al. (2022). Simultaneously transmitting and reflecting intelligent omni-surfaces: modeling and implementation. *IEEE Vehicular Technology Magazine* 17 (2): 46–54. https://doi.org/10.1109/MVT.2022.3157069

88 Mu, X., Liu, Y., Guo, L. et al. (2022). Simultaneously transmitting and reflecting (STAR) RIS aided wireless communications. *IEEE Transactions on Wireless Communications* 21 (5): 3083–3098. https://doi.org/10.1109/TWC.2021.3118225

89 Mu, X., Liu, Y., Xu, J. et al. (2022). Joint beamforming optimization for simultaneously transmitting and reflecting (STAR) RIS aided communications. *55th Asilomar Conference on Signals, Systems, and Computers*, Pacific Grove, CA, USA (31 October–03 November 2021), New York, US: IEEE. https://doi.org/10.1109/IEEECONF53345.2021.9723386

90 Ying, K., Gao, Z., and Lyu, S. (2020). GMD-based hybrid beamforming for large reconfigurable intelligent surface assisted millimeter-wave massive MIMO. *IEEE Access* 8: 19530–19539. https://doi.org/10.1109/ACCESS.2020.2968456

91 Ning, B., Chen, Z., Chen, W., Fang, J. (2020). Beamforming optimization for intelligent reflecting surface assisted MIMO: a sum-path-gain maximization approach. *IEEE Wireless Communications Letters* 9 (7): 1105–1109. https://doi.org/10.1109/LWC.2020.2982140

92 Zhang, S., Zhang, R. (2020). Capacity characterization for intelligent reflecting surface aided MIMO communication. *IEEE Journal on Selected Areas in Communications* 38 (8): 1823–1838. https://doi.org/10.1109/JSAC.2020.3000814

93 Wu, Q., Zhang, R. (2019). Intelligent reflecting surface enhanced wireless network: joint active and passive beamforming design. *IEEE Global Communications Conference (GLOBECOM)*, Abu Dhabi, United Arab Emirates

(09–13 December 2018), New York, US: IEEE. https://doi.org/10.1109/GLOCOM.2018.8647620

94 Yu, X., Xu, D., Schober, R. (2019). MISO wireless communication systems via intelligent reflecting surfaces. *IEEE/CIC International Conference on Communications in China (ICCC)*, Changchun, China (11–13 August 2019), New York, US: IEEE. https://doi.org/10.1109/ICCChina.2019.8855810

95 Yang, Y., Zhang, S., Zhang, R. (2020). IRS-enhanced OFDM: power allocation and passive array optimization. *IEEE Global Communications Conference (GLOBECOM)*, Waikoloa, HI, USA (09–13 December 2019), New York, US: IEEE. https://doi.org/10.1109/GLOBECOM38437.2019.9014204

96 Yuan, J., Liang, Y.-C., Joung, J. et al. (2021). Intelligent reflecting surface-assisted cognitive radio system. *IEEE Transactions on Communications* 69 (1): 675–687. https://doi.org/10.1109/TCOMM.2020.3033006

97 Wang, P., Fang, J., Yuan, X. et al. (2020). Intelligent reflecting surface-assisted millimeter wave communications: joint active and passive precoding design. *IEEE Transactions on Vehicular Technology* 69 (12): 14960–14973. https://doi.org/10.1109/TVT.2020.3031657

98 Saikia, P., Singh, K., Singh, S.K. (2023) Beamforming design in vehicular communication systems with multiple reconfigurable intelligent surfaces: a deep learning approach. *IEEE Access* 11: 100832–100844. https://doi.org/10.1109/ACCESS.2023.3314668

99 Faisal, K.M., Choi, W. (2022). Machine learning approaches for reconfigurable intelligent surfaces: a survey. *IEEE Access* 10: 27343–27367. https://doi.org/10.1109/ACCESS.2022.3157651

100 Shamasundar, B., Daryanavardan, N., Nosratinia, A. (2023). Channel training & estimation for reconfigurable intelligent surfaces: exposition of principles, approaches, and open problems. *IEEE Access* 11: 6717–6734. https://doi.org/10.1109/ACCESS.2023.3236909

101 Wen, E., Yang, X., Sievenpiper, D.F. (2023). Real-data-driven real-time reconfigurable microwave reflective surface. *Nature Communications* 14 (7736): 1–8. https://doi.org/10.1038/s41467-023-43473-y

102 Shen, D., Dai, L. (2021). Channel feedback for reconfigurable intelligent surface assisted wireless communications. *GLOBECOM 2020 - 2020 IEEE Global Communications Conference*, Taipei, Taiwan (07–11 December 2020), New York, US: IEEE. https://doi.org/10.1109/GLOBECOM42002.2020.9347974

103 Usman, M., Rains, J., Cui, T.J. et al. (2022). Intelligent wireless walls for contactless in-home monitoring. *Light: Science & Applications* 11 (212): 1–13. https://doi.org/10.1038/s41377-022-00906-5

104 Han, Y., Tang, W., Li, X. et al. (2023). CSI acquisition in RIS-assisted mobile communication systems. *National Science Review* 10 (8): 1–15. https://doi.org/10.1093/nsr/nwad127

105 Wei, M., Zhao, H., Galdi, V. et al. (2023). Metasurface-enabled smart wireless attacks at the physical layer. *Nature Electronics* 6: 610–618. https://doi.org/10.1038/s41928-023-01011-0

106 Shaikhanov, Z., Hassan, F., Guerboukha, H. et al. (2022). Metasurface-in-the-middle attack: from theory to experiment. *Proceedings of the 15th ACM Conference on Security and Privacy in Wireless and Mobile Networks (WiSec '22)*, San Antonio, TX, USA (May 16–19, 2022), New York, US: ACM. https://doi.org/10.1145/3507657.3528549

107 Wang, Y., Lu, H., Zhao, D. et al. (2022). Wireless communication in the presence of illegal reconfigurable intelligent surface: signal leakage and interference attack. *IEEE Wireless Communications* 29 (3): 131–138. https://doi.org/10.1109/MWC.008.2100560

108 Alexandropoulos, G.C., Katsanos, K.D., Wen, M. et al. (2023). Counteracting eavesdropper attacks through reconfigurable intelligent surfaces: a new threat model and secrecy rate optimization. *IEEE Open Journal of the Communications Society* 4: 1285–1302. https://doi.org/10.1109/OJCOMS.2023.3282814

109 Almasoud, A.M. (2023). Robust anti-jamming technique for UAV data collection in IoT using landing platforms and RIS. *IEEE Access* 11: 70635–70651. https://doi.org/10.1109/ACCESS.2023.3294596

110 Gonçalves Licursi de Mello, R., De Sousa, F.R. (2018). Precise techniques to detect superimposed radar pulses on ESM systems. *IET Radar, Sonar & Navigation* 12 (7): 735–741. https://doi.org/10.1049/iet-rsn.2017.0563

111 Gonçalves Licursi de Mello, R., De Sousa, F.R. (2019). Non-analytical direction-finding method as a key step in pursuing low size, weight, costs, and computational power in the deinterleaving of radar pulses. *IET Radar, Sonar & Navigation* 13 (11): 1876–1882. https://doi.org/10.1049/iet-rsn.2018.5011

112 Mardia, H.K. (1988). Adaptive multidimensional clustering for ESM. *Proceedings of IEE Colloquium on Signal Processing for ESM Systems*, London, UK (26 April 1988), New York, US: IEEE.

113 Mardia, H.K. (1989). New techniques for the deinterleaving of repetitive sequences. *Proceedings of the Institution of Electrical Engineers* 136 (4): 149–154. https://doi.org/10.1049/ip-f-2.1989.0025

114 Xiao, Z., Yan, Z. (2021). Radar emitter identification based on novel time-frequency spectrum and convolutional neural network. *IEEE Communications Letters* 25 (8): 2634–2638. https://doi.org/10.1109/LCOMM.2021.3084043

115 Bhatti, S.G., Bhatti, A.I. (2022). Radar signals intrapulse modulation recognition using phase-based STFT and BiLSTM. *IEEE Access* 10: 80184–80194. https://doi.org/10.1109/ACCESS.2022.3195273

116 Qu, Z., Hou, C., Hou, C. et al. (2020). Radar signal intra-pulse modulation recognition based on convolutional neural network and deep Q-learning network. *IEEE Access* 8: 49125–49136. https://doi.org/10.1109/ACCESS.2020.2980363

117 Alrubeaan, T., Albagami, K., Ragheb, A. et al. (2019). An investigation of LPI radar waveforms classification in RoF channels. *IEEE Access* 7: 124844–124853. https://doi.org/10.1109/ACCESS.2019.2938317

118 Meng, X., Shang, C., and Dong, J. et al. (2021). Automatic modulation classification of noise-like radar intrapulse signals using cascade classifier. *ETRI Journal* 43 (6): 991–1003. https://doi.org/10.4218/etrij.2020-0338

119 Haykin, S. (2006). Cognitive radar: a way of the future. *IEEE Signal Processing Magazine* 23 (1): 30–40. https://doi.org/10.1109/MSP.2006.1593335

120 Smith, G.E., Cammenga, Z., and Mitchell, A. (2016). Experiments with cognitive radar. *IEEE Aerospace and Electronic Systems Magazine* 31 (12): 34–46. https://doi.org/10.1109/MAES.2016.150215

121 Gurbuz, S.Z., Griffiths, H.D., Charlish, A. et al. (2019). An overview of cognitive radar: past, present, and future. *IEEE Aerospace and Electronic Systems Magazine* 34 (12): 6–18. https://doi.org/10.1109/MAES.2019.2953762

10

Summary and Final Remarks

In the last decades, we have witnessed a formidable evolution in EW due to significant advancements in Electronics, Telecommunications, Computer Science, Systems Engineering, and many other domains. The activity, once considered a simple clash between static systems, became a highly complex combat framework fought between EW networks. In this context, researchers and engineers have been pushing the boundaries of both the hardware and software parts of the associated technologies. Recent developments yielded extremely dynamic EW environments made up of reconfigurable systems of systems using flexible radio circuits. All that is driven by the latest AI models to optimize specific tasks on the fly.

In this book, we studied how metasurfaces can impact such scenario by adding extra layers of dynamism at the physical level of the EW problem. The exotic properties provided by metasurfaces allow the tailoring of electromagnetic waves. In this way, first, we may extend the capabilities of legacy hardware and, second, we can make platforms and even the environment itself reconfigurable in the EW scenario. Further, the simplicity, compactness, power efficiency, and low cost, among other features of metasurfaces, are convenient not only to fit the fuselage of aircraft and spacecraft but also to embed extremely constrained applications such as missile seekers or even lightweight, cheap drones.

The latter is of primary importance, considering the doctrinal and technological trends in the latest conflicts. An analysis on the wars of Nagorno-Karabakh, Ukraine, and Gaza makes it evident that the extensive use of lightweight, cheap drones will change the way humans make war once again, rather like the radio did about one century ago. Under such circumstances, we envision highly specialized, autonomous drone swarms to lead, in a near future, a distributed EW instead of its concentration in major platforms.

In Chapter 1, we had the chance to revisit the history of radio technology from a reconfigurability perspective. Then, we contextualized the reader in terms of

Metasurface-Driven Electronic Warfare, First Edition. Rafael Gonçalves Licursi de Mello.

Published 2025 by John Wiley & Sons, Inc.
Companion website: www.wiley.com/go/licursidemello/metasurfacedrivenelectronicwarfare1

metasurface technology, considering the increasingly dynamic nature of the EW scenario.

Chapter 2 provided a pertinent review on electromagnetics fundamentals that are necessary to understand the working principle of metamaterials and metasurfaces in Chapter 3. Everything that a non-expert needs for a good comprehension of the rest of the book was there. On the other hand, metasurface experts and readers that just wanted to know the technology in an application level did not need to follow these chapters.

In Chapter 4, we discussed how we can use metasurfaces to provide passive stealth properties to a platform. We saw different mechanisms to this end, like wave absorption and wave scattering in space, frequency, and polarization. Such chapter also recalled some basic concepts like the radar equation and RCS. Further, we introduced an important concept for the use of metasurfaces coating platforms: the deployment rate.

In Chapter 5, we took a complex application to work on a series of new concepts, namely the active-cancellation stealth. After recalling the LPI radar, we studied the complexity associated with DRFMs and saw how power-amplifying metasurfaces can replace them. This chapter also introduced our EW scenario simulator, a powerful tool to handle the complexity of modern days in the planning of new EW measures.

In Chapter 6, we recalled the role of ES in the EW space, and then we approached the applicable antenna technology. After revisiting the one-way equation, we studied the theory of frequency-independent antennas. Following, we saw how we could associate different reflection schemes using single-band, multiband, frequency-reconfigurable, and inhomogeneous AMCs with wideband antennas in the ES activity.

Chapter 7 addressed EA applications. First, we remembered how a monopulse radar works. Next, we studied the jamming equation and the concept of burn-through range. Then, we illustrated the use of passive and power-amplifying metasurfaces in EA with the cross-polarization jamming and blinking jamming applications. These were the bases to the envision of totally new EA measures, like the generalized EA measure we presented at the end of this chapter.

In Chapter 8, we studied different EP approaches. First, we discussed how traditional FSSs and rasorbers can counter directed-energy weapons, this new element that is each day more and more developed. We also shed some light on the use of polarization-tunable antennas with reconfigurable metasurfaces to counter cross-polarization jamming. After brief comments on radar polarimetry, we studied metasurface-based antennas as an alternative to phased arrays and then delved into the area of multifunctional antennas with metasurfaces.

Chapter 9 is the most visionary chapter of this book. We started by considering the creation of an ES-capable drone swarm with the use of direction-finding metasurfaces. Right after, we entered in the world of multi-agent reinforcement learning by adding the EA capabilities we saw in Chapter 7 to such a swarm. Then, we discussed how RISs can control the spectrum, either to ease the access of friendly forces or to jeopardize that of adversary ones.

The implementation of most of the applications this book proposes is straightforward. For example, metasurface prototypes whose performance was measured already validated the RCS reduction concepts we saw in Chapter 4. Besides, the community verified experimentally the metasurfaces we considered for active-cancellation stealth in Chapter 5. Further, measurements in anechoic chambers confirm the performance of most of the ES-applicable antennas Chapter 6 discussed. The discussion about EA applications in Chapter 7 also relied on the performance of real, implemented metasurfaces, whether for cross-polarization or blinking jamming. Last, experimental verifications concerning the polarization-reconfigurable, the beam-steerable, and the multifunctional antennas using metasurfaces discussed in Chapter 8 already took place too.

We demonstrated the value of the applications we studied in Chapters 4–8 with realistic simulations and exercises using the EW scenario simulator available on the companion website.

The applications Chapter 9 proposed, on the other hand, require a higher maturity level in the associated metasurface technologies before reaching the theater of operation. Nonetheless, once we implement these innovations, the expected impact is unparalleled. We could also see the potential and feasibility of these ideas through the proposed exercises.

In conclusion, we are in the very moment where metasurfaces are becoming popular and commercially successful. If in a civilian context regulatory bodies already elected the technology as a central element for the physical layer of the 6G, in Defense, this book contributes to connect the dots with EW. In any manner, the countless possibilities metasurfaces open up remind us of the maxim:

> God made the bulk. The surface was invented by the devil.
>
> —Wolfgang Pauli, Nobel laureate.

Index

a

active-cancelation stealth 16, 19, 143, 153
ADC 7, 144, 149, 151
AGC 5, 150, 275
aliasing 151
AMC 74, 78, 85, 88, 90, 107, 113, 121, 130, 163, 191, 219, 307, 341, 345
  dual-band 210, 312
  frequency-reconfigurable 209
  inhomogeneous 191, 221
  multiband 191, 210, 311
  multiband reflection strategy 16, 212, 224
  reconfigurable 191, 289
  triple-band 216
  wideband 130, 208
Ampère's law 34, 38, 68, 75, 195, 203
anisotropic media 35
antenna
  active region 199, 204
  AMC-backed 208–210, 216, 219
  Archimedean spiral 194, 204–206, 208, 222
  biconical 196
  bow-tie 197, 208, 209
  cavity-backed 204
  checkerboard 201, 207
  coplanar waveguide-fed 210, 289
  co-polarized pattern 249, 258, 260
  cross-polarized pattern 249, 260
  deformed radiation patterns 303
  effective area 192, 245
  equiangular spiral 199
  fractal 200
  frequency-independent 190, 317
  grooved bow-tie 212, 216, 313
  horn 299, 330
  log-periodic 194, 200, 205, 207, 222
  multifunctional 286, 308
  reflector-backed 205, 305
  sinuous 200
  U-shaped dipole 210

b

Babinet's principle 203
backward waves 11, 13, 33, 54, 57, 63, 67, 71
B.A. Munk 105, 288
beamsteering 85, 286, 290, 292, 294, 307, 314, 329
bit resolution 151–153, 159
blind zones 191, 213, 219
boundary condition 36, 43, 44, 48, 53, 74, 76, 82, 201, 202
bound charges 57, 59, 63, 67

Metasurface-Driven Electronic Warfare, First Edition. Rafael Gonçalves Licursi de Mello.

Published 2025 by John Wiley & Sons, Inc.
Companion website: www.wiley.com/go/licursidemello/metasurfacedrivenelectronicwarfare1

c

cat-and-mouse EW game 5, 99, 204, 219, 239, 285
causality principle 12, 57, 63, 71
C4ISTAR 8
channel
  difference 243
  sum 243, 262
  virtual 243, 249
channel state information 341, 344, 348, 349
cognitive
  EW 7
  radar 7, 8, 163, 239, 274, 278, 309, 334
  radio 7
  system 4
combat cloud 8
complex representation 37, 57, 61, 145, 157
conformal 2, 3, 90, 133, 345
constitutive relation 35, 42, 49, 54, 68, 75
continuity of fields 46, 48, 53, 55
corrugations 296
critical angle 51

d

DAC 7, 144, 149, 151
delay
  phase 71, 72
  time 144, 148, 151, 159, 162, 163, 166, 170
dependence
  angle 3, 89, 122
  frequency 3, 13, 88, 206, 311
  polarization 3, 89, 122, 291
  temporal 275
deployment rate 131, 251, 259, 268
D.F. Sievenpiper 73, 85, 88, 130, 207, 298
directed-energy weapon 9, 105, 286, 287
dissipative paints 106, 127, 143
DRFM 7, 8, 144, 148, 153, 159, 162, 170, 225, 241, 248, 250, 254, 269, 275, 286, 338
  DOA-driven 228
  realistic 179
drone swarm 3, 9, 241, 275, 278, 286, 288, 328, 332, 336, 353
  communication 276
  EA-capable 328, 356
  ES-capable 328, 329, 356
  synchronization 276
D.R. Smith 13, 70
Drude model 68
duality theorem 203
dynamic range 150, 153, 169

e

electronic attack 3, 189, 285, 348, 353
  angle deception jamming 239
  blinking jamming 16, 19, 239, 240, 260, 274
  burn-through range 241, 246, 249, 251, 256, 265, 272, 275
  cognitive jamming 328, 333, 356
  cross-eye jamming 239
  cross-polarization jamming 17, 240, 247, 254, 260, 274, 286, 289
  disadvantage 240
  formation jamming 239
  generalized measure 239, 274, 277
  noise jamming 239
  skirt jamming 239
  terrain bounce 239
electronic intelligence 5
  library 5, 8
electronic protection 3, 9, 105, 189, 285, 353
  at the antenna level 285
  against cross-polarization jammers 289

against directed-energy weapons 287
measure 153, 169, 239
at the processor level 285
radar reaction 249, 251, 258, 265, 270
at the RF-front-end level 285
electronic support 3, 79, 274, 285, 346
advantage 192
antenna 207, 210
architecture 189, 192
deinterleaving 5, 7, 189, 328
dense scenario 328, 354
detection 191, 246
maximum range 192, 193, 207, 208, 217, 221, 225
processing 192
pulse clustering 7, 334
system 144, 147, 162, 170, 189, 277, 334
TOA pattern 7, 334, 355
triangulation 328, 355
energy conservation principle 12, 63, 121
evanescent waves
amplification 11, 13, 33, 55, 57, 63, 67, 71

f

Fabry–Perot
antenna 112, 124
cavity 300
condition 302, 310
resonances 300
resonances, at multiple bands 310
Faraday–Lenz's law 34, 38, 42, 68, 70, 75, 196, 202
ferrite 63, 123, 130, 267
filter 15, 99, 105, 125
Floquet's theorem 71
FMCW 117
FPGA 8, 119, 153, 330
frequency
Doppler 117, 148
hopping 6, 8, 290
Nyquist 151
reconfigurability 72, 210
shift 117, 148, 317
UHF 99, 127, 130, 223
VHF 99, 127, 130
Friis equation
for the noise figure 170
for the transmission of signals 192
FSS 15, 76, 99, 104, 125, 144, 165, 286, 287

g

gain
antenna 193, 292
control 4
co-polarized 113
cross-polarization disadvantage 250, 251, 258, 265, 289
cross-polarized 113
signal processing 148, 189, 192, 241, 249
Gauss's law 34
generalized law 85
of reflection 109, 119, 329
of refraction 297
genetic algorithm 130

h

H.G. Booker 203
high-resolution lenses 72
Huygens' principle 79, 296

i

image theory 205
impact
of complicated RCS patterns 157
of DOA estimation 156, 228
of DRFM nonidealities 277
of gain in radar classification 229

impact (*cont'd*)
of the time delay 159
of the waveform 157
Information Warfare 8
intrinsic impedance 14, 42, 44, 47, 57, 73, 75, 82
invisibility cloak 13, 72, 99
I/Q sampling 151, 171

j
JADC2, 8
jamming-to-signal ratio 240, 246, 249, 251, 264, 265, 273
J.B. Pendry 13, 68, 70
J. Mitola 7

l
lightweight drones 9, 19, 169, 196, 205, 219, 241, 278, 286, 289, 292, 298, 299, 317, 318, 327, 332
linear frequency modulation 117, 156, 159
Lockheed Martin F-22 Raptor 143
Lockheed-Martin F-117 Nighthawk 99, 104
Lockheed SR-71 Blackbird 104
Lorentz model 61, 68
Love's equivalence principle 80, 194, 296
low-noise amplifier 149, 150, 174, 179, 225, 254, 270, 292, 296

m
magnetron 5
malicious applications 3
Matlab 145, 171, 245, 247, 260, 264
metal-organic framework 104
metasurface
absorber 9, 15, 88, 121, 130, 266, 342
asymmetric transmitting 9, 16, 127
checkerboard 106, 115, 119, 130
direction-finding 16, 328, 329
frequency scattering 16
Huygens 84, 87, 312
multifunctional 309
multilayer 89, 90, 111, 117, 122, 126, 130, 299, 330
mushroom-like 73, 85, 207, 296, 298
optically transparent 133
phase-gradient 109, 130, 297
polarization-conversion 9, 16, 112, 250, 286, 290, 292
power-amplifying 16, 144, 163, 252, 268, 276
space–time modulated 16, 118, 328, 329
time-varying 116
MIMO 341
missile seekers 19, 121, 196, 205, 219, 286, 289, 292, 298, 299, 317, 318
MMIC 165, 167, 170, 330
modulation 116
loop sequence 119
pseudorandom sequence 118
multipath propagation 17, 339, 344
multi-walled carbon nanotubes 104

n
N. Engheta 88, 121, 130
network
distributed ES 333
EW 3, 8, 191, 241, 327, 337
jammers 8, 144
radar 8, 121, 144
radio 7
unmanned platform 8
noise figure 150, 163, 165, 166, 169, 170, 174, 179, 225, 254, 270, 330
Northrop B-2 Spirit 143

o
one-way propagation equation 192, 246
OODA loop 333
oversampling 151

p

partially reflective surface 300, 310
perfect electric conductor 16, 44, 73, 76, 90, 106, 194, 201, 205, 211, 223
perfect magnetic conductor 73, 76, 78, 106, 201, 208, 210, 311
permeability/permittivity 35, 52
  arbitrary 12
  negative 11, 63, 67
  relative 35
phase compensation 161, 170
phased array 16, 150, 172, 260, 286
  active electronically scanned 150, 292, 296
  antenna element 150, 172, 174, 254, 260, 293, 305
  passive 297
phase gradient 85
phase velocity 39, 54
  negative 48
photonic techniques 153
physics fundamentals 3, 33
  break of 56
PIN diode 14, 89, 116, 118, 119, 252, 266, 289, 305, 341, 345
polarizability 76, 78, 81, 84
  artificial materials 67
  surface 82, 87
polarization conversion rate 114, 251, 259
polarization vector 58, 63, 67
power amplifier 149, 150, 163, 174, 179, 254, 270, 292, 296
power density 101, 192, 245
Poynting vector 36, 38, 55
probability of intercept 150
  angle 190, 193
  frequency 191, 193, 213, 219
pulse
  chirp 145, 157, 193
  coded 145, 189, 229
  coherence 148, 161, 174, 254
  compression 145, 147, 148, 152, 285
  descriptor word 227, 334
  feature extraction 227
  integration 148, 174, 254
  superimposed 152

q

quantization 159
  effects 158
  levels 4, 8, 178
  noise 152, 163

r

radar
  antenna control system 244, 247, 249, 254, 263, 274
  classification 5, 8, 189, 219, 221, 355
  cognitive 7, 8, 163, 239, 274, 278, 309, 334
  conical-scan 239
  cross-polarized 166
  low-probability-of-intercept 7, 144, 145, 153, 189, 205, 219, 221, 229, 285
  monopulse 239, 241, 253, 260, 269, 278
  monostatic equation 242
  moving target indication 117, 121
  multifunctional 239, 309, 328, 333, 335, 355
  polarimetric 16, 286, 291
  processing 248
  sequential-lobe 239
  surveillance 174, 274, 278, 334
  tracking 176, 239, 247, 254, 274
radar absorbing materials 104, 106, 127, 143
radar cross-section 16
  of complex scatterers 103
  complex square root 100, 154

radar cross-section (*cont'd*)
equivalent area 100, 101
monostatic 104
reduction 72, 99, 103, 105, 106, 111, 118, 143, 155, 159, 174, 265, 274
varying 265
radar equation 99
bistatic 102
maximum range 103
monostatic 102
radomes 105, 127
rasorber 16, 100, 125, 286, 287
RC time constant 117, 210, 332
reconfigurability 6, 8, 13, 33, 89, 209, 219, 250, 252, 269, 289, 339, 343
at the tactical level 18
reconfigurable intelligent surfaces 17, 328, 338
reflectarray 165, 297
reflectionless transmission 84
refraction 47, 51
law of 3
negative 11, 13, 33, 52, 67, 87
reinforcement learning 274, 328, 334
action 335
multi-agent 337
reward 335
state 334
reradiation 100, 101, 103, 163, 252

s

Salisbury lens 121, 127
sampling rate 4, 151, 276
S.A. Schelkunoff 81, 196
satellites 19, 102, 169, 292, 298, 299, 317
scaling principle 127, 194, 317
Schelkunoff's equivalence principle 80
self-complementariness principle 201
signals intelligence 4, 219, 346
skin return 103, 143, 148, 157, 242, 246, 248, 250, 262, 268, 274
bistatic 102
time-sequenced 239
Snell's law 33, 47, 102, 105, 111, 297
software-defined radio 7
spectrum access control 17, 328
open environment 353
steady state 57, 63
stealth 2, 16, 99, 103, 143, 174, 274
Stoke's theorem 201
survivability 143, 274, 277
3D printing 91

t

time-bandwidth product 148
transient analysis 57, 159
transmission line 164
equivalent schematic 45, 50, 76, 105, 122, 125, 164, 288
right-angled 166, 252
truncation principle 197

v

varactor 89, 116, 209, 252, 266, 289, 298, 299, 313, 330, 341
V.H. Rumsey 190, 194, 197, 203, 208
V. Veselago 11, 33, 70
medium 36, 52, 54, 56

w

waveform 8, 147, 157, 189, 250, 254, 340
fidelity 148, 157, 163, 170
noise-like 145
wavelength 39, 51, 73
projection 52
wavenumber 39
negative 55
projection 47
wide-angle impedance matcher 89, 123
metasurface 90
WWII 5, 6, 239

Printed and bound by CPI Group (UK) Ltd, Croydon, CR0 4YY
17/09/2026
14964311-0005